Vacuum Fluctuations

Vacuum Fluctuations

Editors

G. Jordan Maclay
Roberto Passante

Basel • Beijing • Wuhan • Barcelona • Belgrade • Novi Sad • Cluj • Manchester

Editors
G. Jordan Maclay
Quantum Fields LLC
St. Charles
USA

Roberto Passante
Università degli Studi di Palermo
Palermo
Italy

Editorial Office
MDPI
Grosspeteranlage 5
4052 Basel, Switzerland

This is a reprint of articles from the Special Issue published online in the open access journal *Physics* (ISSN 2624-8174) (available at: https://www.mdpi.com/journal/physics/special_issues/vf).

For citation purposes, cite each article independently as indicated on the article page online and as indicated below:

Lastname, A.A.; Lastname, B.B. Article Title. *Journal Name* **Year**, *Volume Number*, Page Range.

ISBN 978-3-7258-1445-9 (Hbk)
ISBN 978-3-7258-1446-6 (PDF)
doi.org/10.3390/books978-3-7258-1446-6

Cover image courtesy of G. Jordan Maclay

© 2024 by the authors. Articles in this book are Open Access and distributed under the Creative Commons Attribution (CC BY) license. The book as a whole is distributed by MDPI under the terms and conditions of the Creative Commons Attribution-NonCommercial-NoDerivs (CC BY-NC-ND) license.

Contents

About the Editors vii

Preface ix

Galina L. Klimchitskaya and Vladimir M. Mostepanenko
Centenary of Alexander Friedmann's Prediction of the Universe Expansion and the Quantum Vacuum
Reprinted from: *Physics* **2022**, *4*, 65, doi:10.3390/physics4030065 1

G. Jordan Maclay
New Insights into the Lamb Shift: The Spectral Density of the Shift
Reprinted from: *Physics* **2022**, *4*, 81, doi:10.3390/physics4040081 15

Michael R. R. Good and Yen Chin Ong
Electron as a Tiny Mirror: Radiation from a Worldline with Asymptotic Inertia
Reprinted from: *Physics* **2023**, *5*, 10, doi:10.3390/physics5010010 40

Gerd Leuchs, Margaret Hawton and Luis L. Sánchez-Soto
Physical Mechanisms Underpinning the Vacuum Permittivity
Reprinted from: *Physics* **2023**, *5*, 14, doi:10.3390/physics5010014 49

Daniel C. Cole
Two New Methods in Stochastic Electrodynamics for Analyzing the SimpleHarmonic Oscillator and Possible Extension to Hydrogen
Reprinted from: *Physics* **2023**, *5*, 18, doi:10.3390/physics5010018 63

A. Salam
van der Waals Dispersion Potential between Excited Chiral Molecules via the Coupling of Induced Dipoles
Reprinted from: *Physics* **2023**, *5*, 19, doi:10.3390/physics5010019 81

Raúl Esquivel-Sirvent
Finite-Size Effects of Casimir–van der Waals Forces in the Self-Assembly of Nanoparticles
Reprinted from: *Physics* **2023**, *5*, 24, doi:10.3390/physics5010024 95

Matthew J. Gorban, William D. Julius, Patrick M. Brown, Jacob A. Matulevich and Gerald B. Cleaver
The Asymmetric Dynamical Casimir Effect
Reprinted from: *Physics* **2023**, *5*, 29, doi:10.3390/physics5020029 104

Matthew Bravo, Jen-Tsung Hsiang and Bei-Lok Hu
Fluctuations-Induced Quantum Radiation and Reaction from an Atom in a Squeezed Quantum Field
Reprinted from: *Physics* **2023**, *5*, 40, doi:10.3390/physics5020040 130

Christophe Hugon and Vladimir Kulikovskiy
Zero-Point Energy Density at the Origin of the Vacuum Permittivity and Photon Propagation Time Fluctuation
Reprinted from: *Physics* **2024**, *6*, 7, doi:10.3390/physics6010007 166

About the Editors

G. Jordan Maclay

G. Jordan Maclay is the Chief Scientist at Quantum Fields LLC, a research organization he founded in 1999 after retiring from the Department of Electrical Engineering, University of Illinois at Chicago in 1998. Prof. Maclay is interested in fundamental quantum phenomena, particularly those related to the quantum fluctuations of the electromagnetic field, such as Casimir forces and radiative shifts, and how these phenomena might relate to macroscopic vacuum fields around matter. He published the first quantum field theoretical calculation of the Casimir force, introducing the Zeta function method, with his thesis advisor, Prof. Lowell S. Brown, in 1969. Prof. Maclay and his students carried out the first work incorporating Casimir forces into microelectromechanical systems (MEMS) in 1995. He has worked extensively in the area of chemical microsensors. Many of his papers are available at Quantum Fields. He received his PhD in physics from Yale University in 1972. In 2021, he edited the Special Reprint volume "Symmetries in Quantum Mechanics" for the MDPI Journal *Symmetry*. He authored the book *Transformations: Poetry and Art*, published in 2022. Currently, he is working on a book on the symmetries of the hydrogen atom.

Roberto Passante

Roberto Passante is a Professor of Physics at the University of Palermo, Italy. His main research interests are currently in the fields of quantum electrodynamics and quantum optics, vacuum fluctuations, Casimir and Casimir–Polder interactions, the Unruh effect, cavity quantum electrodynamics, quantum optomechanics, atomic radiative processes in static or dynamical external environments and backgrounds, the quantum field theory in a curved spacetime, dark energy and dark matter, and axion electrodynamics.

Preface

This Special Issue presents research on the expanding role of quantum vacuum fluctuations in understanding Casimir and van der Waals phenomena in microscopic domains (for example, the use of dispersion forces to manipulate atoms) and in macroscopic domains (for example, the computation of the vacuum's permittivity in terms of virtual electron–positron pairs). Vacuum fluctuations form the fabric of the universe and their manipulation, particularly due to the control of boundary conditions, has led to many discoveries and applications. This Special Issue discusses a historical account of Friedmann's first proposal of an expanding universe due to vacuum energy. Moreover, it presents a review of the changing concepts of the ether. It also contains papers on Casimir forces between chiral molecules, the Lamb shift, the asymmetric dynamical Casimir effect, stochastic electrodynamics, radiation from atoms in squeezed quantum fields, and a model of the electron as a mirror with constant acceleration. This volume is targeted at researchers wanting to enhance their understanding of these phenomena that arise from vacuum fluctuations.

G. Jordan Maclay and Roberto Passante
Editors

Review

Centenary of Alexander Friedmann's Prediction of the Universe Expansion and the Quantum Vacuum

Galina L. Klimchitskaya [1,2] and Vladimir M. Mostepanenko [1,2,3,*]

1   Central Astronomical Observatory at Pulkovo of the Russian Academy of Sciences, 196140 Saint Petersburg, Russia
2   Peter the Great Saint Petersburg Polytechnic University, 195251 Saint Petersburg, Russia
3   Kazan Federal University, 420008 Kazan, Russia
*   Correspondence: vmostepa@gmail.com

Abstract: We review the main scientific pictures of the universe developed from ancient times to Albert Einstein and underline that all of them treated the universe as a stationary system with unchanged physical properties. In contrast to this, 100 years ago Alexander Friedmann predicted that the universe expands starting from the point of infinitely large energy density. We briefly discuss the physical meaning of this prediction and its experimental confirmation consisting of the discovery of redshift in the spectra of remote galaxies and relic radiation. After mentioning the horizon problem in the theory of the hot universe, the inflationary model is considered in connection with the concept of quantum vacuum as an alternative to the inflaton field. The accelerated expansion of the universe is discussed as powered by the cosmological constant originating from the quantum vacuum. The conclusion is made that since Alexander Friedmann's prediction of the universe expansion radically altered our picture of the world in comparison with the previous epochs, his name should be put on a par with the names of Ptolemy and Copernicus.

Keywords: quantum vacuum; Friedmann universe; general theory of relativity

1. Introduction

According to Immanuel Kant [1], the starry heavens is one of two things which "fill the mind with ever new and increasing admiration and awe..." Questions about what our universe is, how it was created and how long it will exist have always aroused great interest. In the pre-scientific era, answers to these questions were usually given on the basis of various myths, religions, and philosophical systems. In ancient Greece, for the first time in the history of mankind, the foundations of a scientific approach to the study and attempts to answer these questions were laid.

In this review, we briefly list the main scientific pictures of the universe, developed during the period of time from ancient Greece to Albert Einstein, and emphasize one characteristic feature common to all of them. This common feature is that the universe has always been thought as a stationary system. The concept of a stationary universe was questioned only 100 years ago in 1922 when Alexander Friedmann, on the basis of the general theory of relativity, demonstrated that our universe expands with time. This prediction was confirmed experimentally very soon and became the basis of modern cosmology given every reason to include the name of Alexander Friedmann on a par with the names of greatest scientists who completely revised our understanding of the world around us.

Another important point is that the Friedmann universe exists for a finite time after its creation at a point "from nothing". This initial state of the universe called the "cosmological singularity" makes a link between the Friedmann discovery of expanding universe and the concept of quantum vacuum. According to modern views, the first moments of the evolution of the universe were governed by quantum theory. In the framework of a

semiclassical model, where the gravitational field remains classical but the fields of matter are quantum, it is possible to consider the stress-energy tensor of the vacuum of quantized fields as the source of gravitational field. Under the influence of quantum vacuum the universe expands exponentially fast which is known as the cosmic inflation. At slightly later time, inflation gives way to the Friedmann expansion following the power law. At the present stage, an expansion of the Friedmann universe is accelerating under the impact of dark energy. One of the most popular explanations of this mysterious substance is again given by the quantum vacuum which leads to a nonzero cosmological constant.

The paper is organized as follows. After a discussion of various pictures of the universe from ancient times to Einstein in Section 2, we briefly consider the main facts of Alexander Friedmann's scientific biography in Section 3. Section 4 is devoted to Friedmann's prediction of the universe expansion made 100 years ago. The experimental facts confirming that the universe is really expanding are presented in Section 5. Section 6 is devoted to the cosmic inflation and creation of the universe from the quantum vacuum. In Section 7, we consider the accelerated expansion of the universe and its explanation in terms of dark energy originating from the vacuum of quantized fields. In Section 8, the reader will find the discussion, and the paper ends with conclusions in Section 9.

2. Pictures of the Universe—From Ancient Times to Albert Einstein

The first scientific picture of the universe based on observations was created by Claudius Ptolemy in the first century AD. The Ptolemy system was geocentric which means that the Earth was placed at the center of the world. All remaining celestial bodies, i.e., the Moon, the Sun and five planets known at that time (Mercury, Venus, Mars, Jupiter, and Saturn), rotated around the Earth in circular orbits. According to Ptolemy's system of the world, beyond Saturn there is a firmament to which the fixed stars are attached. It was assumed that the stars and the firmament do not obey the same physical laws as all bodies on the Earth. Despite the presence of some nonscientific elements, the Ptolemy system gave the possibility to perform calculation of both future and past positions of the Moon, Sun and all five planets with rather high accuracy. In fact, this system was successfully used until the 16th century. Needless to say, Ptolemy's picture of the universe was stationary. It did not vary with time.

Important change in our picture of the world has been made by investigations of Nicolaus Copernicus, Johannes Kepler, and Galileo Galilei, performed in the 16th and 17th centuries. They developed on a scientific basis and supported by observations the long-proposed hypothesis that our universe is in fact heliocentric. According to the heliocentric system, the Earth and all five planets orbit the Sun whereas the Moon orbits the Earth. Johannes Kepler made an important discovery that the orbits of planets are not circles but ellipses with the Sun at one of the ellipse's focuses. Both Copernicus and Kepler believed Ptolemy's idea that the stars are fixed points attached to the firmament. However, Galilei elaborated methods to determine the shape of stars and made estimations of their radii. The radically new picture of the world established by Copernicus, Kepler, and Galilei retains its validity in the scales of Solar system up to the present. However, they persisted in the belief that the universe is stationary in a sense that planets followed and will always follow the same predetermined orbits.

The next dramatic step in our understanding of the universe was made by Isaac Newton who developed the first physical theory, Newtonian mechanics, and laid foundations of the mechanical picture of the world. In his book [2] *Mathematical Principles of Natural Philosophy* published in 1687, Newton formulated the three laws of mechanics and the law of gravitation which must be obeyed by all material bodies on the Earth and in the sky. Newton arrived to the fundamental conclusion that the inertial mass, m_i, of each material body is equal to its gravitational mass, m_g, which is responsible for the gravitational attraction. This was the first formulation of the equivalence principle used by Einstein as the basis of general relativity theory 230 years later.

With the second law of mechanics and the Newton law of gravitation, one finds that the force acting between the test mass $m_i = m_g$ and the Earth of the mass M and radius R can be expressed in two ways:

$$F = m_i a = \frac{G m_g M}{R^2}, \qquad (1)$$

where G is the gravitational constant and a is the acceleration of the test mass. Then, using the equivalence principle, one obtains from Equation (1):

$$a = \frac{GM}{R^2}, \qquad (2)$$

i.e., the conclusion is that all bodies in the vicinity of Earth surface fall down with the same acceleration independently of their mass. The law that light and heavy bodies falling to the ground from the same height reach the ground at the same time was experimentally discovered by Galileo Galilei. Newton derived it theoretically. This fundamental law of Nature, which defies common sense, was destined to play a huge role in elucidating the structure and evolution of our universe.

Newton's concept of the universe pushed its boundaries far beyond the solar system. According to Newton, the universe is infinitely large in volume and contains infinitely many stars. The space of the universe is homogeneous (i.e., all points are equivalent) and isotropic (i.e., all directions are also equivalent). However, keeping unchanged an important element of the previous pictures of the universe, Newton believed that our universe is of an infinitely large age and it will exist forever. In this sense he considered the universe to be stationary.

Newton's picture of the universe was universally accepted until the early 20th century despite some unresolved problems. For instance, according to Olbers paradox proposed in 1823, in the case of an infinitely large universe, in every direction one looks, one should see a star. As a result, the entire night sky would shine like a surface of a star, which is not true. One more difficulty is the problem of the heat death of the universe discussed by Bailly in 1777 and elaborated on by Lord Kelvin in 1851 on the basis of the laws of thermodynamics. Since the Newtonian universe exists for an infinitely large time, it should already have reached a state where all energy is evenly distributed and all dynamical processes are terminated. Thus, the observed temperature differences are in contradiction to an assumption that the universe is infinitely old.

A new era in the study of the universe began in 1915 when Albert Einstein created the general theory of relativity starting from the equivalence principle. According to this theory, there is no gravitational force which attracts material bodies to each other. All bodies move freely along the shortest (geodesic) lines in the Riemann curved space-time of the universe which becomes curved under the impact of energy and momentum of these bodies. Thus, the general theory of relativity describes the self-consistent system where the space-time curvature is determined by the material bodies whereas their motion is caused by the character of this curvature.

The description of the universe as a whole in the framework of general theory of relativity is based on Einstein field equations,

$$R_{ik} - \frac{1}{2} R g_{ik} - \Lambda g_{ik} = \frac{8\pi G}{c^4} T_{ik}, \qquad (3)$$

where R_{ik} is the Ricci tensor characterizing the space–time curvature, $R = g^{ik} R_{ik}$ is the scalar curvature, g_{ik} is the metrical tensor, Λ is the cosmological constant, c is the speed of light, and T_{ik} is the stress-energy tensor of matter. The indices i and k here take the values $0, 1, 2, 3$ and there is a summation over the repeated indices.

The term Λg_{ik} was absent in the original Einstein's equations, published in 1915 [3] (see [4] for English translation). Einstein introduced it when applying his field equations to the universe as a whole in order to compensate the effect of gravity and make the universe stationary [5] (see [6] for English translation). This means that he shared the opinion of Ptolemy, Copernicus, Galilei and Newton that the universe does not vary with time. Using the basic concepts of Newton's picture of the world, Einstein also assumed that the 3-dimensional space of the universe is homogeneous and isotropic. Based on this assumptions, he obtained the model of the stationary universe of finite spatial volume. This universe exists forever. It has never been created. There is a finite number of stars in the Einstein universe.

Thus, all the greatest scientists from Ptolemy to Einstein, who determined the views of mankind on the universe for two millennia believed that it is stationary. A new era in the understanding of the universe began with the paper by Alexander Friedmann [7], published 100 years ago, in 1922, in which he first proved that the universe expands. Before considering Friedmann's discovery, we briefly present the main facts of his scientific biography, making it clear how he came to such a radical conclusion.

3. Brief Scientific Biography of Alexander Friedmann

Alexander Friedmann (see his photo in Figure 1) was born on 6 June, 1888 in Saint Petersburg (the capital of Russian Empire) in the artistic family (a full description of his life can be found in [8,9]). Alexander Friedmann's father's name was also Alexander. He was an artist at the Court Ballet of the Imperial Theater and a ballet composer. Alexander Friedman's mother's maiden name was Lyudmila Voyachek, she was a pianist. She graduated from the Saint Petersburg Conservatoire. Nothing indicated that a child born in such a family will become an eminent mathematician and physicist. After the divorce of Friedmann's parents in 1897, he lived with his father. In the same year, he started to study at the Second Saint Petersburg High School which was known for the highly qualified teachers in the field of mathematics and physics.

Figure 1. Alexander A. Friedmann (6 June 1988–16 September 1925), the founder of modern cosmology.

While still a schoolboy, Alexander Friedmann, in collaboration with his schoolmate Yakov Tamarkin (in the future, a famous mathematician), wrote his first paper, devoted to Bernulli numbers. In 1906, this paper was published in *Mathematische Annalen* by the recommendation of David Hilbert [10].

Alexander Friedmann graduated from High School in 1906 with a gold medal and was admitted to the first course of the Department of Mathematics belonging to the Faculty of Physics and Mathematics at the Saint Petersburg University. During his student years at the university, Alexander Friedmann obtained an intimate knowledge in different fields of mathematics and physics and his successes were always evaluated as "excellent".

After a graduation from the Department of Mathematics in 1910, Alexander Friedmann stayed at the same department as a Postgraduate Researcher and to prepare to the Professor position. His supervisor was the famous mathematician academician Vladimir Steklov. From 1910 to 1913, Alexander Friedmann solved several complicated problems in mathematical physics, published many papers and delivered lectures in mathematics for students. Although he successfully passed examinations for a Master degree, he formally defended the Master thesis only in 1922. By that time, he was already Full Professor at the Perm University (1918–1920), at the Petrograd University, Petrograd Polytechnic Institute, and at the Institute of Railway Engineering (Petrograd, formerly Saint Petersburg, 1920–1925).

In 1913, Alexander Friedmann was employed by the Saint Petersburg Physical (later renamed in Geophysical) Observatory. During the work at this institution, he obtained several fundamental results in dynamical meteorology, hydrodynamics, and aerodynamics. His results retain their significance to the present day, and Friedmann's name is well known to everyone working in these fields. Several important results were obtained by him during the visit to Leipzig University in the first part of 1914.

During the three years of the World War I, from 1914 to 1917, Friedmann served in the Air Force of the Russian Empire. During this period of his life, he personally piloted airplanes, organized the aerological service, and created the mathematical theory of bombing. His service during the war was marked with several military awards.

The period of Alexander Friedmann's life from 1920 to 1925 was especially productive. During these years, he published several books and obtained outstanding results in the field of dynamical meteorology. As a recognition of his scientific merits, in 1925 Alexander Friedmann was appointed Director of the Geophysical Observatory of the Russian Academy of Sciences. Just during this period of time, he published two papers [7,11] containing an extraordinary prediction that our universe expands. Although on 16 September 1925 Alexander Friedmann tragically died of typhus at the age of 37, these papers made his name immortal. In the next section, we briefly discuss the essence of the obtained results and how the distinctive features of Friedmann's scientific career helped him make a discovery that even the great Einstein himself missed.

4. Friedmann's Prediction of Expanding Universe

In his approach to the description of the universe in the framework of general relativity theory, Friedmann assumed that the three-dimensional space is homogeneous and isotropic. In this regard, he followed his predecessors Newton and Einstein. The assumption of homogeneity and isotropy alone gives the possibility to find the metric, i.e., the distance, ds, between two infinitesimally close space-time points, x^i and x^k, using the standard mathematical methods:

$$ds^2 = g_{ik}dx^i dx^k = c^2 dt^2 - a^2(t)[d\chi^2 + f^2(\chi)(d\theta^2 + \sin^2\theta d\varphi^2)]. \qquad (4)$$

Here, t is the time coordinate whereas χ, θ, and φ are analogous to the spherical coordinates in the three-dimensional space. In doing so, the usual Cartesian coordinates and the radial coordinate are expressed as

$$x^1 = r(t)\sin\theta\cos\varphi, \qquad x^2 = r(t)\sin\theta\sin\varphi, \qquad x^3 = r(t)\cos\theta,$$
$$r(t) = a(t)f(\chi). \qquad (5)$$

The function $f(\chi) = \sin\chi$, 0, and $\sinh\chi$ depending on whether the constant curvature of the three-space is equal to $\kappa = 1, 0$, and -1, respectively.

The function $a(t)$ is called the "scale factor". It has the dimension of length. In the case of $f(\chi) = \sin\chi$ (the space of positive curvature), $a(t)$ has the meaning of the radius of curvature. The space of positive curvature has the finite volume $V = 2\pi^2 a^3(t)$. The spaces of zero and negative curvature have an infinitely large volume. In his first paper [7], Friedmann considered the space of positive curvature, $\kappa = 1$, whereas his second paper [11] is devoted to the space of negative curvature $\kappa = -1$.

In the Friedmann approach to the problem, it is important that he was a mathematician who used the rigorous analytic methods. He wished to see what is contained in the fundamental Einstein's equations (3) in the case of a homogeneous isotropic metric (4) independently of our historical and methodological preferences. This approach, which proved to be very fruitful in all Friedmann's diverse scientific activities, was based on the long-standing traditions of the Saint Petersburg mathematical school.

Substituting Equation (4) in Equation (3) and calculating R_{ik} and R by the standard expressions of Riemann geometry, Alexander Friedmann obtained two equations, which were later named after him:

$$\frac{d^2a}{dt^2} = -\frac{4\pi G}{3c^2}a(\varepsilon + 3P) + \frac{1}{3}c^2 a\Lambda,$$

$$\left(\frac{da}{dt}\right)^2 = \frac{8\pi G}{3c^2}a^2\varepsilon - \kappa c^2 + \frac{1}{3}c^2 a^2 \Lambda. \tag{6}$$

In these equations, it was taken into account that in the homogeneous isotropic space the stress-energy tensor, T_{ik}, is diagonal and that its component $T_0^0 = \varepsilon$ has the meaning of the energy density of matter, whereas its components, $T_1^1 = T_2^2 = T_3^3 = -P$, describe the pressure P of matter. Note also that in his papers [7,11] Friedmann considered the so-called "dust matter" for which the relative velocities of its constituents are small as compared to the speed of light. This leads to the zero pressure, $P = 0$, but does not affect any of the fundamental conclusions following from Equation (6).

Friedmann found that for $\kappa = 1$ Equation (6) admits the stationary solution in the special case when

$$\varepsilon + 3P = \frac{c^4 \Lambda}{4\pi G}, \qquad \frac{4\pi G a^2}{c^4}(\varepsilon + P) = \kappa. \tag{7}$$

In this case, from Equation (6) one has:

$$\frac{d^2a}{dt^2} = \frac{da}{dt} = 0, \qquad a = a_0 = \text{const.} \tag{8}$$

This is the stationary universe obtained earlier by Einstein [5,6]. The stationary solution exists only in the case $\kappa = 1$, $\Lambda \neq 0$. If $\Lambda = 0$, the first equality in Equation (6) leads to $d^2a/dt^2 < 0$ because for the usual matter it holds $\varepsilon + 3P > 0$. For $\kappa = -1$ the stationary universe containing matter with $\varepsilon > 0$ is impossible [11]. Thus, the cosmological solution found by Einstein is an exceptional case, whereas in all other cases (i.e., with $\Lambda = 0$ or $\Lambda \neq 0$ but with conditions (7) violated), the universe is nonstationary so that the scale factor $a(t)$ depends on time.

Friedmann derived one more specific solution of Equation (6) for which the scale factor a depends on time but the scalar curvature R is constant. This is the solution previously found by de Sitter [12]. It is most simple to illustrate the de Sitter solution, which is determined by only the cosmological constant in the absence of usual matter, $\varepsilon = P = 0$, for the case $\kappa = 0$. Then Equation (6) is simplified to one equation:

$$\frac{da}{dt} = \sqrt{\frac{\Lambda}{3}} ca, \tag{9}$$

which has the solution

$$a(t) = a_0 \exp\left(c\sqrt{\frac{\Lambda}{3}}t\right). \tag{10}$$

The scalar curvature of the homogeneous isotropic spaces is given by

$$R = -\frac{6}{a^2}\left\{\frac{1}{c^2}\left[\left(\frac{da}{dt}\right)^2 + a\frac{d^2a}{dt^2}\right] + \kappa\right\}. \tag{11}$$

Substituting Equation (10) in Equation (11) for $\kappa = 0$, one obtains the constant scalar curvature of the de Sitter space–time $R = -4\Lambda$. We will return to the consideration of the de Sitter scale factor (10) in Section 6 in connection with the quantum vacuum.

As to the general solution of Friedmann equation (6), it is characterized by the zero initial value of the scale factor $a(0) = 0$ and by the infinitely large values of the initial scalar curvature $R(0)$ and energy density $\varepsilon(0)$. Thus, at the initial moment the space of the universe was compressed to a point and the time period from the creation of the world to the present moment is finite [7]. The initial state of the universe is called the cosmological singularity; see [13,14] for details about the solutions of Friedmann equation (6) for different types of matter and corresponding equations of state.

It should be noted that the first reaction of Einstein to Alexander Friedmann's results was completely negative. Shortly after the publication of Friedmann's paper [7], Einstein published a note [15] (see English translation in [16]) claiming that the cosmological solutions, found by Friedmann, do not satisfy the field equations of general relativity theory. In response to this criticism, Friedmann wrote the explanation letter [17] (for English translation, see [18]), which was put in Einstein's hands by Yurii A. Krutkov during his visit to Germany. This letter contained exhaustive explanations which cannot be ignored. As a result, Einstein published one more note where he recognized that his criticism "was based on an error in calculations" [19] (see [20] for English transaltion). Even after Einstein had realized that his original criticism was incorrect, this did not make him a supporter of the concept of an expanding universe. For a long time Friedmann's discovery went largely unnoticed. Its importance became obvious only after the publications by Georges Lemaêtre and others (see below).

In fact, Alexander Friedmann did not construct a new physical theory for the description of the universe. This was performed by Albert Einstein, who created the general theory of relativity. The greatness of Alexander Friedmann lies in the fact that he was the first to believe that the mathematical solution of Einstein equations, corresponding to an expanding universe, can indeed apply to the real world at the time when this idea was not considered plausible. In doing this, he changed our picture of the universe and gave start to grandiose cosmological investigations of the last century.

Although Friedmann himself believed that the observational data at our disposal are completely insufficient for choosing the solution of his equations that describes our universe [7], the first experimental confirmation of the universe expansion came very soon.

5. Experimental Confirmations of the Universe Expansion

For all nonstationary solutions of the Friedmann equation (6) distances between any two remote bodies in the observed universe increase with time. This is seen from Equation (4), where the spatial distance is proportional to the scale factor $a(t)$. As a result, if the universe expands, all observable galaxies should move away from the Earth. The galaxies are observed due to the light emitted by them. According to the Doppler law, the frequency of an electromagnetic wave emitted by a source moving away from the observer is decreased. This is the so-called "redsqhift" of the emitted light to the red end of the visible spectrum.

Actually, the first observation of the redshift of Andromeda Nebula was made by Slipher in 1913 [21], i.e., before the development of the general theory of relativity by Einstein. He interpreted this observation in the spirit of Doppler effect that the Andromeda Nebula moves away from the Earth. The universal law which connects the redshift in the spectra of remote galaxies with the universe expansion was experimentally discovered in 1927 by Georges Lemaêtre [22] and in 1929 by Edwin Hubble [23] who finally validated that the nebulas are the galaxies outside the Milky Way. According to this law, the velocity of a remote galaxy is proportional to a distance to it, $v = HD$, where H is the Hubble constant which can be expressed via the scale factor as

$$H = H(t) = \frac{1}{a(t)} \frac{da(t)}{dt}, \qquad (12)$$

i.e., it is in fact time-dependent.

The discovery of the redshifts in the spectra of remote galaxies was the first experimental confirmation of the universe expansion predicted by Alexander Friedmann. The next important step was made by George Gamow who elaborated the theory of a hot universe which provided a possibility to explain the creation of chemical elements and the formation of galaxies [24]. The basic point of Gamow's theory was an assumption that the early universe was dominated not by the dust matter but by radiation with the equation of state $P(t) = \varepsilon(t)/3$. In the framework of Gamow's theory of hot universe, Ralph Alpher and Robert Herman predicted the existence of the background relic radiation [25]. The discovery of this radiation served as the second most important experimental confirmation of the expansion of the universe.

The cosmic microwave background electromagnetic radiation, called also the "relic radiation", was discovered in 1965 by Arno Penzias and Robert Wilson [26]. It fills all space and was created in the epoch of formation of first atoms. The observation of relic radiation confirmed the origin of the universe from the cosmological singularity predicted by Alexander Friedmann as a result of the so-called "Big Bang". Based on the theory of the hot universe, it became possible to describe the various eras in the universe evolution starting from the Electroweak Era followed by the Particle Era, the Era of Nucleosynthesis, Eras of Nuclei, Atoms, and, finally, by the Era of Galaxies. This covers the period of the universe evolution from approximately 10^{-33} s after the cosmological singularity to about 14 billion years which is the present age of the universe. As to the very early stage from 0 to 10^{-33} s, it remained a mystery and could not be explained on the basis of the general theory of relativity.

6. Cosmic Inflation and Creation of the Universe from Quantum Vacuum

As was noted above, the basic assumption, used in Friedmann's cosmology and in the theory of hot universe, is that the space is homogeneous and isotropic. This assumption was confirmed by the approximately homogeneous and isotropic large-scale distribution of galaxies and, more importantly, by the properties of relic radiation. It was found that the relic radiation has a blackbody thermal spectrum at $T = 2.726 \pm 0.001$ K average temperature and the variations of this temperature measured from different directions in the sky do not exceed $\Delta T/T \sim 10^{-5}$.

As discussed in Section 5, at the early stages of its evolution the universe was filled with radiation possessing the equation of state $P = \varepsilon/3$. In this case the solution of Friedmann equation (6) is given by $a(t) \sim \sqrt{t}$ and the respective energy density behaves as $\varepsilon(t) \sim 1/t^2$ when t goes to zero.

These behaviors, however, create a problem. The point is that if t is decreased down to the Planck time defined as

$$t_{Pl} = \sqrt{\frac{\hbar G}{c^5}} = 5.39 \times 10^{-44} \text{ s}, \qquad (13)$$

the size of the universe turned out to be unexpectedly large $a(t_{Pl}) \sim 10^{-3}$ cm as compared to the Planck length,

$$l_{Pl} = t_{Pl} c = \sqrt{\frac{\hbar G}{c^3}} = 1.62 \times 10^{-33} \text{ cm}, \qquad (14)$$

traveled by light during the Planck time.

Thus, if the above scale factor were applicable down to $t = 0$, at Planck time the universe would consist of the 10^{89} causally disconnected parts. This is in contradiction with the fact that the relic radiation in all places and all directions in the sky has the same temperature—the so-called "horizon problem".

The horizon problem cannot be solved in the framework of the general theory of relativity. The point is that this is the classical theory and the space-time scales of the order of Planck length and Planck time are outside the region of its applicability. In the absence of quantum theory of gravitation, which is still unavailable in spite or repeated attempts to develop it undertaken during several decades, some semiclassical approaches are believed to lead to at least a partial solution of the problem.

In 1981, Alan Guth [27] found that the symmetry breaking caused by the scalar fields introduced in particle physics can cause the period of exponentially fast expansion of the universe. During this period, the scale factor varies as $\exp(t)$ rather than $\sqrt{t}$. As a result, at the Planck time the universe has the Planck size which solves the horizon problem. It was Guth who introduced the term "inflation" for the exponentially fast expanding universe. The scalar field responsible for the inflation process was called the "inflaton field". The theory of inflation was further developed by Andrei Linde [28]. Actually, the possibility of an early exponential expansion of the universe was predicted before Guth by Sergey Mamaev and one of the authors of this paper [29] and, independently, by Alexei Starobinsky [30] based on the semiclassical Einstein's equations (see below). In doing so, the scale factor was expressed either via the proper synchronous time t [30] or, equivalently, via the conformal time η [29].

Due to a very fast expansion of the universe during the inflationary stage, the energy density of matter becomes very low. The conversion of the energy density of oscillating inflaton field into that of usual matter is called "reheating" after inflation. The theory of reheating was developed by Lev Kofman, Anfrei Linde, and Alexei Starobinsky [31] using the effect of resonant particle production in the time-periodic external field revealed earlier by one of the authors and Valentin Frolov [32].

The weak point of the theory of inflation is that the physical nature of inflaton field remains unclear. In this situation, the question arises of whether there are other possibilities for obtaining the period of exponentially fast expansion in the evolution of the early universe. The possible answer to this question was given by the theory of quantum matter fields in curved space-time [33,34]. This theory is applicable under the condition that the gravitational field can be considered as the classical background, i.e., at $t \gg t_{Pl}$, which is already well satisfied at $t \geqslant 10^{-40}$ s. One can assume that at $t \sim 10^{-40}$ s all quantum matter fields are in the vacuum state $|0\rangle$, i.e., the number of particles of different kinds is zero. This does not mean, however, that the vacuum energy density and pressure are zero because vacuum is polarized by the external gravitational field.

The vacuum stress-energy tensor of quantum matter fields with different spins in the homogeneous isotropic space was calculated in the 1970s by several groups of authors. It is common knowledge that the quantities $\langle 0|T_{ik}|0\rangle$ contain the ultraviolet divergences. These divergences can be interpreted in terms of the bare cosmological constant which is connected with the infinitely large energy density of the zero-point oscillations, the bare gravitational constant, and the bare constants in front of the invariant quadratic combinations of the components of Ricci tensor. The finite expressions for the renormalized values of $\langle 0|T_{ik}|0\rangle_{\text{ren}}$, obtained after the removal of divergencies, can be found in [33,34]. Based on these results, the so-called "self-consistent" Einstein equations were considered:

$$R_{ik} - \frac{1}{2}Rg_{ik} = \frac{8\pi G}{c^4}\langle 0|T_{ik}|0\rangle_{\text{ren}}. \tag{15}$$

In these equations, the vacuum of quantized matter fields $|0\rangle$ is polarized by the gravitational field of the homogeneous isotropic space with metric (4) determining the left-hand side of Equation (15). In free Minkowski space-time, the physical energy density and pressure in the vacuum state obtained after discarding of infinities are equal to zero.

However, in an external field (regardless of whether it is electromagnetic or gravitational) after discarding of infinities the vacuum state is polarized, such as a dielectric in an electric field, i.e., it is characterized by some nonzero stress-energy tensor. On the other hand, the gravitational field described by the left-hand side of Equation (15) is created as a source by the vacuum energy density and pressure on the right-hand side. By solving the self-consistent equations (15), one can find the scale factor of the homogeneous isotropic space determined by the quantum vacuum of the matter fields.

It should be stressed that the theoretical approach based on Equation (15) is semi-classical. This means that the gravitational field and the corresponding metrical tensor in Equation (4) are still treated as the classical ones. It is assumed that only the fields of matter (scalar, spinor, vector, etc.) exhibit a quantum behavior. Recall that this approach is applicable at $t \geqslant 10^{-40}$ s. At earlier moments down to the Planck time and to the domain of singularity in the solution of the classical general theory of relativity, one should take into account the effects of quantization of space-time, i.e., the effects of quantum gravity.

The solutions of Equation (15) for the massless matter fields were obtained in [29,30]. As an example, for a scalar field in the space of positive curvature the self-consistent scale factor is given by

$$a(t) = \sqrt{\frac{\hbar G}{360\pi c^3}} \cosh\left(t\sqrt{\frac{360\pi c^5}{\hbar G}}\right). \qquad (16)$$

Using Equations (13) and (14), one can see that for $t > t_{Pl}$ the scale factor (16) takes the form

$$a(t) = \frac{l_{Pl}}{\sqrt{360\pi}} \exp\left(\frac{\sqrt{360\pi}\, t}{t_{Pl}}\right), \qquad (17)$$

i.e., it is the exponentially increasing with time scale factor of the de Sitter space (cf. with Equation (10)). This is the scale factor describing the cosmic inflation obtained on the fundamental grounds of quantum field theory without introducing the inflaton field.

Actually, in this approach the inflationary universe is spontaneously created from the quantum vacuum. It should be noted that the expressions for $\langle 0|T_{ik}|0\rangle_{\text{ren}}$ contain the third and fourth derivatives of the scale factor, which lead to scalar and tensor instabilities. As a result, the de Sitter solution becomes unstable relative to the spatially homogeneous massive scalar modes (scalarons). Using this fact, Alexei Starobinsky [30] constructed the nonsingular cosmological model where the de Sitter universe describing inflation is spontaneously created from the quantum vacuum. Then, due to the generation of scalarons and the decay of scalarons into usual particles, the exponentially fast expansion of the inflationary stage is replaced by the power-type expansion law $a(t) \sim \sqrt{t}$ of the theory of hot universe. According to current concepts, the inflationary stage of the universe expansion lasts from $t \sim 10^{-36}$ s to $\sim 10^{-33}$ s. More precise measurements of the spectrum of relic radiation, planned for the near future, should provide information concerning the gravitational radiation generated during the exponentially fast expansion. This will help to conclusively establish the main properties of the inflationary stage of the universe evolution.

7. Accelerated Expansion of the Universe, Dark Energy and the Quantum Vacuum

According to the commonly accepted views formed by the end of the 20th century, the present stage of the universe evolution is described by the Friedmann equations (6) with $\Lambda = 0$ and a predominance of dust matter having the energy density $\varepsilon(t) = \rho(t)c^2$, where $\rho(t)$ is the mass of matter per unit volume, and zero pressure $P = 0$. In doing so, the fraction of visible matter is by a factor of 5.4 smaller than that of invisible matter, which possesses the same properties as the visible one and is called the "dark matter". We know about the existence of dark matter due to its gravitational action on visible bodies. In this situation, it was expected that due to the gravitational attraction of matter the expansion of the universe should be decelerating.

It was quite unexpected, however, when in 1998 two research teams (Supernova Cosmology Project and High-Z Supernova Search Team) observed clear evidence that, by contrast, the expansion of the universe is accelerating (see the references and discussion below). This means that the velocities of remote galaxies moving away from the Earth increase with time. Actually, the inflationary stage of the universe evolution was also the period of accelerated expansion which, however, lasted for an infinitesimally short period of time. As to the accelerated expansion observed at present, it has already been going on for several billion years.

The matter which causes acceleration of the universe expansion was called "dark energy". Unlike the dark matter, which is not seen but gravitates like ordinary matter, dark energy acts against the gravitational attraction, i.e., it is characterized by the negative pressure. The observed acceleration rate of the universe expansion requires that 68% of the universe should consist of dark energy. As to dark matter and usual visible matter, they constitute approximately 27% and 5% of the universe's energy, respectively. There were many attempts in the literature to understand the physical nature of dark energy by introducing some new hypothetical particles with unusual properties [35]. However, the most popular explanation of the accelerated expansion returns us back to the concept of cosmological constant and quantum vacuum.

As discussed in Section 4, in the absence of background matter, $\varepsilon = P = 0$, the cosmological term Λg_{ik} in the Einstein's equations (3) determines the de Sitter space-time with a scale factor (10). In the homogeneous isotropic space, the presence of this term just corresponds to the energy density and pressure,

$$\varepsilon_\Lambda = \frac{c^4 \Lambda}{8\pi G} > 0, \qquad P_\Lambda = -\frac{c^4 \Lambda}{8\pi G} = -\varepsilon_\Lambda, \qquad (18)$$

i.e., results in some effective negative pressure. When the cosmological term is considered along with the stress-energy tensor of the ordinary matter T_{ik}, it just leads to the required acceleration of the universe expansion. Calculations show that an agreement with the observed rate of acceleration is reached for $\Lambda \approx 2 \times 10^{-52}$ m^{-2} and the corresponding energy density $\varepsilon_\Lambda \approx 10^{-9}$ J/m^3 [36,37].

The question arises what is the nature of the cosmological constant. Actually, it can be considered as one more fundamental constant closely related to the concept of the quantum vacuum [38]. This statement is based on the geometric structure of the vacuum stress-energy tensor of quantized fields,

$$\langle 0|T_{ik}(x)|0\rangle = I g_{ik}, \qquad (19)$$

where I is the infinitely large constant depending on the number, masses, and spins of quantum fields (see the pedagogical derivation of this equation by the method of dimensional regularization in [39]). The structure of Equation (19) is the same as the cosmological term in Einstein's equation (3).

The only difficulty is that the constant I is diverging. By making the cutoff at the Planck momentum $p_{\text{Pl}} = (\hbar c^3/G)^{1/2}$, one obtains the enormously large value of $I \approx 2 \times 10^{68}$ m^{-2} which exceeds the observed cosmological constant Λ by 120 orders of magnitude [36,40]. This discrepancy was called the "vacuum catastrophe" [37].

One can argue, however, that the large value of I is determined by the contribution of virtual particles and, thus, is of no immediate physical meaning. The energy density $c^4 I/(8\pi G)$ determined by the constant I does not gravitate as the bare (non-renormalized) electric charge in quantum electrodynamics is not a source of the measurable electric field. This is also in some analogy to the Casimir effect [41] where only a difference between two infinite energy densities in the presence and in the absence of plates is the source of gravitational interaction [42,43] and gives rise to the measurable force. The quantity I takes the physical (renormalized) value of $\Lambda \approx 2 \times 10^{-52}$ m^{-2} only after the renormalization procedure. These considerations have already received some substantiation in the framework of quantum field theory in curved space-time, but could obtain the fully rigorous justification only after a construction of the quantum theory of gravitation.

There is a lot of different approaches to the problem of cosmological constants in connection with the origin and physical nature of dark energy. The number of articles devoted to this subject is large and we do not aim to review them here (see [44] for an introduction to the field).

8. Discussion

In previous Sections, we reviewed the main scientific concepts regarding the structure of the universe developed in the history of mankind from Ptolemy to Einstein. All of these concepts imply that the universe is stationary and its properties do not vary with time. Although Copernicus and Newton's pictures of the universe are significantly different from Ptolemy's picture, all of these pictures are similar in one basic point: each is time-invariant. This characteristic point was remained untouched by Einstein, who has had in his mind the predetermined aim to obtain the stationary cosmological model in the framework of the general theory of relativity developed by him. Actually, the mathematical formalism of this theory presumed a much more broad spectrum of cosmological models describing the expanding universe. However, the power of tradition was so strong that even a great innovator like Einstein chose to modify his equations by introducing the cosmological term for the sole purpose of keeping the universe stationary.

This gives us an insight into the fundamental importance of the scientific results obtained by Alexander Friedmann 100 years ago. By solving the same equations of the general relativity theory as Einstein, Alexander Friedmann demonstrated that their general cosmological solution describes the expanding universe whereas the stationary solution is only a particular case. This result was so unexpected that Einstein rejected it as a mathematical error, and only after a detailed written explanations passed to him by Friedmann, Einstein had to recognize that, in fact, he himself made an error.

We considered the main facts of Alexander Friedmann's scientific biography which provides an explanation why a mathematician, educated in the traditions of Saint Petersburg mathematical school, was able to make such a fundamental discovery in the field of theoretical physics. After the brief exposition of the properties of expanding universe based on the Friedmann equations, we discussed the main experimental confirmations of the universe expansion, i.e., the discoveries of redshifts in the spectra of remote galaxies and the relic radiation.

Although the theory of the hot universe raised the possibility to describe the main stages of its evolution, the problem of cosmological singularity and the problem of horizon remained unsolved. The solution of these problems suggested by the theory of cosmic inflation links them to the concept of the quantum vacuum. There are reasons to believe that the inflationary stage of the universe expansion is caused by the vacuum quantum effects of fields of matter rather than by some special inflaton field. This point of view may find confirmation in further developments of the quantum theory of gravitation, on the one hand, and by measurements of the spectrum of relic gravitational radiation, on the other hand.

The discovery of the acceleration in the universe expansion resumed an interest to the cosmological term in the Einstein equations originally introduced with a single aim to make the universe stationary. The point is that this term has the same geometrical form as the vacuum stress-energy tensor of quantized matter fields and provides a possible explanation of the observed acceleration of the universe expansion with some definite value of the cosmological constant. There is a great discrepancy between this value and theoretical predictions of quantum field theory which created a discussion in the literature. However, independently of the resolution of this issue, one can argue that the quantum vacuum bears a direct relation to both the earliest and modern stages of the evolution of our universe.

9. Conclusions

To conclude, Alexander Friedmann made a prediction of the universe expansion which radically altered our scientific picture of the world as compared to the previous epochs.

Later, this prediction was confirmed experimentally, and Friedmann equations became the basis of modern cosmology. Because of this, the Friedmann name should be put on a par with the names of Ptolemy and Copernicus, who created the previous, stationary, pictures of the world around us.

Friedmann's discovery was based on the classical general relativity theory and could not take into account the quantum effects. Nowadays, we know that the quantum vacuum plays an important role in the problem of the origin of Friedmann's universe from the initial singularity, governs the process of cosmic inflation, and can be considered as a possible explanation of the observed acceleration of the universe expansion. Future development of quantum gravity will make our current knowledge more complete but the concept of expanding universe created by Alexander Friedmann will forever remain the cornerstone of our picture of the world.

Author Contributions: Investigation, G.L.K. and V.M.M.; writing, G.L.K. and V.M.M. All authors have read and agreed to the published version of the manuscript.

Funding: This work was supported by the Peter the Great Saint Petersburg Polytechnic University in the framework of the Russian State Assignment for Basic Research (Project No. FSEG-2020-0024). The work of V.M.M. was supported by the Kazan Federal University Strategic Academic Leadership Program.

Data Availability Statement: Not applicable.

Conflicts of Interest: The authors declare no conflict of interest.

References

1. Kant, I. *Critique of Practical Reason*; Cambridge University Press: Cambridge, UK, 1997. [CrossRef]
2. Newton, I. *The Mathematical Principles of Natural Philosophy*; Cambridge University Press: Cambridge, UK, 2021.
3. Einstein, A. Zur allgemeinen Relativitätstheorie. *Sitzungsber. Königlich Preuss. Akad. Wiss. (Berlin)* **1915**, *44*, 778–786. Available online: https://archive.org/details/sitzungsberichte1915deut/page/778/ (accessed on 13 August 2022).
4. Einstein, A. On the general theory of relativity. In *The Collected Papers of Albert Einstein. Volume 6: The Berlin Years: Writings, 1914–1917 (English Translation Supplement)*; Klein, M.J., Kox, A.J., Schulman, R., Eds.; Princeton University Press: Princeton, NJ, USA, 1997; pp. 98–106. Available online: http://einsteinpapers.press.princeton.edu/vol6-trans/110 (accessed on 13 August 2022).
5. Einstein, A. Kosmologische Betrachtungen zur allgemeinen Relativitätstheorie. *Sitzungsber. Königlich Preuss. Akad. Wiss.* **1917**, *6*, 142–152. Available online: https://archive.org/details/sitzungsberichte1917deut/page/142/ (accessed on 13 August 2022).
6. Einstein, A. Cosmological considerations in the general theory of relativity. In *The Collected Papers of Albert Einstein. Volume 6: The Berlin Years: Writings, 1914–1917 (English Translation Supplement)*; Klein, M.J., Kox, A.J., Schulman, R., Eds.; Princeton University Press: Princeton, NJ, USA, 1997; pp. 421–432. Available online: http://einsteinpapers.press.princeton.edu/vol6-trans/433 (accessed on 13 August 2022).
7. Friedmann, A.A. Über die Krümmung des Raumes. *Z. Phys.* **1922**, *10*, 377–386. [CrossRef]
8. Frenkel', V.Y. Aleksandr Aleksandrovich Fridman (Friedmann): A biographical essay. *Uspekhi Fiz. Nauk* **1988**, *155*, 481–516; Translated: *Sov. Phys. Usp.* **1988**, *31*, 645–665. [CrossRef]
9. Tropp, E.A.; Frenkel, V.Y.; Chernin, A.D. *Alexander A. Friedmann: The Man Who Made the Universe Expand*; Cambridge University Press: Cambridge, UK, 2006.
10. Tamarkine, J.; Friedmann, A. Sur les congruences du second degré et les nombres de Bernoulli. *Math. Ann.* **1906**, *62*, 409–412. [CrossRef]
11. Friedmann, A.A. Über die Möglichkeit einer Welt mit konstanter negativer Krümmung des Raumes. *Z. Phys.* **1924**, *21*, 326–332. [CrossRef]
12. de Sitter, W. On Einstein's theory of gravitation and its astronomical consequences. First paper. *Mon. Not. R. Astron. Soc.* **1916**, *76*, 699–728. [CrossRef]
13. Landau, E.M.; Lifshitz, E.M. *The Classical Theory of Fields*; Pergamon: Oxford, UK, 1971.
14. Zeldovich, Y.B.; Novikov, I.D. *The Structure and Evolution of the Universe*; University of Chicago Press: Chicago, IL, USA, 1983.
15. Einstein, A. Bemerkung zu der Arbeit von A. Friedmann "Über die Krümmung des Raumes". *Z. Phys.* **1922**, *11*, 326. [CrossRef]
16. Einstein, A. Comment on A. Friedmann's paper: "On the curvature of space". In *The Collected Papers of Albert Einstein. Volume 13: The Berlin Years: Writings & Correspondence, January 1922–March 1923 (English Translation Supplement)*; Kormos Buchwald, D., Illy, J., Rosenkranz, Z., Sauer, T., Eds.; Princeton University Press: Princeton, NJ, USA, 2013; pp. 271–272. Available online: http://einsteinpapers.press.princeton.edu/vol13-trans/301 (accessed on 13 August 2022).

17. Friedmann, A. From Alexander Friedmann. In *The Collected Papers of Albert Einstein. Volume 13: The Berlin Years: Writings & Correspondence, January 1922–March 1923*; Kormos Buchwald, D., Illy, J., Rosenkranz, Z., Sauer, T., Eds.; Princeton University Press: Princeton, NJ, USA, 2013; pp. 601–604. Available online: http://einsteinpapers.press.princeton.edu/vol13-doc/697 (accessed on 13 August 2022). (In German)
18. Friedmann, A. From Alexander Friedmann. In *The Collected Papers of Albert Einstein. Volume 13: The Berlin Years: Writings & Correspondence, January 1922 – March 1923 (English Translation Supplement)*; Kormos Buchwald, D., Illy, J., Rosenkranz, Z., Sauer, T., Eds.; Princeton University Press: Princeton, NJ, USA, 2013; pp. 333–335. Available online: http://einsteinpapers.press.princeton.edu/vol13-trans/363 (accessed on 13 August 2022).
19. Einstein, A. Notiz zu der Arbeit von A. Friedmann "Über die Krümmung des Raumes". *Z. Phys.* **1923**, *16*, 228. [CrossRef]
20. Einstein, A. Note to the paper by A. Friedmann "On the curvature of space". In *The Collected Papers of Albert Einstein. Volume 14: The Berlin Years: Writings & Correspondence, April 1923–May 1925 (English Translation Supplement)*; Kormos Buchwald, D., Illy, J., Rosenkranz, Z., Sauer, T., Eds.; Princeton University Press: Princeton, NJ, USA, 2015; p. 47. Available online: http://einsteinpapers.press.princeton.edu/vol14-trans/77 (accessed on 13 August 2022).
21. Slipher, V.M. The radial velocity of the Andromeda Nebula. *Lowell Observat. Bull.* **1913**, *2*, 56–57. Available online: https://articles.adsabs.harvard.edu/pdf/1913LowOB...2...56S (accessed on 13 August 2022).
22. Lemaître, G. Un univers homogène de masse constante et de rayon croissant, rendant compte de la vitesse radiale des nébuleuses extra-galactiques. *Ann. Soc. Sci. Brux. A* **1927**, *47*, 49–59. Available online: https://adsabs.harvard.edu/full/1927ASSB...47...49L (accessed on 13 August 2022).
23. Hubble, E. A relation between distance and radial velocity among extra-galactic nebulae. *Proc. Nat. Acad. Sci. USA* **1929**, *15*, 168–173. [CrossRef] [PubMed]
24. Gamow, G. Expanding universe and the origin of elements. *Phys. Rev.* **1946**, *70*, 572–573. [CrossRef]
25. Alpher, R.A.; Herman, R.C. Evolution of the Universe. *Nature* **1948**, *162*, 774–775. [CrossRef]
26. Penzias, A.A.; Wilson, R.W. A Measurement of Excess Antenna Temperature at 4080 Mc/s. *Astrophys. J. Lett.* **1965**, *142*, 419–421. [CrossRef]
27. Guth, A.H. Inflationary universe: A possible solution to the horizon and flatness problems. *Phys. Rev. D* **1981**, *23*, 347–356. [CrossRef]
28. Linde, A.D. *Particle Physics and Inflationary Cosmology*; Harwood: Chur, Switzerland, 1990. Available online: https://arxiv.org/abs/hep-th/0503203 (accessed on 13 August 2022).
29. Mamayev, S.G.; Mostepanenko V.M. Isotropic cosmological models determined by the vacuum quantum effects. *Zh. Eksp. Teor. Fiz.* **1980**, *78*, 20–27; Translated: *Sov. Phys. JETP* **1980**, *51*, 9–13. Available online: http://jetp.ras.ru/cgi-bin/e/index/e/51/1/p9?a=list (accessed on 13 August 2022).
30. Starobinsky, A.A. A new type of isotropic cosmological models without singularity. *Phys. Lett. A* **1980**, *91*, 99–102. [CrossRef]
31. Kofman, L.; Linde, A.D.; Starobinsky, A.A. Towards the theory of reheating after inflation. *Phys. Rev. D* **1997**, *56*, 3258–3295. [CrossRef]
32. Mostepanenko, V.M.; Frolov, V.M. Production of particles from vacuum by a uniform electric-field with periodic time-dependence. *Yad. Fiz.* **1974**, *19*, 885–896; Translated: *Sov. J. Nucl. Phys.* **1974**, *19*, 451–456.
33. Birrell, N.D.; Davies, P.C.D. *Quantum Fields in Curved Space*; Cambridge University Press: Cambridge, UK, 1982. [CrossRef]
34. Grib, A.A.; Mamayev, S.G.; Mostepanenko, V.M. *Vacuum Quantum Effects in Strong Fields*; Friedmann Laboratory Publishing: St. Petersburg, Russia, 1994.
35. Klimchitskaya, G.L. Constraints on theoretical predictions beyond the Standard Model from the Casimir effect and some other tabletop physics. *Universe* **2021**, *7*, 47. [CrossRef]
36. Frieman, J.A.; Turner, M.S.; Huterer, D. Dark energy and the accelerating universe. *Annu. Rev. Astron. Astrophys.* **2008**, *46*, 385–432. [CrossRef]
37. Adler, R.J.; Casey, B.; Jacob, O.C. Vacuum catastrophe: An elementary exposition of the cosmological constant problem. *Am. J. Phys.* **1995**, *63*, 620–626. [CrossRef]
38. Zel'dovich, Y.B. The cosmological constant and the theory of elementary particles. *Uspekhi Fiz. Nauk* **1968**, *95*, 209–230; Translated: *Sov. Phys. Usp.* **1968**, *11*, 381–393. [CrossRef]
39. Mostepanenko, V.M.; Klimchitskaya, G.L. Whether an enormously large energy density of the quantum vacuum is catastrophic. *Symmetry* **2019**, *11*, 314. [CrossRef]
40. Weinberg, S. The cosmological constant problem. *Rev. Mod. Phys.* **1989**, *61*, 1–23. [CrossRef]
41. Bordag, M.; Klimchitskaya, G.L.; Mohideen, U.; Mostepanenko, V.M. *Advances in the Casimir Effect*; Oxford University Press: Oxford, UK, 2015. [CrossRef]
42. Bimonte, G.; Calloni, E.; Esposito, G.; Rosa, L. Energy-momentum tensor for a Casimir apparatus in a weak gravitational field. *Phys. Rev. D* **2006**, *74*, 085011. [CrossRef]
43. Bimonte, G.; Calloni, E.; Esposito, G.; Rosa, L. Relativistic mechanics of Casimir apparatuses in a weak gravitational field. *Phys. Rev. D* **2007**, *76*, 025008. [CrossRef]
44. Peebles, P.J.E.; Ratra, B. The cosmological constant and dark energy. *Rev. Mod. Phys.* **2003**, *75*, 559–606. [CrossRef]

Article

New Insights into the Lamb Shift: The Spectral Density of the Shift

G. Jordan Maclay

Quantum Fields LLC, 147 Hunt Club Drive, St. Charles, IL 60174, USA; jordanmaclay@quantumfields.com

Abstract: In an atom, the interaction of a bound electron with the vacuum fluctuations of the electromagnetic field leads to complex shifts in the energy levels of the electron, with the real part of the shift corresponding to a shift in the energy level and the imaginary part to the width of the energy level. The most celebrated radiative shift is the Lamb shift between the $2s_{1/2}$ and the $2p_{1/2}$ levels of the hydrogen atom. The measurement of this shift in 1947 by Willis Lamb Jr. proved that the prediction by Dirac theory that the energy levels were degenerate was incorrect. Hans Bethe's non-relativistic calculation of the shift using second-order perturbation theory demonstrated the renormalization process required to deal with the divergences plaguing the existing theories and led to the understanding that it was essential for theory to include interactions with the zero-point quantum vacuum field. This was the birth of modern quantum electrodynamics (QED). Numerous calculations of the Lamb shift followed including relativistic and covariant calculations, all of which contain a nonrelativistic contribution equal to that computed by Bethe. The semi-quantitative models for the radiative shift of Welton and Power, which were developed in an effort to demonstrate physical mechanisms by which vacuum fluctuations lead to the shift, are also considered here. This paper describes a calculation of the shift using a group theoretical approach which gives the shift as an integral over frequency of a function, which is called the "spectral density of the shift." The energy shift computed by group theory is equivalent to that derived by Bethe yet, unlike in other calculations of the non-relativistic radiative shift, no sum over a complete set of states is required. The spectral density, which is obtained by a relatively simple computation, reveals how different frequencies of vacuum fluctuations contribute to the total energy shift. The analysis shows, for example, that half the radiative shift for the ground state 1S level in H comes from virtual photon energies below 9700 eV, and that the expressions of Power and Welton have the correct high-frequency behavior, but not the correct low-frequency behavior, although they do give approximately the correct value for the total shift.

Keywords: Bethe; radiative shift; shift spectral density; spectral density; vacuum fluctuations; vacuum field; mass renormalization; Lamb shift; QED; radiative reaction; zero point fluctuations; hydrogen atom

Citation: Maclay, G.J. New Insights into the Lamb Shift: The Spectral Density of the Shift. *Physics* **2022**, *4*, 1253–1277. https://doi.org/10.3390/physics4040081

Received: 22 April 2022
Accepted: 31 August 2022
Published: 19 October 2022

Publisher's Note: MDPI stays neutral with regard to jurisdictional claims in published maps and institutional affiliations.

Copyright: © 2022 by the author. Licensee MDPI, Basel, Switzerland. This article is an open access article distributed under the terms and conditions of the Creative Commons Attribution (CC BY) license (https://creativecommons.org/licenses/by/4.0/).

1. Introduction

In astronomy, in quantum theory, and in quantum electrodynamics (QED), there have been periods of great progress in which solutions to challenging problems have been obtained, and the fields have moved forward. However, in some cases getting the right answers can still leave fundamental questions unanswered. The Big Bang explained the origin of cosmic background radiation but left the problem of why the universe appears to be made of matter and not equal amounts of matter and antimatter [1]. In quantum theory, physicists can compute the behavior of atoms yet cannot describe a measurement in a self-consistent way [2], or make sense of the collapse of a photon wavefunction from a near infinite volume to a point [3]. In quantum electrodynamics, we can compute the Lamb shift of the H atom to 15 decimal places [4], yet we are left with the paradox of using perturbation theory to remove infinite terms or to understand a quantum vacuum with infinite energy. In this paper, several different approaches to the computation of the non-relativistic Lamb shift are examined. For these approaches, the Lamb shift can be expressed in different ways as an integral over the frequency of a spectral density. This paper analyzes of the differences in the non-relativistic spectral densities for the different approaches as a function of frequency and compares the spectral densities to those obtained using a group theoretical analysis. The integral of the spectral density over all frequencies gives the corresponding value of the non-relativistic Lamb shift.

Feynman called the three-page long 1947 non-relativistic Lamb shift calculation by Hans Bethe the most important calculation in quantum electrodynamics because it tamed the infinities plaguing earlier attempts [5,6]. When the sum over all states is evaluated numerically, it gives a finite prediction that agreed with experiment [7,8]. Dirac said it "fundamentally changed the nature of theoretical physics" [6]. Quoting the classic text of *Quantum Electrodynamics*, Volume 4 in the Landau and Lifshiftz series on Theoretical Physics: "This work provided the initial stimulus for the whole subsequent development of quantum electrodynamics" [5]. Yet when this calculation is explored more deeply, questions arise about it and about other fundamentally different calculations of the Lamb shift, for example, those by Welton [9] and Power [10], that employ different conceptual approaches that have similar high-frequency but different low-frequency behavior from Bethe's result yet give approximately the same value for the level shift [6]. These three approaches to the Lamb shift and the corresponding vacuum energy densities have also been considered in [11].

There is an intimate relationship between radiative shifts and vacuum fluctuations. The shift can be interpreted as arising from virtual transitions induced by the quantum fluctuations of the electromagnetic field. Since the vacuum field contains all frequencies, virtual transitions to all states, bound and scattering, are possible. These short-lived virtual transition result in a slight shift in the average energy of the atom, a shift which is the Lamb shift [12]. The Lamb shift can also be described as an interaction of the electron with its own radiation field, yielding the same results as the vacuum field [6].

Shortly after Bethe's calculation, Dyson published, as a problem assigned by Bethe, a calculation of the Lamb shift for a spinless electron [13]. Formal and rigorous relativistic perturbation theory calculations to first-order in the radiation field and to fourth order in the ratio of the velocity of the atomic electron to the velocity of light, based on the Dirac equation, including spin and relativistic effects, were carried out independently in 1949 by J. French and V. Weisskopf [14] and N. Kroll and W. Lamb [15]. They used a high-frequency cutoff for the non-relativistic contribution which also acted as a low-frequency cutoff for the relativistic portion, and when both contributions were added together, the sum was independent of the cutoff, and equaled a term corresponding to the Bethe log plus constants corresponding to relativistic corrections. Although these calculations were difficult and cumbersome, they have stood the test of time [16]. These relativistic calculations gave a value of 1052 MHz for the Lamb shift, compared to Bethe's result of 1040 MHz, about a 1% difference. The determination of the second-order corrections for the Lamb shift is quite complicated. The most complete tabulation and systematic derivation was carried out by Erickson and Yennie [17]. Today the Lamb shift is computed precisely to over 10 decimal places to be about 1057 MHz, the largest error being due to the uncertainty in the radius of the proton [4].

At about the same time as the first relativistic calculations were published, Schwinger published a general covariant approach to quantum electrodynamics which he applied to the Lamb shift computation [18]. Within a year, three different approaches to quantum electrodynamics were independently developed that were relativistic and could deal with divergences with some success. Schwinger, Tomonaga, and Feynman each had proposed a manifestly covariant method, and shown its capability to address a broader range of QED problems than just the energy levels of the H atom [19]. Freeman Dyson showed that these three methods had essential similarities and were mutually consistent [20].

The classic text, *Quantum Electrodynamics*, Volume 4 of the Landau and Lifshitz series on Theoretical Physics, second edition published in 1979, does the Lamb shift calculation by dividing the vacuum energy spectrum into a low and high-frequency region with a cutoff [5]. For the high-energy region, they use scattering theory based on Schwinger's approach. For the low-frequency region they use non-relativistic kinematics and start with the expression for dipole radiation, which they transform with some mathematical operations on delta function; then they integrate over frequency and obtain essentially the same result as Bethe plus constants corresponding to vacuum polarization and other corrections that are not considered here. When the high and low-frequency results are added, the cutoff cancels. The final result contains the Bethe log. A slightly different approach is taken but with similar results in [21]. Another important text, published in 1980 by Itzykson and Zuber, starts with a modified Dirac equation with a fictitious photon mass, and that includes vacuum polarization and vertex corrections for an electron in a slowly varying external field [22]. This equation yields a level shift with two terms that correspond to the high-energy region and low-energy region. After evaluation, the sum of the terms is taken, the fictitious mass cancels, and the result is the Bethe log and a constant.

Modern QED texts use similar approaches. Weinberg uses a fictitious photon mass in the photon propagator which leads to a high and low-energy term [16]. When added, the fictitious mass cancels, yielding a constant and the Bethe log term. The evolution of Lamb shift calculations is outlined in [12] and the detailed status is in [23].

For the purposes of the present paper, which is a non-relativistic calculation of the radiative Lamb shift, not including vacuum polarization, it is important to note that Bethe's non-relativistic result and the numerically evaluated Bethe log have been obtained in virtually all Lamb shift calculations, including all relativistic calculations [5,6,16,22]. The Bethe log represents the non-relativistic radiative contribution in all these diverse calculations. The Lamb shift spectral density obtained here can be integrated to obtain the non-relativistic Lamb shift, a quantity given by the Bethe logarithm. Furthermore, the integral can be taken to the desired energy to be consistent with a particular relativistic calculation. Thus, the spectral density computed is a standalone fundamental quantity that relates to all Lamb shift calculations described in the literature. Consequently, there is no need to consider the numerous relativistic calculations further in this paper.

Bethe's non-relativistic calculation to order α in the vacuum field (one virtual photon) was based on second-order perturbation theory applied to the minimal coupling of the atom with the vacuum field, $(e/mc)\mathbf{A} \cdot \mathbf{p}$, and a dipole approximation; here m, e, and $\mathbf{p}$ are the mass, charge and momentum of the electron, respectively, c is the speed of light in a vacuum, and $\mathbf{A}$ is the vector potential. This interaction leads to the emission and absorption of virtual photons corresponding to virtual transitions. The shift is expressed as a sum over the infinite number of intermediate states, bound and scattering, reached by virtual transitions. The predicted shift is divergent, but Bethe subtracted the term that corresponded to the linearly divergent vacuum energy shift for a free bare electron, essentially doing a mass renormalization to remove this higher-order divergence in the spectral density for the shift. For S states, the resulting spectral density has a 1/frequency behavior for frequencies above about 1000 eV/$\hbar$, (with $\hbar$ the reduced Planck's constant) giving a logarithmic divergence in the shift. Bethe used a high-frequency cutoff of $\omega = mc^2/\hbar$.

The models of Welton and Power embody different perspectives on the role of vacuum fluctuations than the Bethe calculation. Itzykson and Zuber note: "Charged particles interact with the fluctuations of the quantized electromagnetic field ... we may give, following Welton, a qualitative description of the main effect: the Lamb shift" [22]. Welton's semi-quantitative model for computing the Lamb shift was based on the perturbation of the motion of a bound electron in the H atom due to the quantum vacuum fluctuations altering the location of the electron, which resulted in a slight shift of the bound state energy [6,9,12]. This simplified intuitive model predicts a spectral density proportional to 1/frequency for all frequencies and a shift only for S states. The approach of Feynman [24], interpreted by Power [10], considers a large box containing H atoms and is based on the shift in the energy in the quantum vacuum field due to the change in the index of refraction arising from the presence of H atoms. This approach predicts that the shift in the energy in the vacuum field around the H atoms exactly equals the radiative shift [6,11]. It gives a spectral density with the same high-frequency dependence as Bethe, but a different low-frequency dependence. A similar calculation to Power's models the Lamb shift as a Stark shift [6].

The Lamb shift has been previously computed using O(4) symmetry [25] and by using SO(4,2) symmetry [26] using a different approach from that presented here. In both these cases, the Bethe log is obtained for low-lying states as a converging series. The results presented here describe a calculation of the Lamb shift for all states that is based on a SO(4,2) group theoretical analysis of the H atom that allows us to determine the dependence of the shift on the frequency with no sum over states [27]. The spectral shift due to all virtual transitions is computed analytically.

The degeneracy group of the non-relativistic H atom is O(4), with generators angular momentum operator L and Runge-Lenz vector A. A representation of O(4) of dimension n^2 exists for each value of the principal quantum number n, where the angular momentum L has values from 0 to $n-1$, and there are $2L+1$ possible values of $L_z = m$, the azimuthal quantum number. If this group is extended by adding a 4-vector of generators the non-invariance group SO(4,1) results, which has representations that include all states of different n and L and operators that connect states with different principal quantum numbers. Adding a 5-vector of additional generators produces the group SO(4,2) and allows us to express Schrödinger's equation in terms of the new generators, and to make effective group theoretical calculations [27]. The basis states used permit both bound and scattering states to be included seamlessly [28] and no sum over states appears in the final expression for the spectral density. One advantage of this approach is that for each energy level it is possible to readily compute a spectral density for the shift whose integral over frequency from 0 to $mc^2/2\pi\hbar$ is the radiative shift that includes transitions to all possible states. This reveals how different frequencies of the vacuum field contribute to the radiative shift.

In this paper, the different results of Bethe, Welton and Power are compared to the group theoretical spectral density for the non-relativistic Lamb shift for the 1S ground state, and the 2S and 2P levels. With this new picture of the Lamb shift, differences between the various approaches are seen. Knowing the spectral density of the shift provides new insights into understanding the Lamb shift.

2. Background of Radiative Shift Calculations

The first calculation of the Lamb shift of a hydrogen atom was carried out by Bethe in 1947 [8], who assumed the shift was due to the interaction of the atom with the vacuum field. He calculated the shift using second-order perturbation theory, assuming that there was minimal coupling in the Hamiltonian

$$H_{\text{int}} = -\frac{e}{mc}\mathbf{A}\cdot\mathbf{p} + \frac{e^2}{2mc^2}\mathbf{A}^2, \qquad (1)$$

where $\mathbf{A}$ is the vector potential in the dipole approximation for the vacuum field in a large quantization volume V:

$$\mathbf{A} = \sum_{\mathbf{k},\lambda}\left(\frac{2\pi\hbar}{\omega_k V}\right)^{1/2}(a_{\mathbf{k},\lambda} + a^{\dagger}_{\mathbf{k},\lambda})\mathbf{e}_{\mathbf{k}\lambda}, \qquad (2)$$

where the sum is over the virtual photon frequency ω_k, and the polarization λ; $a_{\mathbf{k},\lambda}$ and $a^{\dagger}_{\mathbf{k},\lambda}$ are the annihilation and creation operators, and $\mathbf{e}_{\mathbf{k}\lambda}$ is a unit vector in the direction of polarization of the electric field. The shift from the $\mathbf{A}^2$ term is independent of the state of the atom and is therefore neglected. The total shift $\Delta E_{n\text{Tot}}$ for energy level n of the atom in state $|n\rangle$ is given by second-order perturbation theory as

$$\Delta E_{n\text{Tot}} = -\frac{2}{3\pi^2}\frac{\alpha}{m^2c^2}\sum_{m}|\mathbf{P}_{mn}|^2\int\frac{E\,dE}{E_m - E_n + E}, \qquad (3)$$

where the quantum vacuum field energy is $E = \hbar\omega$ and the momentum matrix elements are $|\mathbf{p}_{mn}| = |\langle m|\mathbf{p}|n\rangle|$. The sum is over all intermediate states $|m\rangle$, scattering and bound, where $m \neq n$ and the integral is over the energy E of the vacuum field. The fine structure constant is $\alpha = e^2/\hbar c$.

The integrand in Equation (3) has a linear divergence. Bethe observed that this divergence corresponded to the integral that occurs when the binding energy vanishes so $(E_m - E_n) \to 0$ and the electrons are free:

$$\Delta E_{\text{free}} = -\frac{2}{3\pi^2}\frac{\alpha}{m^2 c^2}\sum_m |\mathbf{p}_{mn}|^2 \int dE. \quad (4)$$

He subtracted this divergent term ΔE_{free} from the total shift $\Delta E_{n\text{Tot}}$,

$$\Delta E_{nL} = \Delta E_{n\text{Tot}} - \Delta E_{\text{free}}, \quad (5)$$

to obtain a finite observable shift ΔE_{nL} for the state $|nL\rangle$:

$$\Delta E_{nL} = \frac{2\alpha}{3\pi(mc)^2}\sum_m^s |\mathbf{p}_{mn}|^2 \int_0^{\hbar\omega_c} dE\, \frac{(E_m - E_n)}{E_m - E_n + E - i\epsilon}, \quad (6)$$

where ω_C is a cutoff frequency for the integration that Bethe took as $\hbar\omega_c = mc^2$. Using an idea from Kramers [6,29], Bethe did this renormalization, taking the difference between the terms with a potential present and without a potential present, essentially performing the free electron mass renormalization. He reasoned that relativistic retardation could be neglected and the radiative shift could be reasonably approximated using a non-relativistic approach and he cut the integration off at an energy corresponding to the mass of the electron. He obtained a finite result that required a numerical calculation over all states, bound and scattering, that gave good agreement with measurements [7,8,30].

The spectral density in the Bethe formalism, which is analysed in this paper, is the quantity in Equation (6) being integrated over E. It includes the sum over states m and the constants. The term for m represents the contribution to the Lamb shift for the virtual transition from state n to state m. Note that since the ground state is the lowest state, all intermediate states have higher energies so the ground state shift has to be positive.

For the purposes of comparison to the other calculations of the Lamb shift, it is helpful to show the next steps Bethe took to evaluate the shift ΔE_n for S states, which have the largest shifts. Note that the spectral density in Equation (6) that will be analyzed is not affected by the subsequent approximations Bethe made to evaluate the integral. First, the $E-$ integration is carried out:

$$\Delta E_n^{\text{Bethe}} = \frac{2\alpha}{3\pi}\left(\frac{1}{mc}\right)^2 \sum_m |\mathbf{p}_{nm}|^2 (E_m - E_n) \ln\frac{(mc^2 + E_m - E_n)}{|E_m - E_n|}. \quad (7)$$

To simplify the evaluation Bethe assumed $|E_m - E_n| \ll mc^2$ in the logarithm and that the logarithm would vary slowly with the index m so it could be replaced by an average value,

$$\widehat{\Delta E}_n^{\text{Bethe}} = \frac{2\alpha}{3\pi}\left(\frac{1}{mc}\right)^2 \ln\frac{mc^2}{|E_m - E_n|_{\text{ave}}} \sum_m |\mathbf{p}_{nm}|^2 (E_m - E_n), \quad (8)$$

where the hat over the ΔE indicates this is an approximation to Equation (7). Now that the E-integration is done, the spectral density is no longer manifest. The summation can be evaluated using the dipole sum rule:

$$2\sum_m^s |\mathbf{p}_{nm}|^2 (E_m - E_n) = \hbar^2 \langle n|\nabla^2 V|n\rangle. \quad (9)$$

where $\langle ... \rangle$ indicates an expectation value. The Laplacian with a Coulomb potential $V = -Ze^2/r$ is $\nabla^2 V(r) = 4\pi Ze^2 \delta(\mathbf{r})$ which gives

$$\left\langle n \left| \nabla^2 V \right| n \right\rangle = 4\pi Ze^2 |\psi_n(0)|^2, \tag{10}$$

where Z is the positive charge of the nucleus, $\psi(r)$ is the wave function for the Coulomb potential and $|\psi_n(0)|^2$ is zero except for S states,

$$|\psi_n(0)|^2 = \frac{1}{\pi} \left(\frac{Z\alpha mc}{n\hbar} \right)^3. \tag{11}$$

For S states, this gives an energy shift equal to [6]

$$\widehat{\Delta E}_n^{\text{Bethe}} = \frac{4mc^2}{3\pi} \alpha(Z\alpha)^4 \frac{1}{n^3} \ln \frac{mc^2}{|E_m - E_n|_{\text{ave}}}. \tag{12}$$

The so-called "Bethe log" for an S state with principal quantum number n is

$$\ln \frac{mc^2}{|E_m - E_n|_{\text{ave}}} = \frac{\sum_m |\mathbf{p}_{nm}|^2 (E_m - E_n) \ln \frac{mc^2}{|E_m - E_n|}}{\sum_m |\mathbf{p}_{nm}|^2 (E_m - E_n)}, \tag{13}$$

where the sum is over all states, bound and scattering. Bethe also has extended the formalism for shifts for states that are not S states [30].

Regarding the approximations Bethe made to obtain Equation (8) from Equation (7) and the use of the Bethe log Equation (13) he commented: "The important values of $|E_m - E_n|$ will be of the order of the ground state binding energy for a hydrogenic atom. This energy is very small compared to mc^2 so the log [in Equation (7) here] is very large and not sensitive to the exact value of $(E_m - E_n)$. In the numerator we neglect $(E_m - E_n)$ altogether and replace it by an average energy [30]". This study shows that Bethe was correct that the relative contribution from energies of the order of the ground state is very important, but the contribution from higher energy scattering states is quite significant, and therefore the approximation $|E_m - E_n| \ll mc^2$ is not valid for higher energy scattering states for which $\bar{E}_m$ increases to the value mc^2. I am not aware of any quantitative estimates of the error in the approximation. The difference, 0.3%, between the value obtained here for the total 1S shift and that of Bethe may be due to this approximation, although this has not been verified. On the other hand, Bethe's approximation may have made his non-relativistic approach viable.

To provide a more intuitive and qualitative physical picture of the shift, Welton considered [9] the effect of a zero-point vacuum field on the motion of an electron bound in a Coulomb potential $V(\mathbf{r})$ at a location $\mathbf{r}$. The perturbation $\vec{\xi}=(\xi_x,\xi_y,\xi_z)$ in the position of the bound electron due to the random zero-point vacuum field $\mathbf{E}_0$ causes a variation in the potential energy,

$$V(\mathbf{r}+\vec{\xi}) = V(\mathbf{r}) + \vec{\xi} \cdot \nabla V(\mathbf{r}) + \frac{1}{2}(\vec{\xi} \cdot \nabla)^2 V(\mathbf{r}) + \cdots. \tag{14}$$

Because of the harmonic time dependence of the vacuum field $\langle \vec{\xi} \rangle$ vanishes and the radiative shift is given approximately by the vacuum expectation value of the last term:

$$\Delta E_n^{\text{Welton}} = \frac{\langle \vec{\xi}^2 \rangle}{6} \left\langle \nabla^2 V(\vec{r}) \right\rangle_n. \tag{15}$$

Since the potential has spherical symmetry $\langle \xi_x^2 \rangle = \langle \xi_y^2 \rangle = \langle \xi_z^2 \rangle = \langle \vec{\xi}^2/3 \rangle$. Equation (15) gives $\Delta E_n^{\text{Welton}}$ as the product of two factors, the first depending on the nature of the fluctuations in the position of the bound electron due to the vacuum field and the second depending on the structure of the system. ζ is determined by $m\ddot{\zeta}=e\mathbf{E}_0$, where top double-dot denotes the second order time derivative.

With a Fourier decomposition of E_0 and ζ, and integrating over the frequency distribution of the vacuum field, the vacuum expectation value can be computed[6,12]:

$$\langle(\vec{\zeta})^2\rangle = \frac{2\alpha}{\pi}\left(\frac{\hbar}{mc}\right)^2 \int_0^{mc^2} \frac{dE}{E}. \tag{16}$$

With the results in Equations (10) and (11) the Laplacian in Equation (15) can be evaluated giving a shift for S states equal to [6]:

$$\Delta E_n^{\text{Welton}} = \frac{4mc^2}{3\pi}\alpha(Z\alpha)^4 \frac{1}{n^3} \int_0^{mc^2} \frac{dE}{E}. \tag{17}$$

Equation (17) shows that the spectral density for the Welton approach is proportional to $1/E$. For the upper limit of integration, we assume the maximum wavelength equals the Compton wavelength corresponding to an energy mc^2, which Bethe used. The lower limit of 0 gives a divergent shift, clearly showing that the model of the bound state is deficient at low energies. To give an approximate lower limit, note that the large wavelength modes are sensitive to low-lying electronic states, which suggests a wavelength cutoff about equal to the radius a of the ground state so $E = \hbar c/a = \alpha mc^2 = 4.01$ keV. For the $n = 2$ level of hydrogen, this gives a shift of 660 MHz, about 60% of the observed shift [22], which confirms the model is reasonable. On the other hand, if one happens to compare Equation (17) to Equation (12) it is clear that with the lower limit $|E_m - E_n|_{\text{ave}}$, as defined in the Bethe log Equation (13), Welton's model gives exactly the same total S state shift as in the approximate Bethe formalism Equation (12). With these limits, the RMS (root mean square) amplitude of oscillation of the electron bound in the Coulomb potential, $\sqrt{\langle(\vec{\zeta})^2\rangle}$, is about 60 fm, which is about 0.037% of the mean radius of the 2S electron orbit.

Feynman proposed another approach for computing the Lamb shift based on a fundamental observation about the interaction of matter and the vacuum field [24]. He considered a large box containing a low density of atoms in the quantum vacuum. The atoms cause a change in the index of refraction, which leads to changes in the frequencies of the vacuum field. The wavelengths remain the same. Feynman maintained that the change in the energy of the zero point vacuum field in the box due to the frequency changes resulting from a weak perturbing background of atoms acting as a refracting medium would correspond to the self energy of the atoms, which is precisely the Lamb shift.

Power, based on the suggestion by Feynman, considered the change in vacuum energy when N hydrogen atoms are placed in a volume V using the Kramers-Heisenberg expression for the index of refraction $n(\omega_k)$ [6,10]. The H atoms cause a change in the index of refraction and therefore a change in the frequencies of the vacuum fluctuations present. The corresponding change in vacuum energy ΔE is

$$\Delta E = \sum_k \frac{1}{n(\omega_k)}\frac{1}{2}\hbar\omega_k - \frac{1}{2}\hbar\omega_k, \tag{18}$$

where the sum is over all frequencies ω_k present. For a dilute gas of atoms in a level n, the index of refraction is [6]

$$n(\omega_k) = 1 + \frac{4\pi N}{3\hbar}\sum_m \frac{\omega_{mn}|\mathbf{d}|_{mn}^2}{\omega_{mn}^2 - \omega_k^2}, \tag{19}$$

where $\omega_{mn} = (E_m - E_n)/\hbar$ and $\mathbf{d}_{mn} = e\mathbf{x}_{mn}$, the transition dipole moment. Substituting $n(\omega_k)$ into Equation (18) gives a divergent result for the energy shift. Following Bethe's approach, Power subtracted from ΔE the energy shift for the N free electrons, which equals the shift when $\omega_{mn} \to 0$, with no binding energy. After making this subtraction and converting the sum over ω_k to an integral over ω, and letting $NV \to 1$ the observable shift in energy is obtained [6]:

$$\Delta E_n^{\text{Power}} = -\frac{2}{3\pi c^3} \sum_m \omega_{mn}^3 |\mathbf{d}_{mn}|^2 \int_0^{mc^2/\hbar} \frac{\omega \, d\omega}{\omega_{mn}^2 - \omega^2}. \tag{20}$$

Noting that

$$\langle m|\frac{\mathbf{P}}{m}|n\rangle = \frac{i}{\hbar}\langle m|[H,\mathbf{x}]|n\rangle = \frac{i}{\hbar}(E_m - E_n)\langle m|\mathbf{x}|n\rangle, \tag{21}$$

it follows

$$|\mathbf{P}_{mn}|^2 = m^2 \omega_{mn}^2 |\mathbf{x}_{mn}|^2 = \frac{m^2 \omega_{mn}^2}{e^2}|\mathbf{d}_{mn}|^2. \tag{22}$$

This allows us to write Power's result Equation (20) as

$$\Delta E_n^{\text{Power}} = -\frac{2e^2}{3\pi m^2 c^3} \sum_m \omega_{mn} |\mathbf{P}_{mn}|^2 \int_0^{mc^2/\hbar} \frac{\omega \, d\omega}{\omega_{mn}^2 - \omega^2}. \tag{23}$$

Writing this equation in terms of $E = \hbar \omega$ instead of ω yields:

$$\Delta E_n^{\text{Power}} = -\frac{2\alpha}{3\pi}\left(\frac{1}{mc}\right)^2 \sum_m |\mathbf{P}_{mn}|^2 (E_m - E_n) \int_0^{mc^2} \frac{E\,dE}{(E_m - E_n)^2 - E^2}. \tag{24}$$

Below, in Section 5, this equation will be used to analyze the spectral density for Power's method, showing the spectral density is different from Bethe's at low frequencies but the same at high frequencies. When Equation (24) is integrated with respect to E, taking the principal value, one obtains

$$\Delta E_n^{\text{Power}} = \frac{2\alpha}{3\pi}\left(\frac{1}{mc}\right)^2 \sum_m |\mathbf{P}_{mn}|^2 (E_m - E_n) \ln\left[\frac{mc^2 + (E_m - E_n)}{E_m - E_n} \cdot \frac{mc^2 - (E_m - E_n)}{E_m - E_n}\right]^{1/2}. \tag{25}$$

Except for the argument in the ln function, which corresponds to the upper limit of integration, this is the same as Bethe's expression (7) for the shift. Assuming $mc^2 \gg E_m - E_n$, as Bethe did, then both expressions for the total shift are identical. It is clear, however, that this approximation is not valid at high energies since the second factor in the ln function in Equation (25) may even become less than one making the ln term negative. Feynman's approach highlights the changes in the vacuum field energy due to the interactions with the H atoms.

One assumption in the computation by Power is that the index of refraction in the box containing the atoms is spatially uniform. In Section 6, this assumption will be revisited and a model suggested that predicts for a single atom the changes in the vacuum field energy as a function of position for each spectral component of the radiative shift.

3. Spectral Density of the Lamb Shift

Our goal is to develop an expression for the energy shift of a level, in terms of the generators of the group SO(4,2), that is an integral over frequency. Then the integrand will be the spectral density of the shift, and group theoretical techniques can be used to evaluate it [27]. This approach yields a generating function for the shifts for all levels. The initial focus is on the ground state 1S level as an illustration of the results. At ordinary temperatures and pressures, most atoms are in the ground state. The radiative shift for the 1S level is [27]

$$\Delta E_1 = \frac{4mc^2 \alpha (Z\alpha)^4}{3\pi} \int_0^{\phi_c} d\phi e^\phi \sinh\phi \int_0^\infty ds e^{se^{-\phi}} \frac{d}{ds} \frac{1}{(\coth\frac{s}{2} + \cosh\phi)^2}, \tag{26}$$

where the dimensionless normalized frequency variable ϕ is defined as

$$\phi = \frac{1}{2}\ln\left[1 + \frac{\hbar\omega}{|E_1|}\right], \tag{27}$$

and E_1 is the ground state energy of the H atom -13.6 eV. The cutoff ϕ_c corresponds to $E = \hbar\omega_c = mc^2 = 511$ keV corresponding to the electron mass.

The group theoretical expression for the Lamb shift (26) is directly derived from the Klein-Gordon equations of motion using a non-relativistic dipole approximation, assuming infinite proton mass, and minimal coupling with the vacuum field. Basis states of $(1/Z\alpha)$ are used since they have no scattering states and have the same quantum numbers as the usual bound energy eigenstates [27]. The level shift is obtained as the difference between the mass renormalization for a spinless meson bound in the desired state and the mass renormalization for a free meson. Second-order perturbation theory is not used. Near the end of the derivation, an equation that is equivalent to Bethe's result (6) for the radiative shift can be derived by inserting a complete set of Schrödinger energy eigenstates. Thus, we expect the fundamental results from Bethe's spectral density (with no approximations) and the group's theoretical spectral density to be in agreement [12,27]. For convenience, in Appendix A an explanation of the basis states used to derive Equation (26) is given, and in Appendix B, the derivation of Equation (26) is given, since the derivation in [27] is spread in steps throughout the paper as the group theory methods are developed.

Equation (26) can be written as an integral over $E = \hbar\omega$, which is the energy of the vacuum field in eV. The definite integral over s can be evaluated analytically for different values of ϕ or $E = \hbar\omega$. The ground state Lamb shift ΔE_1 is measured in eV so the spectral density of the shift $d\Delta E_1/dE$ is measured in eV/eV which is dimensionless:

$$\Delta E_1 = \int_0^{mc^2} \frac{d\Delta E_1}{dE} dE, \tag{28}$$

where the ground state spectral density from Equation (26) is

$$\frac{d\Delta E_1}{dE} = \frac{4\alpha^3}{3\pi} e^{-2\phi} \sinh\phi \int_0^\infty ds\, e^{se^{-\phi}} \frac{1}{\sinh^2(\frac{s}{2})} \frac{1}{(\coth\frac{s}{2} + \cosh\phi)^3}. \tag{29}$$

Figure 1 shows a logarithmic plot (ordinate is a log, abscissa is linear) of the spectral density $\frac{d\Delta E_1}{dE}$ of the ground state Lamb shift with $Z = 1$ over the entire range of energy E computed from Equation (29) using Mathematica. The spectral density is largest at the lowest energies, and decreases monotonically by about 4 orders of magnitude as the energy increases to 511 keV. The ground state shift is the integral of the spectral density from energy 0 to 511 keV.

Figure 2 is a log-log plot of the same information. The use of the log-log plot expands the energy range for each decade, revealing that for energy above about 1000 eV the slope is approximately -1, indicating that the spectral density is nearly proportional to $1/E$. For energy below about 10 eV, the spectral density in Figure 2 is almost flat, corresponding to a linear decrease with E, with a maximum at the lowest energy computed, as shown in Figure 3. Figure 2 shows that there are essentially two different behaviors of the spectral density, one for values of the energy E of the vacuum field that are less than 10 eV, which is less than $E_m - E_n$, the energy difference for all bound state transitions, and another region where E is much larger than the bound state energies, and the spectral density goes as $1/E$. The origin of this behavior is clear mathematically from the factor $B = 1/(E_m - E_n + E)$. In the expression for the spectral shift (6) for $E < E_m - E_n$, $B \propto 1 - E/(E_m - E_n)$ and for $E > E_m - E_n$, $B \propto 1/E$.

Figure 3 shows linear plots of the spectral density of the shift for the ground state computed from Equation (29) for several lower energy regions. Figure 3a shows a linear increase in the spectral density as the energy decreases over the small energy interval plotted. Figure 3b show a linear increase of about 15% as the energy decreases from 3 eV to

0 eV. Figure 3c shows that the spectral density increases by a factor of about 4 as the energy decreases from 100 eV to 0 eV. In the low-frequency limit, the spectral density increases linearly to a constant value as the energy is reduced.

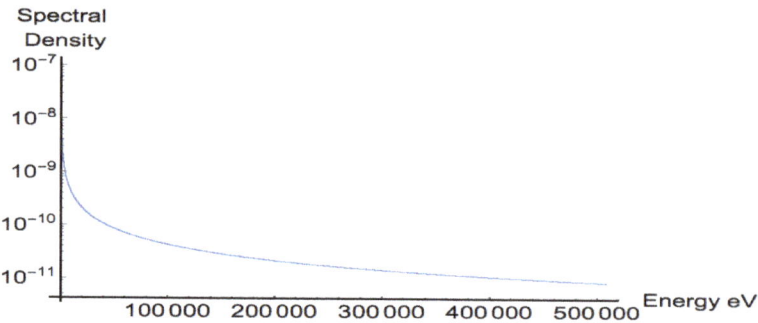

Figure 1. Plot of the log of the spectral density of the ground state Lamb shift from the group theoretical expression (29) versus the energy from 0 to 510 keV.

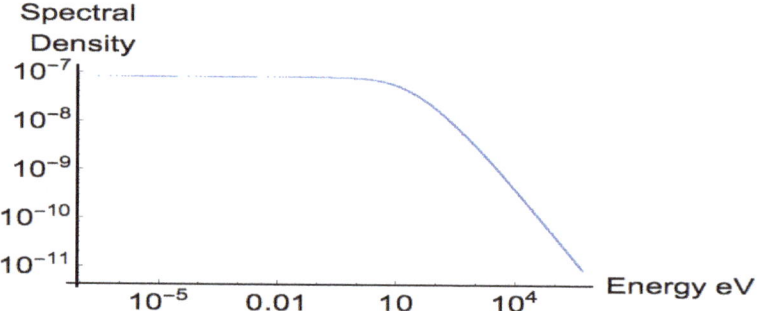

Figure 2. Log-log plot of the spectral density of the ground state shift from the group theoretical expression (29) versus the energy. For energies above about 1000 eV, the behavior is dominated by a 1/energy dependence. From about 10 eV to 0 eV, there is a slow linear increase in the spectral density.

From explicit evaluations, Section 4 shows that for shifts in S states with principal quantum number n, the asymptotic spectral density for large E is proportional to $\alpha(Z\alpha)^4(1/n^3)/E$, and Section 5 shows that as the energy E goes to zero, the spectral density increases linearly, reaching a maximum value that is proportional to $\alpha(Z\alpha)^2(1/n^2)$. An approximate fit to the ground state data in Figure 1 is

$$\frac{d\Delta E_1^{\text{fit}}}{dE} = D\frac{(1+e^{-HE})}{(E+C)}, \qquad (30)$$

where $D = 4.4008 \times 10^{-6}$, $H = 0.08445/$ eV, $C = 106.79$ eV. The fit is quite good at the asymptotes and within 10% over the entire energy range.

The spectral density shown in Figures 1 or 2 can be used to determine the contribution to the total ground state shift from different energy regions. Integrating the spectral density from 0 eV to energy E gives the value of the partial shift $\Delta_1(E)$ that these energies (0 eV to E eV) contribute to the total shift ΔE_1 for the ground state. In Figure 4, $\Delta_1(E)/\Delta E_1$, which is the fraction of the total shift ΔE_1 due to the contributions from energies below E, is plotted as a function of E. Figure 4a shows that almost 80% of the shift comes from energies below about 100,000 eV. Figure 4b shows that about half the total shift is from energies below 9050 eV. Figure 4c shows that energies below 100 eV contribute about 10% of the total shift. Energies below 13.6 eV contribute about 2.5% while energies below 1 eV contribute about 1/4% of the total. As Figure 4c shows, the fraction of the total shift increases linearly for $E < 10$ eV, corresponding to the nearly horizontal portion of the shift density for $E < 10$ eV, as shown in Figure 2. The contribution to the total 1S shift for the visible spectral interval 400–700 nm (1.770 eV to 3.10 eV) is about 1.00342×10^{-7} eV or about 3/10 % of the total shift.

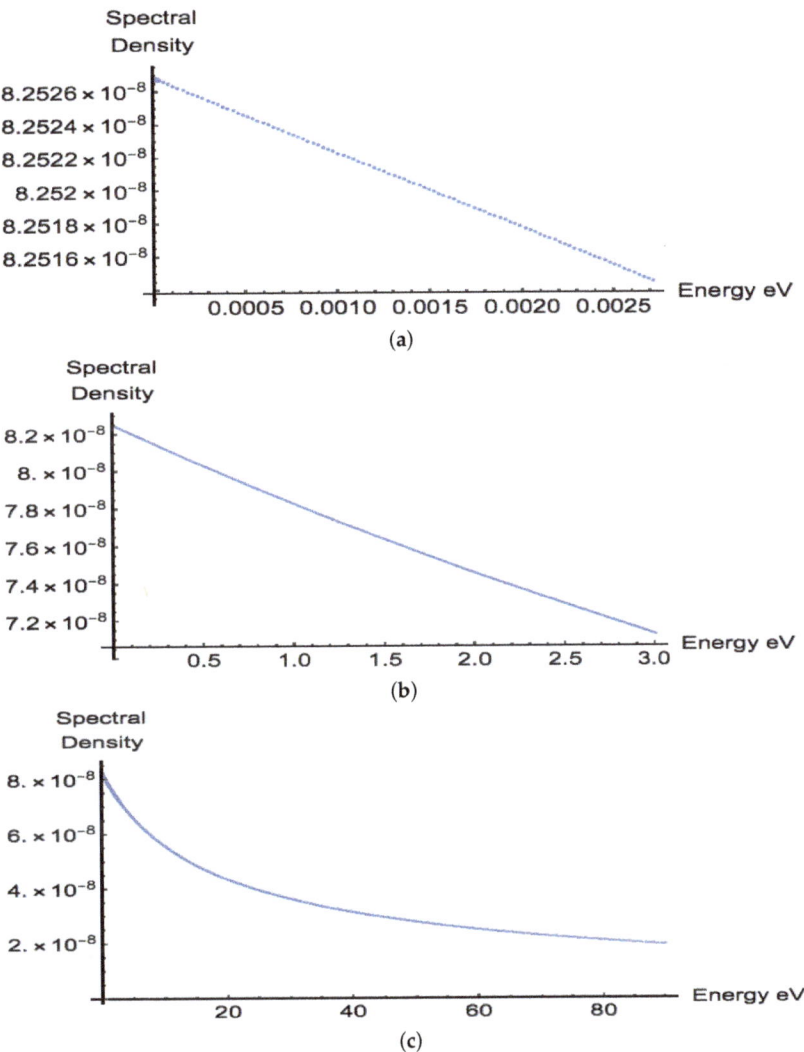

Figure 3. Linear plots of the ground state spectral density calculated from group theory as a function of energy for low and mid energies. From about 10 eV to 0 eV, the spectral density increases linearly to its maximum value. The value of the abscissa at the origin is 0 eV for all graphs. (**a**): Linear change in ground state spectral density at very low energies. (**b**): Near linear change in ground state spectral density for visible and near infra-red energies. The contribution to the total shift for energies below 3 eV is about 0.7%. (**c**): Ground state spectral density calculated for energies below 80 eV, which contribute about 8.6% to the total shift.

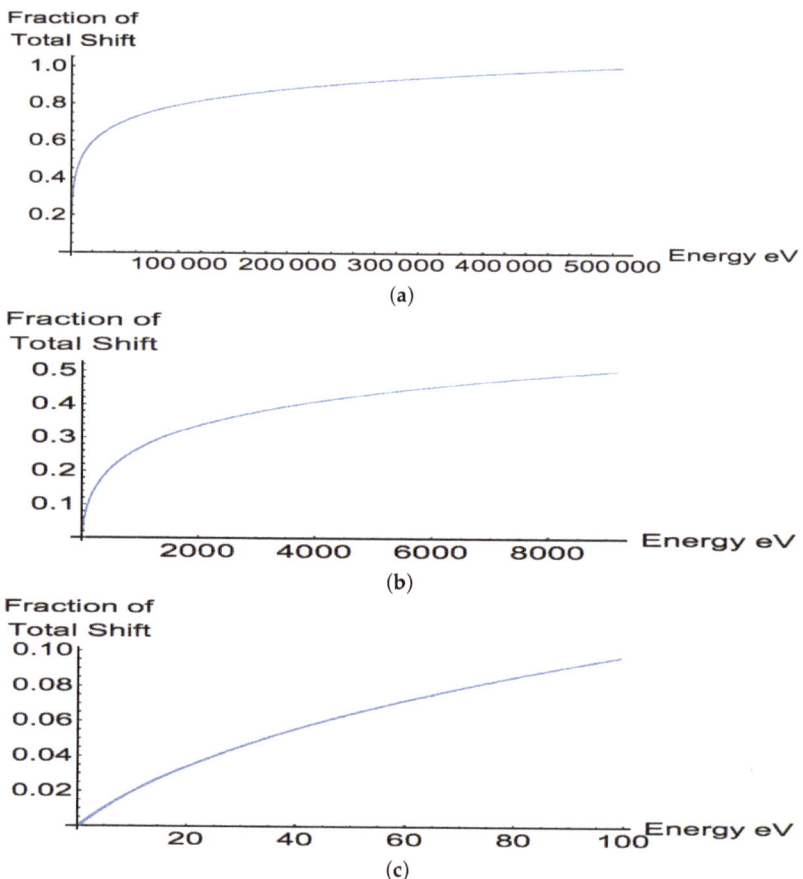

Figure 4. The ordinate is the fraction of the ground state shift ΔE_1 due to vacuum field energies between 0 and E, plotted as a function of E. This plot is obtained by integration of the spectral density Equation (29), shown in Figure 1. The plot is linear in the ordinate and abscissa. The origin corresponds to (0,0) for all plots. (**a**): Fraction of the 1S shift due to energies from 0 to E for $0 < E < 510$ keV. (**b**): Fraction of the 1S shift due to energies below E, for $0 < E < 9000$ eV. (**c**): Fraction of the 1S shift due to energies from 0 to E, for $0 < E < 100$ eV. Energies below 30 eV account for about 0.05 of the total shift. The variation is linear for $E < 10$ eV.

The relative contribution to the total shift per eV is much greater for lower energies. For example, half the 1S shift corresponds to energies 0 to 9000 eV, but only about 0.2% corresponds to 500,000 to 509,000 eV. The largest contribution to the shift per eV is at the lowest energies, which have the steepest slope of the spectral density curve in Figure 1, about 1000 times greater than the slope for the largest values of the energy. However, the total range for the large energies, from 9050 to 510,000 eV is so large that the absolute contribution to the total shift for large energies is significant.

For the ground state, Figure 5 shows how the dominant terms for different m in the Bethe sum over states in Equation (6) contribute to the full spectral density obtained from group theory Equation (29). Each such term in the Bethe sum could be interpreted as corresponding to the shift resulting from virtual transitions from state n to state m occurring due to the vacuum field. Each term shown has a behavior similar to that of the full spectral density, but the magnitudes decrease as the transition probabilities decrease.

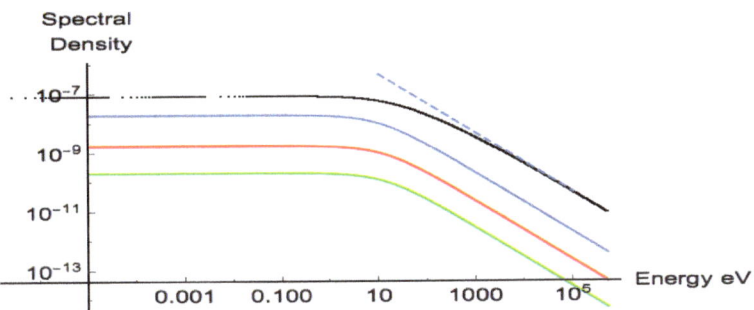

Figure 5. Log-log plot of the 1S spectral density from group theory Equation (29) in black, and the contributions to this shift in the Bethe formalism for the transition $1S \to 2P$ (blue), $1S \to 4P$ (red), $1S \to 8P$ (green). The dashed blue line shows the high-frequency $1/E$ asymptote. The black line is the complete spectral density which is the summation of the contributions from all transitions.

Figure 6 shows the spectral densities for 1S (black) and 2S (orange) shifts. The shapes are similar but the spectral density for the 1S shift is about eight times as large at high frequencies and about four times as large at low frequencies, factors that are derived explicitly in Sections 4 and 5 by considering the asymptotic forms of the spectra density for S states with different principal quantum numbers. Both S states have a $1/E$ high-frequency behavior. The s-integration in the group theoretical calculation for the 2S state diverges for energies below 10.2 eV due to a non-relativistic approximation, but the spectral density of the shift can be obtained from a low-energy approximation (48) to the group theory result, which is derived in Section 5.

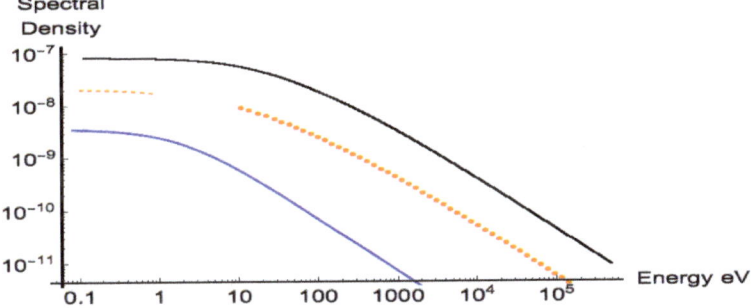

Figure 6. A log-log plot of the group theoretical spectral density for the 1S (black) and 2S (orange) shifts versus energy. The dashed orange curve below 1 eV is a 2S low-energy approximation (47) from group theory or the Bethe formula. The blue is the largest single contribution in the Bethe formalism to the 2S shift for the transition $2S \to 3P$.

The spectral density $d\Delta E_n/dE$ for a state n can be defined in a convenient form suggested by Equation (29),

$$\frac{d\Delta E_n}{dE} = \frac{4\alpha^3}{3\pi} \int_0^\infty ds\, W_n(s, \phi_n), \qquad \phi_n = \ln\left[1 + \frac{E}{|E_n|}\right], \tag{31}$$

where the energy for state n is $E_n = -mc^2(Z\alpha)^2/2n^2$. The group theoretical results give a spectral density for the 2S–2P Lamb shift [27]:

$$W_{2S-2P}(s, \phi_2) = \frac{4e^{(2se^{-\phi_2} + \phi_2)} \sinh^3(\phi_2) \operatorname{csch}^2\left(\frac{s}{2}\right)}{\left[\cosh(\phi_2) + \coth\left(\frac{s}{2}\right)\right]^5}, \tag{32}$$

and for the 2P shift [27]:

$$W_{2P}(s,\phi_2) = -\frac{e^{(2se^{-\phi_2}+\phi_2)}\sinh(\phi_2)\operatorname{csch}^4\left(\frac{s}{2}\right)[\cosh(\phi_2)\sinh(s)+\cosh(s)-3]}{2\left[\cosh(\phi_2)+\coth\left(\frac{s}{2}\right)\right]^5}. \quad (33)$$

The spectral density of the 2P shift has a very different behavior from the spectral density of the 2S shift (Figure 7). It is negative and and it falls off as $1/E^2$. The shift is negative because the dominate contribution to the shift is from virtual transitions from the 2P state to the lower 1S state, with an energy difference of about 10.2 eV. For energies below about 20 eV, the absolute value of the spectral density of the 2P shift increases rapidly in magnitude as the energy is reduced and is much bigger than the spectral density for the 2S shift. The 2S shift cannot have a negative contribution from the lower 1S state since the transition 2S→1S is forbidden by the conservation of angular momentum. The classic Lamb shift arises from the difference between the two spectral densities, so the negative 2P spectral density actually increases the 2S–2P Lamb shift as the energy decreases (Figure 8). In effect, the 2S–2P shift is dominated by vacuum energies below about 100 eV. The total 2P shift is about 0.3% percent of the 2S shift. Bethe also computed a negative contribution for the shift from the 2P state [30].

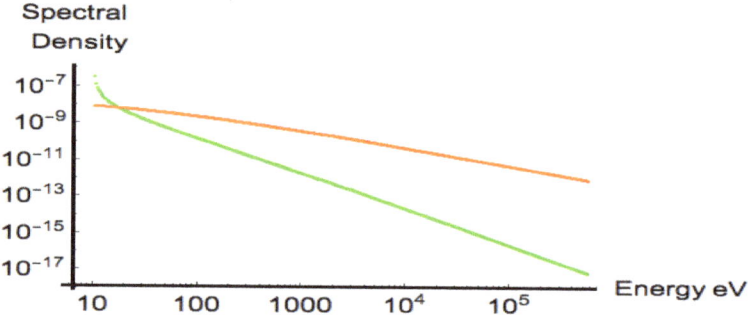

Figure 7. Log-log plot of the absolute value of the spectral density versus the energy for the 2S shift (orange), which goes as $1/E$ for large E, and for the 2P shift (green), which goes as $1/E^2$ for large E. At 511 keV, the 2P spectral density is about 5 orders of magnitude smaller than the 2S spectral density. Below 20 eV, the absolute value of the 2P spectral density is greater than the 2S spectral density. Note that the 2P spectral density is actually negative and the 2S spectral density is positive.

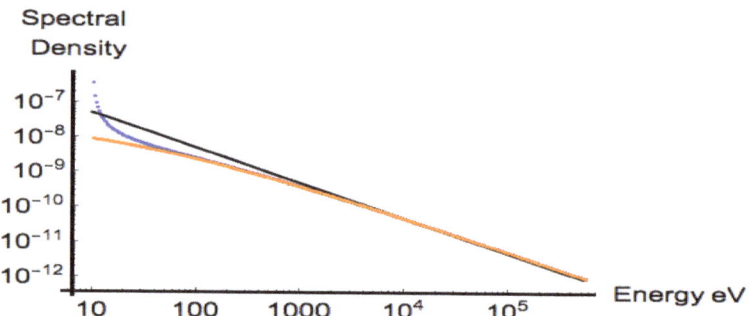

Figure 8. Log-log plot of the spectral density for the 2S shift (orange) and the 2S–2P Lamb shift (blue) versus energy. The solid black line is the $1/E$ asymptote.

Comparing the Ground State Group Theoretical Lamb Shift Calculations to Those of Bethe, Welton, and Feynman

Integrating the group theoretical spectral density Equation(29) from near zero energy (5.4×10^{-7} eV) to 511 keV, about the rest mass energy of the electron, gives the 1S shift of

3.4027 × 10⁻⁵ eV, in agreement with the numerical result of Bethe and Salpeter summing over states and using the Bethe log approximation, 3.392 × 10⁻⁵ eV, to about 0.3% [8].

Bethe and Salpeter [30] reported that the ground state Bethe log, as defined in Equation 13, which is a logarithmically weighted average value of the excitation of the energy levels contributing to the radiative 1S shift, was equivalent to $|E_m - E_n|_{ave}$ of 19.77 Ry or 269 eV [30]. Because of the weighting, it is not clear how to interpret this value, other than it indicates that high-energy photons and scattering states contribute significantly to the shift. The group theoretical method does not provide an equivalent weighted average value for direct comparison.

Although the methods of Bethe, Welton, and Power as defined all give approximately the same value for the 1S shift, which equals the integral of the spectral density, they differ significantly in their frequency dependence, which is examined in Section 4.

4. The Spectral Density of The Lamb Shift at High Frequency

The form for $d\Delta E_n/dE$, which is the Lamb shift spectral density for level n, can be obtained at high energies from (i) the classic calculation by Bethe using second-order perturbation theory; (ii) the calculation by Welton of the Lamb shift; (iii) the calculation by Power of the Lamb shift based on Feynman's approach; and (iv) the group theoretical calculation.

The spectral density for level n can be written from Bethe's expression (6):

$$\frac{\Delta E_n^{Bethe}}{\Delta E} = \frac{2\alpha}{3\pi}\left(\frac{1}{mc}\right)^2 \sum_m |\mathbf{p}_{mn}|^2 (E_n - E_m)\frac{1}{E_n - E_m - E}. \quad (34)$$

For the ground state spectral density $n = 1$, $Z = 1$, $E_1 = -13.613$ eV, and for the bound states with principal quantum number m, $E_m = -13.613$ eV/m^2. For scattering states, E_m is positive. Hence the denominator is negative for all terms in the sum over m and never vanishes, and the spectral density is positive so the ground state shift is positive as it must be. For large values of E, we can make the approximation,

$$\frac{\Delta E_n^{Bethe}}{\Delta E}\bigg|_{E\to\infty} = \frac{2\alpha}{3\pi}\left(\frac{1}{mc}\right)^2 \sum_m |\mathbf{p}_{mn}|^2 (E_m - E_n)\frac{1}{E}. \quad (35)$$

The summation can be evaluated using the dipole sum rule (9)–(11) for the Coulomb S state wavefunction, obtaining the final result for the high-frequency spectral density for S states with the principal quantum number, n:

$$\frac{d\Delta E_n^{Bethe}}{dE}\bigg|_{E\to\infty} = \frac{4mc^2}{3\pi}\alpha(Z\alpha)^4\frac{1}{n^3}\frac{1}{E}. \quad (36)$$

The result highlights the $1/E$ divergence at high frequencies and shows the presence of a coefficient proportional to $1/n^3$. To put a scale on the coefficient, the high-frequency spectral density can be written as $(8/3\pi)(\alpha(Z\alpha)^2/n)(E_n/E)$.

The spectral density for all frequencies from Welton's qualitative model, Equation (17), is identical to this high-frequency limit of Bethe's calculation. Thus, at low frequencies, the spectral density for Welton's semi-quantitative calculation diverges as $1/E$. Because of the expectation value of the Laplacian, Welton's approach predicts a shift only for S states. Its appeal is that it gives a physical picture of the primary role of vacuum fluctuations in the Lamb shift and shows the presence of the $1/E$ characteristic behavior. Its treatment of bound states at low energies is incomplete and inaccurate. To obtain a level shift, it requires providing a low-energy limit for the integration. As noted previously, if the lower limit is Bethe's log average excitation energy, 269 eV for n = 1, and the upper limit mc^2 then Welton's total 1S shift agrees with Bethe's. A choice of this type works since: (i) it does not include any contributions from energies below 269 eV and (ii) it gives a compensating contribution for energies from 269 eV to about 1000 eV that is larger than the actual spectral density, as shown in Figures 4 and (iii) above about 1000 eV, Welton's model gives the same $1/E$ spectral density as Bethe.

The spectral density for Power's model can be obtained from Equation (24):

$$\frac{\Delta E_n^{\text{Power}}}{dE} = -\frac{2\alpha}{3\pi}\frac{1}{(mc)^2}\sum_m |\mathbf{P}_{mn}|^2 (E_m - E_n)\frac{E}{(E_m - E_n)^2 - E^2}. \quad (37)$$

Letting E become large gives a result identical to the high-frequency limit (35) for the Bethe formalism and the Welton model, namely

$$\frac{\Delta E_n^{\text{Power}}}{dE}\bigg|_{E\to\infty} = \frac{4mc^2}{3\pi}\frac{\alpha(Z\alpha)^4}{n^3}\frac{1}{E}. \quad (38)$$

Thus, the models of Bethe, Welton, and Power predict S states with the same $1/E$ dependence of the high-frequency spectral density, corresponding to the logarithmic divergence at high-frequency. The asymptotic theoretical result can be written in a form allowing convenient comparison to the calculated group theoretical spectral density

$$\frac{d\Delta E_n^{\text{Bethe}}}{dE}\bigg|_{E\to\infty} = \frac{4mc^2}{3\pi}\frac{\alpha(Z\alpha)^4}{n^3}\frac{1}{E}. \quad (39)$$

The spectral density goes as $1/n^3$ for S states. For the ground state $n=1$, $Z=1$, this gives:

$$\frac{d\Delta E_1^{\text{Bethe}}}{dE}\bigg|_{E\to\infty} = 4.488\times 10^{-6}\frac{1}{E}. \quad (40)$$

A fit to the last two data points near 510 keV in the group theoretical calculations (GTcalc) gives:

$$\frac{d\Delta E_1^{\text{GTcalc}}}{dE}\bigg|_{E\to\infty} = 4.4008\times 10^{-6}\frac{1}{E}. \quad (41)$$

The coefficients differ by about 2%. Figure 9 is a plot of the ground state group theoretical calculated spectral density (red) from Equation (29) and the theoretical asymptotic behavior from Bethe, Power and Welton, Equation (40) (black), and the difference times of a factor of 10. The asymptotic theoretical result agrees with the full group theoretical calculation from Equation (29) to within about 2% at 511 keV, and to about 6% at 50 KeV. It is notable that the high-frequency form is a reasonable approximation down to 50 keV. Indeed, the Welton qualitative approach is based on this observation, it has the same $1/E$ energy dependence at all energies.

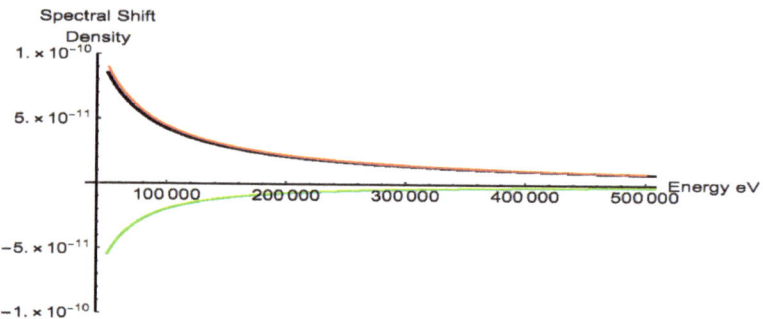

Figure 9. Top red curve is the 1S group theoretical calculated spectral density (29), slightly lower black curve is the $1/E$ asymptotic model Equation (39), and the bottom green curve is the difference times 10, plotted for the interval 50–510 keV. Both axes are linear.

5. Spectral Density of the Lamb Shift at Low Frequency

To obtain a low-frequency limit of the spectral density of the Lamb shift from the Bethe spectral density (34), expand the spectral density to first-order in E, giving

$$\frac{\Delta E_n^{\text{Bethe}}}{dE}\bigg|_{E\to 0} = \frac{2\alpha}{3\pi}\left(\frac{1}{mc}\right)^2 \sum_m |\mathbf{p}_{mn}|^2 \left(1 - \frac{E}{E_m - E_n}\right). \qquad (42)$$

Since the sum is over a complete set of states m, including scattering states, the first term in parenthesis can be evaluated using the sum rule,

$$\sum_m |\mathbf{p}_{mn}|^2 = -2mE_n = (mc)^2 \frac{(Z\alpha)^2}{n^2}. \qquad (43)$$

To evaluate the second term, Equation (22) and the Thomas–Reiche–Kuhn sum rule [31] can be used, giving

$$\sum_m \omega_{mn} |\mathbf{d}_{mn}|^2 = \frac{3e^2\hbar}{2m}. \qquad (44)$$

The final result for $E \to 0$ is

$$\frac{\Delta E_n^{\text{Bethe}}}{dE}\bigg|_{E\to 0} = \frac{2\alpha}{3\pi}\frac{(Z\alpha)^2}{n^2} - \frac{\alpha}{\pi mc^2}E. \qquad (45)$$

The corresponding spectral density for $n=1$, $Z=1$, is

$$\frac{d\Delta E_1^{\text{Bethe}}}{dE}\bigg|_{E\to 0} = \frac{4\alpha \times 13.6}{3\pi mc^2}\left(1 - \frac{3E}{4 \times 13.6}\right) = 8.253 \times 10^{-8}(1 - 0.0551E). \qquad (46)$$

As E decreases to zero, the spectral density increases linearly to a constant value,

$$4\alpha |E_n|/3\pi mc^2 = 2\alpha^3 Z^2/3\pi n^2 = 8.253 \times 10^{-8}/n^2.$$

The intercept goes as $1/n^2$, but the slope is $\alpha/\pi mc^2$, which has a remarkably simple form and is independent of n.

Taking the low-frequency limit of the group theoretical result analytically gives exactly the same result as Equation (45) from the Bethe formulation:

$$\frac{d\Delta E_1^{\text{GTcalc}}}{dE}\bigg|_{E\to 0} = \frac{d\Delta E_1^{\text{Bethe}}}{dE}\bigg|_{E\to 0} = \frac{2\alpha}{3\pi}\frac{(Z\alpha)^2}{n^2} - \frac{\alpha}{\pi mc^2}E. \qquad (47)$$

Figure 3 shows the results of group theoretical calculations of the spectral density of the ground state Lamb shift for different energy regions, showing the near linear increase in the spectral density as the frequency decreases from 80 eV to 10^{-5} eV. For low values of E, the slopes and intercept agree with Equation (47) within about two tenths of a percent.

To explore Power's approach at low frequencies, let E become small in the spectral density Equation (37), giving

$$\frac{\Delta E_n^{\text{Power}}}{dE}\bigg|_{E\to 0} = -\frac{2NV}{3\pi c^3}\sum_m |\mathbf{p}_{mn}|^2 \frac{E}{E_m - E_n}, \qquad (48)$$

which is identical to the second term in the low E approximation to the Bethe result (45) so

$$\frac{\Delta E_n^{\text{Power}}}{dE}\bigg|_{E\to 0} = -\frac{1}{\pi}\frac{\alpha}{mc^2}E. \qquad (49)$$

The result Equation (49) is identical to the frequency-dependent term in Equation (47) which is the low-frequency spectral density from the Bethe approach and from the group theoretical expression. However, in the low-frequency limit based on Power's expression for the spectral density, the constant term that is present in the other approaches does not appear. This is a consequence of the form used for the index of refraction, which assumes that real photons are present that can excite the atom with resonant transitions. More sophisticated implementations of Feynman's proposal may avoid this issue.

6. Comparison of the Spectral Energy Density of the Vacuum Field and the Spectral Density of the Radiative Shift

The theory of Feynman proposes that the vacuum energy density in a large box containing H atoms, which are assumed to be in the 1S ground state, increases uniformly with the addition of the atoms. Feynman maintains that the total vacuum energy in the box increases by the Lamb shift times the number of atoms present [6,10]. If there were only one atom in a very large box, one would not expect the change in energy density to be spatially uniform but more concentrated near the atom. To develop a model of the spatial dependence of the change in energy density for one atom, the close relationship between the vacuum field and the radiative shift can be used. The spectral densities of the ground state shift and of the quantum vacuum with no H atoms present are both known. In the box, the vacuum field density must increase so that the integral gives the 1S Lamb shift. The spectral energy density of the vacuum field with no H atom present is equal to [6]

$$\rho_0(\omega) = \frac{\hbar \omega^3}{2\pi^2 c^3}, \tag{50}$$

where c is in cm/sec and ω is in sec^{-1}. If frequency is measured in eV, so $\hbar\omega = E$, then the vacuum spectral energy density in $1/cc$ is

$$\rho_0(E) = \frac{E^3}{2\pi^2 \hbar^3 c^3}, \tag{51}$$

and $\int_{E_1}^{E_2} \rho_0(E) dE$ would be the energy density eV/cc in the energy interval E_1 to E_2. The question being addressed is: what volume of vacuum energy of density $\rho_0(E)$ is required to supply the amount of energy needed for the radiative shift? The total radiative shift ΔE_1 be expressed as the integral of the vacuum energy density $\rho_0(E)$ over an effective volume $V_1(E)$,

$$\Delta E_1 = \int_0^{mc^2} dE \rho_0(E) V_1(E), \tag{52}$$

where the same upper limit for E is used as in previous calculations. Recall the definition (28) of the spectral shift:

$$\Delta E_1 = \int_0^{mc^2} dE \frac{\Delta E_1}{dE}. \tag{53}$$

Comparing Equations (52) and (53) and insuring the energy balance at each energy E, gives the effective spectral volume $V_1(E)$

$$V_1(E) = \frac{d\Delta E_1}{dE} \frac{1}{\rho_0(E)}. \tag{54}$$

The spectral volume $V_1(E)$ has the dimensions of cc and contains the amount of vacuum energy at energy value E that corresponds to the ground state spectral density at the same energy E. In Figure 10, for the 1S ground state radiative shift, the log of the spectral volume $V_1(E)$, in cubic Angstroms (Å^3), is plotted versus the log of the energy, E.

For energies above about 100 eV, the spectral volume is less than 1 Å^3, approximately the volume of the ground state wavefunction. For an energy of 1 eV, the spectral volume is 11850 Å^3, corresponding to a sphere of radius about 14 Å. This calculation predicts that there is a sphere of positive vacuum energy of radius 14 Å around the atom corresponding to the 1 eV shift spectral density. Figure 11 shows the radius of the spherical spectral volume $V_1(E)$ for energies from 0.05 eV, with spectral radius of 278 Å, to 23 eV, with radius 0.49 Å.

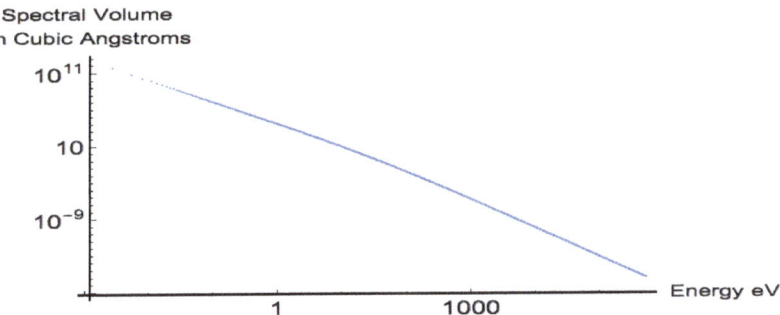

Figure 10. Log-log plot of the spectral volume $V_1(E)$ as a function of E. The spectral volume $V_1(E)$ contains the free field vacuum energy at energy value E that corresponds to the ground state shift spectral density at the same energy E.

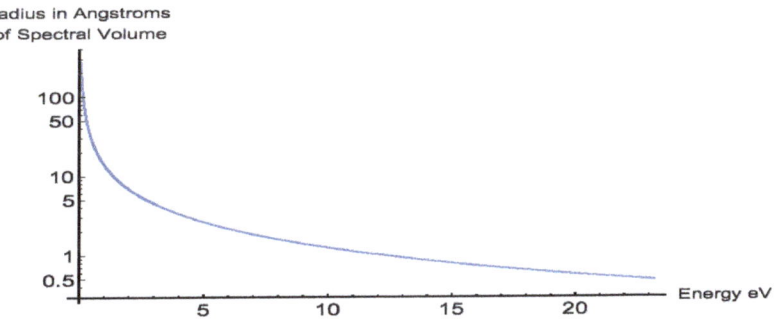

Figure 11. Log of the radius of the spherical spectral volume $V_1(E)$, as a function of the vacuum field energy E, from 0.05 eV to 23 eV.

7. Conclusions

The non-relativistic Lamb shift can be interpreted as due to the interaction of an atom with the fluctuating electromagnetic field of the quantum vacuum. We introduce the concept of a spectral shift density which is a function of frequency ω or energy $E = \hbar\omega$ of the vacuum field. The integral of the spectral density from $E = 0$ to the rest mass energy of an electron, 511 keV, gives the radiative shift. We report on calculations of the spectral density of the level shifts for 1S, 2S and 2P states based on a group theoretical analysis and compare the results to the spectral densities implicit in previous non-relativistic calculations of the Lamb shift by Bethe, Welton, and Power. The group theoretical calculation provides an explicit form for the spectral density over the entire spectral range with no summation over intermediate states. Bethe's approach requires a summation over an infinite number of states, including all bound and all scattering states, to obtain a comparable spectral density. The different approaches for asymptotic cases, for very large and very small energies E, are compared.

The calculations of the shift spectral density provide a new perspective on radiative shifts. The group theory approach as well as the approaches of Bethe, Power, and Welton all show the same $1/E$ high-frequency behavior for S states above about $E = \hbar\omega = 1000$ eV to 511 keV, namely an asymptotic spectral density for S states equal to $(4/3\pi)(\alpha(Z\alpha)^4 mc^2/n^3)(1/E)$ for principal quantum number n. The group theory calculation shows that about 76% of the ground state 1S shift is contributed by E above 1000 eV, which is essentially why all the approaches give approximately the same result for the 1S Lamb shift.

Only the Bethe and group theory calculations have the correct low-frequency behavior. For S states the spectral density increases linearly as E decreases to zero. Its maximum value is at $E = 0$ and for S states equals $(2\alpha/3\pi)(Z\alpha)^2/(n^2)$. This maximum value is about

$1/(Z\alpha)^2$ or about 2×10^4 larger than the high-frequency spectral density at $E = 511$ keV. Thus, low energies contribute much more to the shift for a given spectral interval than high energies. Energies below 13.6 eV contribute about 2.5 %. Because of the huge spectral range contributing to the shift, contributions to the shift from high energies are very important. Half the contribution to the 1S shift is from energies above 9050 eV.

The 2P shift has a very different spectral density from an S state: it is negative and has an asymptotic behavior that goes as $1/E^2$ rather than as $1/E$. Below about 20 eV, the absolute value of the 2P spectral density is much larger than the 2S spectral density and it dominates the 2S–2P shift spectral density, yet the total 2P shift is only about 0.3% of the total 2S shift.

Funding: This research received no external funding.

Data Availability Statement: Not applicable.

Acknowledgments: I thank Peter Milonni for many insightful and enjoyable discussions, particularly about the resonant behavior of the index of refraction and the volume of vacuum energy corresponding to the spectral density, and I thank Lowell S. Brown for his observations, especially about the 1/ frequency asymptotic behavior.

Conflicts of Interest: The author declares no conflicts of interest.

Appendix A. Eigenstates $|nlm;a\rangle$ of $1/Z\alpha$

To obtain an equation for these basis states $|nlm;a\rangle$, write Schrödinger's equation for a charged non-relativistic particle with with momentum p and energy $E = -\frac{a^2}{2m}$ [27,28] in a Coulomb potential:

$$\left[p^2 + a^2 - \frac{2m\hbar c Z\alpha}{r}\right]|a\rangle = 0. \tag{A1}$$

Solutions for $|a\rangle$ exist for certain critical values of the energy $E_n = -\frac{a_n^2}{2m}$ or equivalently for critical values of $a = a_n$ where $\frac{a_n}{mcZ\alpha} = \frac{1}{n}$. These are the usual energy eigenstates which are labeled as $|nlm;a_n\rangle$. Conversely a be fixed in value and $Z\alpha$ may have different values. If it has certain eigenvalues $Z\alpha_n$ then for any value of a there is another set of eigenvectors corresponding to eigenvalues $\frac{a}{mcZ\alpha_n} = \frac{1}{n}$. To demonstrate this we start by inserting factors of $1 = \sqrt{ar}\frac{1}{\sqrt{ar}}$ in Schrödinger's Equation (A1) obtaining

$$\left(\sqrt{ar}(p^2 + a^2)\sqrt{ar} - 2amZ\alpha\right)\frac{1}{\sqrt{ar}}|a\rangle = 0. \tag{A2}$$

We can rewrite this equation, multiplying successively from the left by $\frac{1}{\sqrt{ar}}$, $\frac{1}{p^2+a^2}$, and $\frac{1}{\sqrt{ar}}$, and then multiplying by a^2, and dividing by $mcZ\alpha$, multiplying by $\sqrt{n\hbar}$ obtaining

$$\left[\frac{a}{mcZ\alpha} - K_1(a)\right]\sqrt{\frac{n\hbar}{ar}}|a\rangle = 0, \tag{A3}$$

where

$$K_1(a) = \frac{1}{\sqrt{ar}}\frac{2a^2\hbar}{p^2+a^2}\frac{1}{\sqrt{ar}}. \tag{A4}$$

Solutions exist to this equation for eigenvalues of $1/Z\alpha$ such that $\frac{a}{mcZ\alpha_n} = \frac{1}{n}$:

$$\left(\frac{1}{n} - K_1(a)\right)|nlm;a\rangle = 0, \tag{A5}$$

where

$$\sqrt{\frac{n\hbar}{ar}}|nlm;a\rangle = |nlm;a\rangle.$$

The $n\hbar$ in the square root insures the new states are also normalized to 1. The kernel $K_1(a)$ is bounded and Hermitian with respect to the eigenstates $|nlm;a\rangle$ of $1/Z\alpha$, therefore these eigenstates of $1/Z\alpha$ form a complete orthonormal basis for the hydrogen atom. Because the kernel is bounded, there are no continuum states in this representation. To show they have the same quantum numbers as the usual states, note that when $a = a_n$ the eigenstates of $K_1(a_n)$ becomes $|nlm;a_n)$ and these corresponds to the usual energy eigenstates $|nlm;a_n\rangle$. The value of a in Equation (A5) can be changed to obtain these eigenstates using the dilation operator $D(\lambda) = e^{i S \lambda}$, where the dimensionless operator S, which is also a generator of transformations of SO(4,2), is

$$S = \frac{1}{2\hbar}(\mathbf{p} \cdot \mathbf{r} + \mathbf{r} \cdot \mathbf{p}). \tag{A6}$$

S transforms the canonical variables:

$$D(\lambda)\mathbf{p}D^{-1}(\lambda) = e^{-\lambda}\mathbf{r},$$

$$D(\lambda)\mathbf{r}D^{-1}(\lambda) = e^{\lambda}\mathbf{r}.$$

Operating on $K_1(a)$ with $D(\lambda)$ gives:

$$D(\lambda)K_1(a)D^{-1}(\lambda) = K_1(ae^{\lambda}).$$

If λ is chosen as

$$\lambda_n = \ln(a_n/a),$$

then $ae^{\lambda_n} = a_n$. Thus, operating with $D(\lambda_n)$ on Equation (A5) yields:

$$\left(\frac{1}{n} - K_1(a_n)\right)D(\lambda_n)|nlm;a) = 0. \tag{A7}$$

This is the equation for the usual Schrödinger energy eigenstates, so

$$D(\lambda_n)|nlm;a) = |nlm;a_n) = \sqrt{\frac{n\hbar}{a_n r}}|nlm;a_n\rangle. \tag{A8}$$

The usual Schrödinger energy eigenstates $|nlm;a_n\rangle$ can be expressed in terms of the eigenstates of $1/Z\alpha$ as

$$|nlm;a_n\rangle = \sqrt{\frac{a_n r}{n\hbar}}D(\lambda_n)|nlm;a). \tag{A9}$$

This relationship shows that the complete basis functions $|nlm;a)$ of $1/Z\alpha$ are proportional to the ordinary bound state energy wavefunctions and therefore have the same quantum numbers as the ordinary bound states [27,28]. A comparable set of $1/Z\alpha$ eigenstates useful for momentum space calculations is derived in [27].

Appendix B. Derivation of Group Theoretical Formula for the Shift Spectral Density

The group theoretical approach is based solely on the Schrödinger and Klein-Gordon equations of motion in the non-relativistic dipole approximation. We obtain a result [27]:

$$\Delta E_{NL} = \frac{2\alpha}{3\pi(mc)^2}\int_0^{\hbar\omega_c} dE \langle NL|p_i \frac{H - E_N}{H - (E_N - E) - i\epsilon} p_i|NL\rangle, \tag{A10}$$

where $E = \hbar\omega$, $H = \frac{p^2}{2m} - \frac{Z\alpha\hbar c}{r}$ and the states $|NL\rangle$ are the usual H atom energy eigenstates. ω_C is a cutoff frequency for the integration that is taken as $\hbar\omega_c = mc^2$. Inserting a complete set of states in this expression yields Bethe's result (6) a step avoided with the group theoretical approach. Adding and subtracting E from the numerator in Equation (A10), we find the real part of the shift is

$$\Delta E_{NL} = \frac{2\alpha}{3\pi(mc)^2} Re \int_0^{\hbar\omega_c} dE[\langle NL|p^2|NL\rangle - E\Omega_{NL}],\quad\text{(A11)}$$

where

$$\Omega_{NL} = \langle NL|p_i \frac{1}{H - E_N + \hbar\omega - i\epsilon} p_i|NL\rangle.\quad\text{(A12)}$$

The goal is to convert the matrix element Ω_{NL} to a matrix element of a function of SO(4,2) generators taken between a new set of basis states $|nlm;a\rangle$, which are complete with no scattering states, where $a = \sqrt{2m|E|}$, and n,l,m have their usual meaning and values. The new basis states $|nlm;a\rangle$ are eigenstates of $(Z\alpha)^{-1}$ [27,28]. Sometimes we write them as $|nlm\rangle$ with the a implicit.

A generator of SO(4,1) is $\Gamma_0 = 1/K_1(a)$, defined in Equation (A4), so

$$(\Gamma_0 - n)|nlm;a\rangle = 0.\quad\text{(A13)}$$

This is Schrödinger's equation in the language of SO(4,2).

Several more generators need to be defined. Since the algebra of SO(4,2) generators closes, commutators of generators must also be generators. To find Γ_4 calculate $\Gamma_4 = -i[S,\Gamma_0]$, where the generator S is defined in Equation (A6), obtaining

$$\Gamma_4 = \frac{1}{2\hbar}\left(\frac{\sqrt{r}p^2\sqrt{r}}{a} - ar\right),\quad \Gamma_0 = \frac{1}{2\hbar}\left(\frac{\sqrt{r}p^2\sqrt{r}}{a} + ar\right).\quad\text{(A14)}$$

The generators $(\Gamma_4,S,\Gamma_0) = (j_1,j_2,j_3)$ form a O(2,1) subgroup of SO(4,2) and $S = i[\Gamma_4,\Gamma_0]$, $\Gamma_0 = -i[S,\Gamma_4]$ and for our representations, $\Gamma_0^2 - \Gamma_4^2 - S^2 = \mathbf{L}^2 = l(l+1)$. The scale change S transforms $\Gamma_0 \equiv \Gamma_0(a)$ according to the equation,

$$e^{i\lambda S}\Gamma_0(a)e^{-i\lambda S} = \Gamma_0(e^\lambda a) = \Gamma_0 \cosh\lambda - \Gamma_4 \sinh\lambda,\quad\text{(A15)}$$

and similarly,

$$e^{i\lambda S}\Gamma_4(a)e^{-i\lambda S} = \Gamma_4(e^\lambda a) = \Gamma_4 \cosh\lambda - \Gamma_0 \sinh\lambda.\quad\text{(A16)}$$

Finally define a three vector of generators proportional to the momentum

$$\Gamma_i = \frac{1}{\hbar}\sqrt{r}p_i\sqrt{r}.\quad\text{(A17)}$$

The quantity $\mathbf{\Gamma} = (\Gamma_0,\Gamma_1,\Gamma_2,\Gamma_3,\Gamma_4)$ is a 5-vector of generators under transformations generated by SO(4,2). For the representation of SO(4,2) based on the states $|nlm\rangle$, all generators are Hermitian, and $\mathbf{\Gamma}^2 = \Gamma_A\Gamma^A = -\Gamma_0^2 + \Gamma_1^2 + \Gamma_2^2 + \Gamma_3^2 + \Gamma_4^2 = 1$ for our representation, and $g_{AB} = (-1,1,1,1,1)$ for $A,B = 0,1,2,3,4$. The commutators of the components of the five vector are also generators of $SO(4,2)$ transformations.

Inserting factors of $1 = \sqrt{ar}\frac{1}{\sqrt{ar}}$ and using the definitions of the generators we can transform Equation (A12) to

$$\Omega_{NL} = \frac{mv}{N^2}\langle NL|\Gamma_i \frac{1}{\Gamma n(\xi) - v}\Gamma_i|NL\rangle,\quad\text{(A18)}$$

where

$$n^0(\xi) = \frac{2+\xi}{2\sqrt{1+\xi}} = \cosh\phi,\quad n^i = 0,\quad n^4(\xi) = -\frac{\xi}{2\sqrt{1+\xi}} = -\sinh\phi,\quad\text{(A19)}$$

and

$$\xi = \frac{\hbar\omega}{|E_N|}\quad v = \frac{N}{\sqrt{1+\xi}} = Ne^{-\phi}.\quad\text{(A20)}$$

From the definitions, $\phi = \frac{1}{2}\ln(1+\zeta) > 0$ and $n_A(\zeta)n^A(\zeta) = -1$. The contraction over i in Ω_{NL} may be evaluated using the group theoretical formula [27]:

$$\sum_B \Gamma_B f(n\Gamma)\Gamma^B = \frac{1}{2}(n\Gamma+1)^2 f(n\Gamma+1) + \frac{1}{2}(n\Gamma-1)^2 f(n\Gamma-1) - (n\Gamma)^2 f(n\Gamma). \quad (A21)$$

Applying the contraction formula to the integral representation,

$$f(n\Gamma) = \frac{1}{\Gamma n - \nu} = \int_0^\infty ds\, e^{\nu s} e^{-n\Gamma s}, \quad (A22)$$

gives the result

$$\Gamma_A \frac{1}{\Gamma n - \nu} \Gamma^A = -2\nu \int_0^\infty ds\, e^{\nu s} \frac{d}{ds}\left(\sinh^2 \frac{s}{2} e^{-n\Gamma s}\right). \quad (A23)$$

Applying this to Equation (A18) gives:

$$\Omega_{NL} = -2\frac{m\nu^2}{N^2} \int_0^\infty ds\, e^{\nu s} \frac{d}{ds}\left(\sinh^2 \frac{s}{2} M_{NL}(s)\right) \\ - m\frac{\nu}{N^2}(NL|\Gamma_4 \frac{1}{\Gamma n(\zeta) - \nu} \Gamma_4|NL) + m\frac{\nu}{N^2}(NL|\Gamma_0 \frac{1}{\Gamma n(\zeta) - \nu} \Gamma_0|NL), \quad (A24)$$

where

$$M_{NL}(s) = (NL|e^{-\Gamma n(\zeta)s}|NL). \quad (A25)$$

In order to evaluate the last two terms in Equation (A24) use $\Gamma_0|NL) = N|NL)$ and express the action of Γ_4 on our states as $\Gamma_4 = N - (1/\sinh\phi)(\Gamma n(\zeta) - \nu)$. This expression for Γ_4 is derived from Equations (A18) and (A19): $\Gamma n(\zeta) - \nu = \Gamma_0 \cosh\phi - \Gamma_4 \sinh\phi - \nu$, and then substituting Equation (A20), $\nu = Ne^{-\phi}$. Using the virial theorem $(NLM|p^2|NLM) = a_N^2$, we find that the term in p^2 in Equation (A11) exactly cancels the last two terms in Ω_{NL}, yielding the result for the level shift:

$$\Delta E_{NL} = \frac{4mc^2\alpha(Z\alpha)^4}{3\pi N^4} \int_0^{\phi_c} d\phi\, \sinh\phi e^\phi \int_0^\infty ds\, e^{\nu s} \frac{d}{ds}\left(\sinh^2 \frac{s}{2} M_{NL}(s)\right) \quad (A26)$$

where

$$\phi_c = \frac{1}{2}\ln\left(1 + \frac{\hbar\omega_c}{|E_N|}\right) = \frac{1}{2}\ln\left(1 + \frac{2N^2}{(Z\alpha)^2}\right). \quad (A27)$$

To derive a generating function for the shifts for any eigenstate characterized by N and L multiply Equation (A26) by $N^4 e^{-\beta N}$ and sum over all $N, N \geq L+1$. To simplify the right side of the resulting equation, use the definition (A25) and the fact that Γ_4, S, and Γ_0 form an $O(2,1)$ algebra, obtaining:

$$\sum_{N=L+1}^\infty e^{-\beta N} M_{NL} = \sum_{N=L+1}^\infty (NL|e^{-j\cdot\psi}|NL), \quad (A28)$$

where

$$e^{-j\cdot\psi} \equiv e^{-\beta\Gamma_0} e^{-s\Gamma n(\zeta)}. \quad (A29)$$

Perform a j transformation generated by $e^{i\phi S}$, such that $e^{-j\cdot\psi} \to e^{-j_3\psi} = e^{-\Gamma_0\psi}$. The trace is invariant with respect to this transformation, therefore:

$$\sum_{N=L+1}^\infty e^{-\beta N} M_{NL} = \sum_{N=L+1}^\infty (NL|e^{-j_3\psi}|NL) = \sum_{N=L+1}^\infty e^{-N\psi} = \frac{e^{-\psi(L+1)}}{1 - e^{-\psi}}, \quad (A30)$$

where $(NL|\Gamma_0)|NL) = N$ is used.

In order to find a particular M_{NL}, expand the right hand side of Equation (A30) in powers of $e^{-\beta}$ and equate the coefficients to those on the left hand side. Using the isomorphism between j and the Pauli σ matrices $(\Gamma_4, S, \Gamma_0) \to (j_1, j_2, j_3) \to (\tfrac{i}{2}\sigma_1, \tfrac{i}{2}\sigma_2, \tfrac{1}{2}\sigma_3)$ gives the result:

$$\cosh\frac{\psi}{2} = \cosh\frac{\beta}{2}\cosh\frac{s}{2} + \sinh\frac{\beta}{2}\sinh\frac{s}{2}\cosh\phi. \tag{A31}$$

Rewriting this equation gives:

$$e^{+\frac{1}{2}\psi} = de^{\frac{1}{2}\beta} + be^{-\frac{1}{2}\beta} - e^{-\frac{1}{2}\psi}, \tag{A32}$$

where

$$\begin{aligned} d &= \cosh\tfrac{s}{2} + \sinh\tfrac{s}{2}\cosh\phi, \\ b &= \cosh\tfrac{s}{2} - \sinh\tfrac{s}{2}\cosh\phi. \end{aligned} \tag{A33}$$

Let β become large, which implies large ψ, and iterate the equation for $e^{-\frac{1}{2}\psi}$ to obtain

$$e^{-\psi} = Ae^{-\beta}\left[1 + A_1 e^{-\beta} + A_2 e^{-2\beta} + \cdots\right], \tag{A34}$$

where $A = 1/d^2$ and $A_1 = -(2/d)(b - d^{-1})$. To obtain M_{NL}, expand the right side of Equation (A30) in powers of ψ:

$$\frac{e^{-\psi(L+1)}}{1 - e^{-\psi}} = \sum_{m=1}^{\infty} e^{-\psi(m+L)}. \tag{A35}$$

For large β, it follows from Equations (A30), (A34) and (A35) that

$$\sum_{N=L+1}^{\infty} e^{-\beta N} M_{NL} = \sum_{m=1}^{\infty}\left[e^{-\beta} A(1 + A_1 e^{-\beta} + A_2 e^{-2\beta} + \ldots)\right]^{m+L}. \tag{A36}$$

Using the multinomial theorem [32] the right side of Equation (A36) becomes

$$\sum_{m=1}^{\infty} A^{m+L} \sum_{r,s,t,\ldots} \frac{(m+L)!}{r!s!t!\ldots} A_1^s A_2^t \ldots e^{-\beta(m+L+s+2t+\cdots)}, \tag{A37}$$

where $r + s + t + \cdots = m + L$. To obtain the expression for M_{NL}, we note N is the coefficient of β so $N = m + L + s + 2t + \cdots = r + 2s + 3t + \cdots$ Accordingly we find

$$M_{NL} = \sum_{r,s,t,} A^{(r+s+t+\cdots)} \frac{(r+s+t+\ldots)!}{r!s!t!} A_1^s A_2^t \cdots, \tag{A38}$$

where $r + s + t + \cdots = N$ and $r + s + t + \cdots > L$. As Equation (A26) indicates, the 1S shift corresponds to the matrix element M_{10}, which multiplies $e^{-\beta}$, so $M_{10} = A$. For the 2S shift $M_{20} = A^2 + AA_1$, and for the 2P shift $M_{21} = A^2$. Therefore the radiative shift for the 1S ground state is

$$\mathrm{Re}\Delta E_{10} = \frac{4mc^2\alpha(Z\alpha)^4}{3\pi}\int_0^{\phi_c} d\phi e^{\phi}\sinh\phi \int_0^{\infty} ds e^{se^{-\phi}}\frac{d}{ds}\frac{1}{\left(\coth\frac{s}{2} + \cosh\phi\right)^2}. \tag{A39}$$

The shift for the 2S–2P level is

$$\mathrm{Re}(\Delta E_{20} - \Delta E_{21}) = \frac{m\alpha(Z\alpha)^4}{6\pi}\int_0^{\phi_c} d\phi e^{\phi}\sinh^3\phi \int_0^{\infty} ds e^{2se^{-\phi}}\frac{d}{ds}\frac{1}{\left(\coth\frac{s}{2} + \cosh\phi\right)^4}. \tag{A40}$$

References

1. Choudhuri, A.R. *Astrophysics for Physicists*; Cambridge University Press: Cambridge, UK, 2010. [CrossRef]
2. Carroll, S.M. Addressing the quantum measurement problem. *Phys. Today* **2022**, *75*, 62–64. [CrossRef]
3. D'Espagnat, B. *Veiled Reality: An Analysis of Present-Day Quantum Mechanical Concepts*; CRC Press/Taylor & Francis Group: Boca Raton, FL, USA, 2013. [CrossRef]

4. Beyer, A.; Maisenbacher, L.; Matveev, A.; Pohl, R.; Khabarova, K.; Grinin, A.; Lamour, T.; Yost, D.C.; Hänsch, T.W.; Kolachevsky, N.; et al. The Rydberg constant and proton size from atomic hydrogen. *Science* **2017**, *358*, 79–85. [CrossRef] [PubMed]
5. Berestetskii, V.B.; Lifshitz, E.M.; Pitaevskii, L.P. *Quantum Electrodynamics. Course of Theoretical Physics: Volume 4, Second Edition* Pergamon Press: Oxford, UK, 1982. [CrossRef]
6. Milonni, P.W. *The Quantum Vacuum. An Introduction to Quantum Electrodynamics*; Academic Press, Inc.: San Diego, CA, USA, 1994. [CrossRef]
7. Lamb, W.E.; Retherford, R.C. Fine tructure of the hydrogen atom by a microwave method. *Phys. Rev.* **1947**, *72*, 241–243. [CrossRef]
8. Bethe, H.A. The electromagnetic shift of energy levels. *Phys. Rev.* **1947**, *72*, 339–341. [CrossRef]
9. Welton, T.A. Some observable effects of the quantum-mechanical fluctuations of the electromagnetic field. *Phys. Rev.* **1948**, *74*, 1157–1168. [CrossRef]
10. Power, E.A. Zero-point energy and the Lamb shift. *Am. J. Phys.* **1966**, *34*, 516–518. [CrossRef]
11. Compagno, G., Passante, R., Persico F., *Atom-Field Interactions and Dressed Atoms*; Cambridge University Press: Cambridge, UK, 1995. [CrossRef]
12. Maclay, G.J. History and some aspects of the Lamb shift. *Physics* **2020**, *2*, 105–149. [CrossRef]
13. Dyson, F.J. The electromagnetic shift of energy levels. *Phy. Rev.* **1948**, *73*, 617–626. [CrossRef]
14. French, J.B.; Weisskopf, V.F. The electromagnetic shift of energy levels. *Phys. Rev.* **1949**, *75*, 1240–1248. [CrossRef]
15. Kroll, N.M.; Lamb, W.E., Jr. On the self-energy of a bound electron. *Phys. Rev.* **1949**, *75*, 388–398. [CrossRef]
16. Weinberg, S. *The Quantum Theory of Fields. Volume 1: Foundations*; Cambridge University Press: Cambridge, UK, 1995. [CrossRef]
17. Erickson, G.W.; Yennie, D.R. Radiative level shifts, I. Formulation and lowest order Lamb shift. *Ann. Phys.* **1965**, *35*, 271–313. [CrossRef]
18. Schwinger, J. Quantum electrodynamics. I. A covariant formulation. *Phy. Rev.* **1948**, *74*, 1439–1461. [CrossRef]
19. Feynman, R.P. The development of the space-time view of quantum electrodynamics. *Phys. Today* **1966**, *19*, 31–44. [CrossRef]
20. Dyson, F.J. The radiation theories of Tomonaga, Schwinger, and Feynman. *Phys. Rev.* **1949**, *75*, 486–502. [CrossRef]
21. Akhiezer, A.I.; Berestetskiĭ, V.B. *Quantum Electrodynamics*; Interscience Publishers/John Wiley and Sons, Inc.: New York, NY, USA, 1965. Available online: https://archive.org/details/akhiezer-berestetskii-quantum-electrodynamics/page/n7/mode/2up (accessed on 30 August 2022).
22. Itzykson, C.; Zuber, J.-B. *Quantum Field Theory*; McGraw-Hill Inc.: New York, NY, USA, 1980.
23. Grotch, H. Status of the theory of the Hydrogen Lamb shift. *Found. Phys.* **1994**, *24*, 249–272. [CrossRef]
24. Feynman, R.P. The present status of quantum electrodynamics. In Proceedings of the 12th La Théorie Quantique des Chomps. Douziéme Conseil de Physique, Brussels, Belgium, 9–14 October 1961; Stoops, R., Ed.; Interscience Publishers/John Wiley and Sons, Inc.: New York, NY, USA, 1962; pp. 61–99. Available online: http://www.solvayinstitutes.be/html/solvayconf_physics.html (accessed on 30 August 2022).
25. Lieber, M. O(4) symmetry of the hydrogen atom and the Lamb shift. *Phy. Rev.* **1968**, *174*, 2037–2054. [CrossRef]
26. Huff, R.W. Simplified calculation of Lamb shift using algebraic techniques. *Phys. Rev.* **1969**, *186*, 1367–1379. [CrossRef]
27. Maclay, G.J., Dynamical symmetries of the H atom, one of the most important tools of modern physics: SO(4) to SO(4,2), background, theory, and use in calculating radiative shifts. *Symmetry* **2020**, *12*, 1323. [CrossRef]
28. Brown, L.S. Bounds on screening corrections in beta decay. *Phys. Rev.* **1964**, *135*, B314–B319. [CrossRef]
29. Schweber, S.S., *QED and the men who made it: Dyson, Feynman, Schwinger, and Tomonaga*; Princeton University Press: Princeton, NJ, USA, 1994
30. Bethe, H.A.; Salpeter, E.E., *The Quantum Mechanics of One- and Two-Electron Atoms*; Springer-Verlag: Berlin/Heidelberg, Germany, 1957. [CrossRef]
31. Sakurai, J.J. *Modern Quantum Mechanics*; Addison-Wesley Publishing Company, Inc.: New York, NY, USA, 1994. Available online: https://www.fisica.net/mecanica-quantica/Sakurai%20-%20Modern%20Quantum%20Mechanics.pdf (accessed online on 30 August 2022).
32. Morse, P.M.; Feshbach, H. *Methods of Theoretical Physics*; McGraw-Hill Book Company, Inc.: New York, NY, USA, 1953; Volume 2.

Communication

Electron as a Tiny Mirror: Radiation from a Worldline with Asymptotic Inertia

Michael R. R. Good [1,2,*] and Yen Chin Ong [3,4]

[1] Department of Physics & Energetic Cosmos Laboratory, Nazarbayev University, Astana 010000, Qazaqstan
[2] Leung Center for Cosmology & Particle Astrophysics, National Taiwan University, Taipei 10617, Taiwan
[3] Center for Gravitation and Cosmology, College of Physical Science and Technology, Yangzhou University, Yangzhou 225002, China
[4] Shanghai Frontier Science Center for Gravitational Wave Detection, School of Aeronautics and Astronautics, Shanghai Jiao Tong University, Shanghai 200240, China
* Correspondence: michael.good@nu.edu.kz

Abstract: We present a moving mirror analog of the electron, whose worldline possesses asymptotic constant velocity with corresponding Bogoliubov β coefficients that are consistent with finite total emitted energy. Furthermore, the quantum analog model is in agreement with the total energy obtained by integrating the classical Larmor power.

Keywords: acceleration radiation; moving mirrors; radiation by moving charges; quantum aspects of black holes; Davies-Fulling-Unruh effect

---

Citation: Good, M.R.R.; Ong, Y.C. Electron as a Tiny Mirror: Radiation from a Worldline with Asymptotic Inertia. *Physics* **2023**, *5*, 131–139. https://doi.org/10.3390/physics5010010

Received: 21 November 2022
Revised: 17 December 2022
Accepted: 3 January 2023
Published: 28 January 2023

Copyright: © 2023 by the authors. Licensee MDPI, Basel, Switzerland. This article is an open access article distributed under the terms and conditions of the Creative Commons Attribution (CC BY) license (https://creativecommons.org/licenses/by/4.0/).

1. Introduction: Fixed Radiation

Uniform acceleration, while attractive and simple enough is not globally physical. Consider the problem of infinite radiation energy from an eternal uniformly accelerated charge. The physics of eternal unlimited motions is not only the cause of misunderstandings, but also the starting point of incorrect physical interpretations, especially when considering global calculable quantities, such as the total radiation emitted of a moving charge. Infinite radiation energy afflicts the accurate scrutiny of physical connections between acceleration, temperature, and particle creation.

More collective consideration should be prescribed to straighten out the issue. One path forward is the use of limited non-uniform accelerated trajectories that are capable of rendering finite global total radiation energy. The trade-off with these trajectories is usually the lack of simplicity or tractability in determining the radiation spectrum in the first place.

In this short paper, we present a solution for finite radiation energy and its corresponding spectrum. Limited solutions of this type are rare and can be employed to investigate the physics associated with contexts where a globally continuous equation of motion is desired. For instance, the solution is suited for applications such as the harvesting of entropy from a globally defined trajectory of an Unruh-DeWitt detector, the non-equilibrium thermodynamics of the non-uniform Davies-Fulling-Unruh effect, or the dynamical Casimir effect [1], and the particle production of the moving mirror model [2–4].

Providing a straightforward conceptual and quantitative analog application to understanding the radiation emitted by an electron, we demonstrate the existence of a correspondence (see similar correspondences in Refs. [5–11]) between the electron and the moving mirror. At the very least, this functional coincidence is general enough to be applied to any tractably integrable rectilinear classical trajectory that emits finite radiation energy. Here, we analytically compute the relevant integrable quantities for the specific solution and demonstrate full consistency. The analog approach treats the electron as a tiny moving mirror, somewhat similar to the Schwarzschild [12], Reissner–Nordström [13], and Kerr [14] black mirror analogies, but with the asymptotic inertia of a limited acceleration trajectory. Interestingly, the analog reveals previously unknown electron acceleration radiation spectra, thus helping to develop general but precise links between acceleration, gravity, and thermodynamics.

2. Elements of Electrodynamics: Energy From Moving Electrons

In electrodynamics [15–17], the relativistically covariant Larmor formula (the speed of light, c, the electron charge, q_e, and vacuum permittivity, ϵ_0, are set to unity),

$$P = \frac{\alpha^2}{6\pi}, \tag{1}$$

is used to calculate the total power radiated by an accelerating point charge [15]. The Larmor formula's usefulness is due in part to Lorentz invariance and proper acceleration, α, is intuitive, being what an accelerometer measures in the accelerometer's own instantaneous rest frame [18].

When any charged particle accelerates, energy is radiated in the form of electromagnetic waves, and the total energy of these waves is found by integrating over coordinate time. That is, the time-integral,

$$E = \int_{-\infty}^{\infty} P \, dt, \tag{2}$$

demonstrates that the Larmor power (1) immediately tells an observer the total energy emitted by a point charge along the point's entire time-like worldline. This includes trajectories that lack horizons; see, e.g., [19]. This result is finite only when the proper acceleration is asymptotically zero, i.e., the worldline must be asymptotically inertial.

The force of radiation resistance, whose magnitude is given relativistically as the proper time, τ, derivative (notified by the prime) of the proper acceleration,

$$F = \frac{\alpha'(\tau)}{6\pi}, \tag{3}$$

is known as the magnitude of the Lorentz–Abraham–Dirac (LAD) force, see, e.g., [20]. The power, $F \cdot v$, associated with this force can be called the Feynman power [21]; here, v denotes the speed of the point. The total energy emitted is also consistent with the Feynman power, where one checks:

$$E = -\int_{-\infty}^{\infty} F \cdot v \, dt. \tag{4}$$

The negative sign demonstrates that the total work against the LAD force represents the total energy loss. That is, the total energy loss from radiation resistance due to Feynman power must equal the total energy radiated by the Larmor power (1). Larmor and Feynman powers are not the same, but the magnitude of the total energy from both are identical, at least for rectilinear trajectories that are asymptotically inertial.

Interestingly, the above results also hold in a quantum analog model of a moving mirror. A central novelty of this paper is to explicitly connect the quantum moving mirror radiation spectra with classical moving point charge radiation spectra. Traditionally (e.g., [2–4,22–24]) and recently (e.g., [25–27]), moving mirror models in $(1+1)$ dimensions are employed to study the properties of Hawking radiation for black holes. Here, we show that it is also useful to model the spectral finite energy of electron radiation. In particular, a suitably constructed mirror trajectory (which is quite natural) can produce the same total energy consistent with Equation (4) via the Bogoliubov energy, (see, e.g., [22])

$$E = \int_0^\infty \int_0^\infty \omega |\beta_{\omega\omega'}|^2 \, d\omega \, d\omega'. \tag{5}$$

Here, β are the Bogoliubov β coefficients and ω' and ω are the two sets of incoming and outgoing mode frequencies, respectively.

The final drifting speed, s, of the mirror or electron will be less than the speed of light: $0 < s < 1$. We also denote $a := \omega(1+s) + \omega'(1-s)$, $b := \omega(1-s) + \omega'(1+s)$, $c := a+b$, and $d := a - b$; note that then $c = 2(\omega + \omega')$.

3. GO Trajectory for Finite Energy Emission

We consider a globally defined, continuous worldline, which is rectilinear, time-like, and possesses asymptotic zero velocity in the far past, while travelling to an asymptotically constant velocity in the far future (but asymptotically inertial both to the past and the future). It radiates a finite amount of positive energy and has Bogoliubov β coefficients that are analytically tractable. The Good-Ong (GO) trajectory, as defined by the authors [28], proceeds as follows:

$$z(t) = \frac{s}{2\kappa} \ln(e^{2\kappa t} + 1), \qquad (6)$$

where κ is an acceleration parameter. The GO trajectory's total power, applying the Larmor formula (1), is

$$P = \frac{2\kappa^2}{3\pi} \frac{s^2 e^{-4\kappa t}\left(1 + e^{-2\kappa t}\right)^2}{\left[(1 + e^{-2\kappa t})^2 - s^2\right]^3}. \qquad (7)$$

Let us note that the power is always positive and it asymptotically drops to zero, as Figure 1 illustrates.

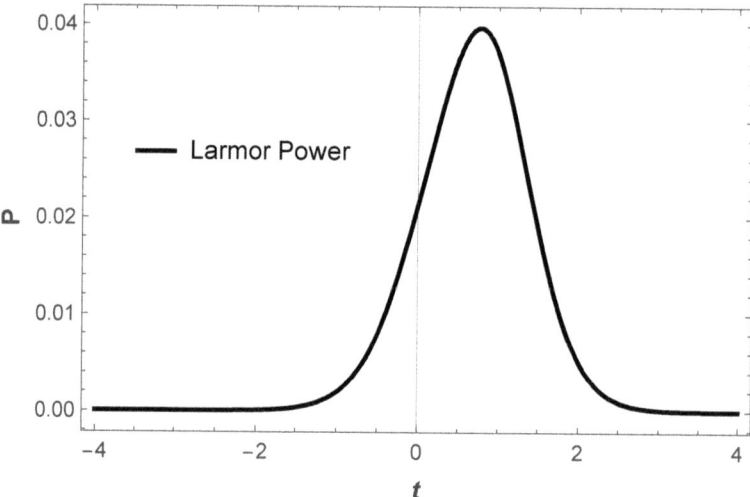

Figure 1. The Larmor power (1) of the Good-Ong (GO) trajectory (6) as a function of time, t, and at final constant speed, $s = 0.9$, i.e., Equation (7), with acceleration parameter, $\kappa = 1$. This plot illustrates that the Larmor power never emits negative energy flux (NEF) and asymptotically dies off, consistent with a physically finite amount of total radiation energy (2).

The Feynman power, $F \cdot v$, associated with the self-force (3), is

$$F \cdot v = \frac{2\kappa^2 s^2 e^{4\kappa t}\left(j_1 e^{6\kappa t} + j_2 e^{4\kappa t} + e^{2\kappa t} + 1\right)}{3\pi\left(-j_1 e^{4\kappa t} + 2e^{2\kappa t} + 1\right)^3} \qquad (8)$$

where $j_1 = s^2 - 1$ and $j_2 = 2s^2 - 1$. Similar to the Larmor power (7), the Feynman power (8) asymptotically dies off, but unlike the Larmor power, the Feynman power has a period of negative radiation reaction, as illustrated in Figure 2.

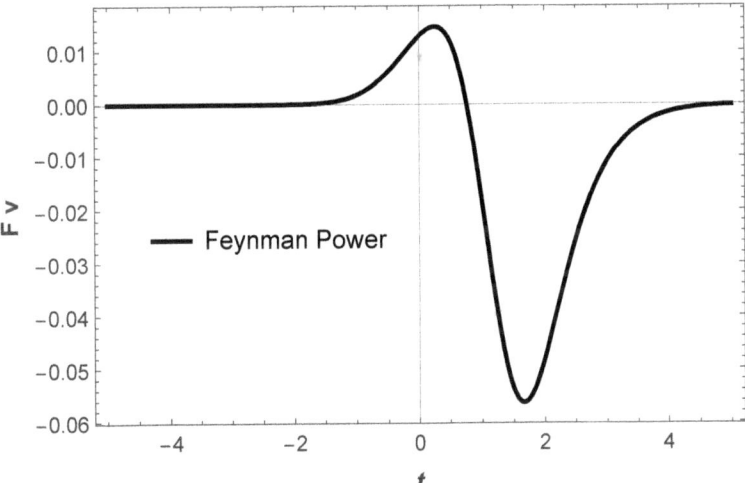

Figure 2. The Feynman power, $F \cdot v$, associated with the self-force, F (3), of the GO trajectory (6), as a function of time and at final constant speed, $s = 0.9$, i.e., Equation (8), with $\kappa = 1$. This plot illustrates the Feynman power dies off asymptotically, has a period of negative radiation reaction, and is also consistent with a physically finite amount of total radiation energy (4).

Let us now compute the total energy using either the Larmor power or Feynman power, and integrating over time. In terms of the rapidity, $\eta = \tanh^{-1} v$, and Lorentz factor, γ, the total energy is given by

$$E = \frac{\kappa}{24\pi}\left[\left(\gamma^2 - 1\right) + \left(\frac{\eta}{s} - 1\right)\right]. \tag{9}$$

The second term, $(\frac{\eta}{s} - 1) = \frac{1}{2s}\ln\frac{1+s}{1-s} - 1$, is proportional to the lowest-order soft energy of inner bremsstrahlung in the case of beta decay (see Equation (3) in Ref. [11]), which is the deep infra-red contribution. One can see that Equation (9) is finite for all $0 < s < 1$ and consistent with both the Larmor power and Feynman power. Below, after we compute the Bogoliubov β-spectrum (and plot it in Figure 3), we compute the Bogoliubov total energy (and plot it in Figure 4). We call Equation (9) the Larmor energy to differentiate it from the Bogoliubov energy (5), while substituting Equation (13) below. The energy is a function of the final constant speed, s.

Finally, the spectrum given by the Bogoliubov coefficients is best found by first considering the presence of a mirror in vacuum, e.g., [29,30]. The mode functions that correspond to the in-vacuum state,

$$\phi_{\omega'}^{\text{in}} = \frac{1}{\sqrt{4\pi\omega'}}\left[e^{-i\omega' v} - e^{-i\omega' p(u)}\right], \tag{10}$$

and mode functions that correspond to the out-vacuum state,

$$\phi_\omega^{\text{out}} = \frac{1}{\sqrt{4\pi\omega}}\left[e^{-i\omega f(v)} - e^{-i\omega u}\right], \tag{11}$$

comprise the two sets of incoming and outgoing modes needed for the Bogoliubov coefficients. The $f(v)$ and $p(u)$ functions express the trajectory (6) of the mirror, but in null coordinates, $u = t - z$ and $v = t + z$. In spacetime coordinates congruent with Equation (6), one form of the β-integral [19] for one side of the mirror is

$$\beta_{\omega\omega'} = \int_{-\infty}^{\infty} dz \frac{e^{i\omega_n z - i\omega_p t(z)}}{4\pi\sqrt{\omega\omega'}}\left[\omega_p - \omega_n t'(z)\right], \tag{12}$$

where $\omega_p = \omega + \omega'$ and $\omega_n = \omega - \omega'$. Combining the results for each side of the mirror [31] by adding the squares of the Bogoliubov β coefficients ensures that one accounts for all of the radiation emitted by the mirror [32]. The overall count per double-mode is

$$|\beta_{\omega\omega'}|^2 = \frac{s^2 \omega \omega' Z}{2\pi abcd\kappa}\left(\frac{e^{\frac{\pi d}{4\kappa}} - 1}{e^{\frac{\pi c}{4\kappa}} - 1}\right) e^{\frac{\pi b}{4\kappa}}, \qquad (13)$$

where $Z = b\,\mathrm{csch}\left(\frac{\pi a}{4\kappa}\right) + a\,\mathrm{csch}\left(\frac{\pi b}{4\kappa}\right)$. Equation (13) combines the squares, $|\beta_R|^2 + |\beta_L|^2$, of the coefficients for left (L) and right (R) sides of mirror [28]. Figure 3 shows a plot of the symmetry between the modes ω and ω'.

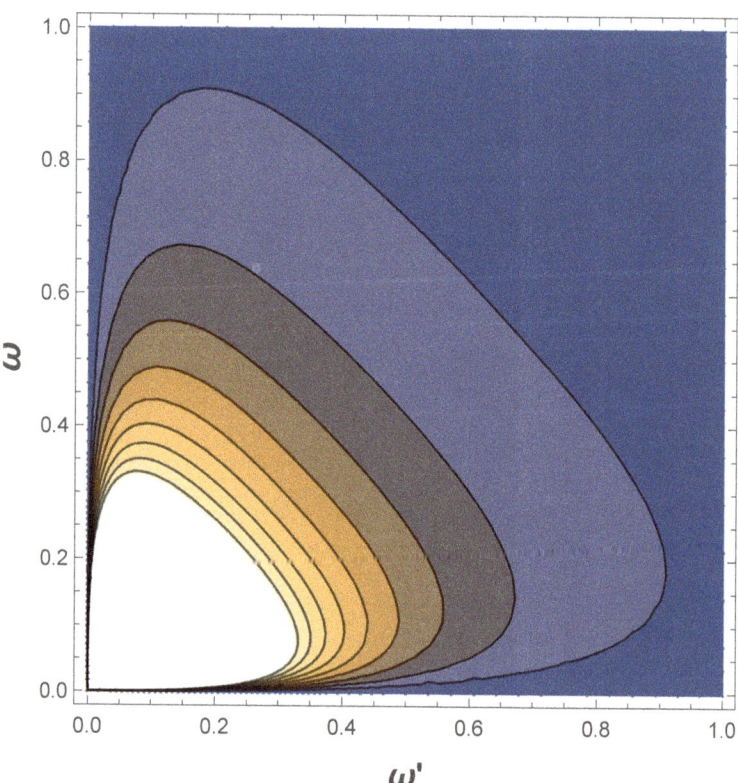

Figure 3. A contour plot of the coefficients (13) as a function of in and out modes, ω' and ω, where final constant speed, $s = 0.444$ and $\kappa = 1$. The color gradient darkens for lower values of the count. This plot underscores the symmetry of the modes in the particle per mode squared distribution spectrum of the Bogoliubov β coefficients (13).

It is then straightforward to verify that the total energy obtained by integrating the power is the same as by using the Bogoliubov β integral (5). We cannot prove this analytically, but a numerical integral is quite convincing; see Figure 4 for a plot of the Larmor and Bogoliubov energies as a function of final constant speed.

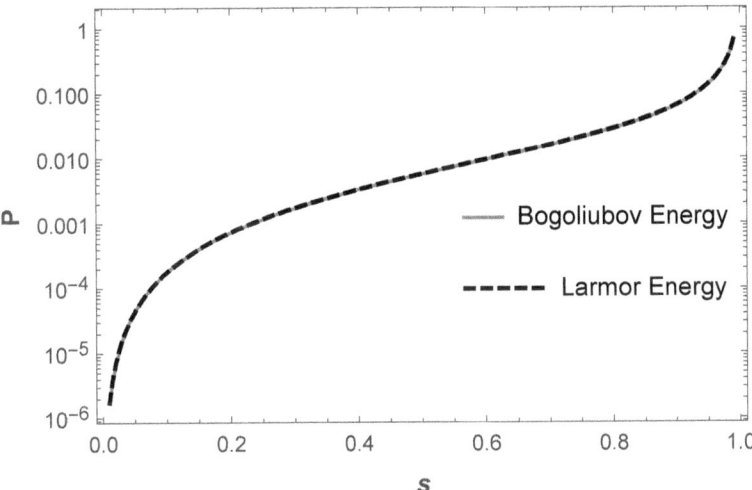

Figure 4. The Larmor energy (9) and Bogoliubov energy (5) as a function of final constant speed, s. Here, $0 < s < 0.99$ and $\kappa = 1$. This plot confirms that Larmor and Bogoliubov energies are equivalent, substantiating the double-sided moving mirror as an analog model of the electron.

4. Discussions: Mirrors, Electrons, and Black Holes

Prior studies of accelerated electrons and their relationship to mirrors are few; however, several papers, e.g., [33,34], connect electrons to the general Davies-Fulling-Unruh effect (for example, perhaps the most known one is by Bell and Leinaas [35], which considered the possibility of using accelerated electrons as thermometers to demonstrate the relationship between acceleration and temperature). Nevertheless, perhaps an early clue a that functional identity existed was made in 1982 by Ford and Vilenkin [24], who found that the LAD self-force was the same form for both mirrors and electrons. In 1995, Nikishov and Ritus [5] asserted the spectral symmetry and found that the LAD radiation reaction has a term that corresponds to the negative energy flux (NEF) from moving mirrors. Ritus examined [6–8] the correspondence connecting the radiation from both the electron and mirror systems, claiming not only a deep symmetry between the two, but a fundamental identity related to bare charge quantization [10]. Recently, the duality was extended to Larmor power [32] and deep infrared radiation [11]. The approach has pedagogical application; for instance, it was used to demonstrate the physical difference between radiation power loss and kinetic power loss [9].

The GO moving mirror was initially constructed to model the evaporation of black holes that exhibit a "death gasp" [36–38]—an emission of NEF due to unitarity being preserved. Therefore, it is a sturdy result that the total finite energy emitted from the double-sided mirror matches the result from the Larmor formula for an electron. In this sense, the GO mirror trajectory represents a crude but functional analog of a drifting electron that starts at zero velocity and speeds away to some constant velocity. A single-sided moving mirror does not account for all of the radiation emitted by an electron, and as such, the single-sided mirror radiation spectrum differs from the electron spectrum. A notable difference: there is no known NEF radiated from an electron.

In the literature, one finds some properties that are shared by black holes and electrons. For example, the ratio of the magnetic moment of an electron (of mass m and charge e) to its spin angular momentum is $ge/2m$ with $g = 2$, which is twice the value of the gyromagnetic ratio for classical rotating charged bodies ($g = 1$). Curiously, as Carter has shown [39], a Kerr–Newman black hole also has $g = 2$. This has led to some speculations as to whether the electron is a Kerr–Newman singularity (the angular momentum and charge of the electron are too large for a black hole of the electron's mass, so there is no horizon) [40] (see also [41,42]). The no-hair property of black holes is also similar to elementary particle indistinguishably: all electrons look the same. Certainly, the mirror model considered here looks too simple to seek certain further connections between particle physics and black holes; in particular, it does not involve any charge or angular momentum. Nevertheless, precisely because of this, it is surprising the total emitted energy in the model should be given by the integral of the Larmor formula.

Near-term possible theoretical applications of electron-mirror correspondence include extension to non-rectilinear trajectories; notably, the uniform accelerated worldlines of Letaw [43], which have Unruh-like temperatures [44] and power distributions [45]. Applying the general study of Kothawala and Padmanabhan [46] to electrons moving along time-dependent accelerations, and comparing the effect to an Unruh-DeWitt detector may prove to be fruitful for understanding the thermal response. Moreover, moving mirror models can be useful in cosmology [47], in particular, in modeling particle production due to the expansion of space [48]. This expansion is accelerated due to an unknown dark energy, which may not be a cosmological constant and thus can decay [49,50]. If dark energy is some kind of vacuum energy, it might be subject to further study from mirror analogs just like Casimir energy. (Actually, dark energy could be Casimir-like [51].)

Near-term possible experimental applications of electron-mirror holography include leveraging the correspondence to disentangle effects in experiments like in the Analog Black Hole Evaporation via Lasers (AnaBHEL) [52] and the RDK II [53,54] experiments (see also [55]). The former exploits the accelerating relativistic moving mirror as a probe of the spectrum of quantum vacuum radiation [56,57], and the latter measures the photon spectrum with high precision as the electron-mirror is subjected to extreme accelerations during the process of radiative neutron beta decay.

Author Contributions: Both authors have contributed equally to the work. All authors have read and agreed to the published version of the manuscript.

Funding: M.R.R.G. thanks the FY2021-SGP-1-STMM Faculty Development Competitive Research Grant No. 021220FD3951 at Nazarbayev University (Astana, Qazaqstan). Y.C.O. thanks the National Natural Science Foundation of China (No. 11922508) for funding support.

Data Availability Statement: Not applicable

Conflicts of Interest: The authors declare no conflict of interest.

References

1. Moore, G.T. Quantum theory of the electromagnetic field in a variable-length one-dimensional cavity. *J. Math. Phys.* **1970**, *11*, 2679–2691. [CrossRef]
2. DeWitt, B.S. Quantum field theory in curved space-time. *Phys. Rep.* **1975**, *19*, 295–357. [CrossRef]
3. Fulling, S.A.; Davies, P.C.W. Radiation from a moving mirror in two dimensional space-time: Conformal anomaly. *Proc. R. Soc. Lond. A* **1976**, *348*, 393–414. [CrossRef]
4. Davies, P.C.W.; Fulling, S.A. Radiation from moving mirrors and from black holes. *Proc. R. Soc. Lond. A* **1977**, *356*, 237–257. [CrossRef]
5. Nikishov, A.; Ritus, V. Emission of scalar photons by an accelerated mirror in (1+1) space and its relation to the radiation from an electrical charge in classical electrodynamics. *J. Exp. Theor. Phys.* **1995**, *81*, 615–624. Available online: http://jetp.ras.ru/cgi-bin/e/index/e/81/4/p615?a=list (accessed on 30 December 2022).
6. Ritus, V. The Symmetry, inferable from Bogoliubov transformation, between the processes induced by the mirror in two-dimensional and the charge in four-dimensional space-time. *J. Exp. Theor. Phys.* **2003**, *97*, 10–23. [CrossRef]
7. Ritus, V. Vacuum-vacuum amplitudes in the theory of quantum radiation by mirrors in 1+1-space and charges in 3+1-space. *Int. J. Mod. Phys. A* **2002**, *17*, 1033–1040. [CrossRef]
8. Ritus, V. Symmetries and causes of the coincidence of the radiation spectra of mirrors and charges in (1+1) and (3+1) spaces. *J. Exp. Theor. Phys.* **1998**, *87*, 25–34. [CrossRef]
9. Good, M.R.R.; Singha, C.; Zarikas, V. Extreme electron acceleration with fixed radiation energy. *Entropy* **2022**, *24*, 1570. [CrossRef]
10. Ritus, V.I. Finite value of the bare charge and the relation of the fine structure constant ratio for physical and bare charges to zero-point oscillations of the electromagnetic field in the vacuum. *Phys.-Usp.* **2022**, *65*, 468–486. [CrossRef]
11. Good, M.R.R.; Davies, P.C.W. Infrared acceleration radiation. *arXiv* **2022**, arXiv:2206.07291. [CrossRef]
12. Good, M.R.R.; Anderson, P.R.; Evans, C.R. Mirror reflections of a black hole. *Phys. Rev. D* **2016**, *94*, 065010. [CrossRef]

13. Good, M.R.R.; Ong, Y.C. Particle spectrum of the Reissner–Nordström black hole. *Eur. Phys. J. C* **2020**, *80*, 1169. [CrossRef]
14. Good, M.R.R.; Foo, J.; Linder, E.V. Accelerating boundary analog of a Kerr black hole. *Class. Quantum Grav.* **2021**, *38*, 085011. [CrossRef]
15. Jackson, J.D. *Classical Electrodynamics*; John Wiley & Sons, Inc.: Hoboken, NJ, USA, 1999. Available online: https://www.scribd.com/doc/200995304/Classical-Electrodynamics-3rd-Ed-1999-John-David-Jackson (accessed on 30 December 2022).
16. Zangwill, A. *Modern Electrodynamics*; Cambridge University Press: Cambridge, UK, 2012. Available online: https://faculty.kashanu.ac.ir/file/download/page/1604994434-modern-electrodynamics.pdf (accessed on 30 December 2022).
17. Griffiths, D.J. *Introduction to Electrodynamics*; Cambridge University Press: New York, NY, USA, 2017. [CrossRef]
18. Rindler, W.A. *Essential Relativity: Special, General and Cosmological*; Springer-Verlag: Berlin/Heidelber, Germany, 1977. [CrossRef]
19. Good, M.R.R.; Yelshibekov, K.; Ong, Y.C. On horizonless temperature with an accelerating mirror. *J. High Energy Phys.* **2017**, *3*, 13. [CrossRef]
20. Myrzakul, A.; Xiong, C.; Good, M.R.R. CGHS black hole analog moving mirror and its relativistic quantum information as radiation reaction. *Entropy* **2021**, *23*, 1664. [CrossRef]
21. Feynman, R.P. *Feynman Lectures on Gravitation*; CRC Press/Taylor & Francis Group: Boca Raton, FL, USA, 2003. [CrossRef]
22. Walker, W.R. Particle and energy creation by moving mirrors. *Phys. Rev. D* **1985**, *31*, 767–774. [CrossRef]
23. Carlitz, R.D.; Willey, R.S. Reflections on moving mirrors. *Phys. Rev. D* **1987**, *36*, 2327–2335. [CrossRef]
24. Ford, L.H.; Vilenkin, A. Quantum radiation by moving mirrors. *Phys. Rev. D* **1982**, *25*, 2569–2575. [CrossRef]
25. Good, M.R.; Linder, E.V.; Wilczek, F. Moving mirror model for quasithermal radiation fields. *Phys. Rev. D* **2020**, *101*, 025012. [CrossRef]
26. Moreno-Ruiz, A.; Bermudez, D. Optical analogue of the Schwarzschild–Planck metric. *Class. Quant. Grav.* **2022**, *39*, 145001. [CrossRef]
27. Good, M.R.R.; Linder, E.V. Modified Schwarzschild metric from a unitary accelerating mirror analog. *New J. Phys.* **2021**, *23*, 043007. [CrossRef]
28. Good, M.R.R.; Ong, Y.C. Signatures of energy flux in particle production: A black hole birth cry and death gasp. *J. High Energy Phys.* **2015**, *7*, 145. [CrossRef]
29. Birrell, N.D.; Davies, P.C.W. *Quantum Fields in Curved Space*; Cambridge Univercity Press: New York, NY, USA, 1982. [CrossRef]
30. Good, M.R.R. Extremal Hawking radiation. *Phys. Rev. D* **2020**, *101*, 104050. [CrossRef]
31. Good, M.R.R. Reflecting at the speed of light. In *Memorial Volume for Kerson Huang*; Phua, K.K., Low, H.B., Xiong, C., Eds.; World Scientific Co., Ltd.: Singapore, 2017; pp. 113–116. [CrossRef]
32. Zhakenuly, A.; Temirkhan, M.; Good, M.R.R.; Chen, P. Quantum power distribution of relativistic acceleration radiation: Classical electrodynamic analogies with perfectly reflecting moving mirrors. *Symmetry* **2021**, *13*, 653. [CrossRef]
33. Myhrvold, N.P. Thermal radiation from accelerated electrons. *Ann. Phys.* **1985**, *160*, 102–113. [CrossRef]
34. Paithankar, K.; Kolekar, S. Role of the Unruh effect in Bremsstrahlung. *Phys. Rev. D* **2020**, *101*, 065012. [CrossRef]
35. Bell, J.S.; Leinaas, J.M. Electrons as accelerated thermometers. *Nucl. Phys. B* **1983**, *212*, 131–150. [CrossRef]
36. Bianchi, E.; Smerlak, M. Last gasp of a black hole: Unitary evaporation implies non-monotonic mass loss. *Gen. Rel. Grav.* **2014**, *46*, 1809. [CrossRef]
37. Bianchi, E.; Smerlak, M. Entanglement entropy and negative energy in two dimensions. *Phys. Rev. D* **2014**, *90*, 041904. [CrossRef]
38. Abdolrahimi, S.; Page, D.N. Hawking radiation energy and entropy from a Bianchi-Smerlak semiclassical black hole. *Phys. Rev. D* **2015**, *92*, 083005. [CrossRef]
39. Carter, B. Global structure of the Kerr family of gravitational fields. *Phys. Rev.* **1968**, *174*, 1559–1571. [CrossRef]
40. Burinskii, A. The Dirac-Kerr electron. *Grav. Cosmol.* **2008**, *14*, 109–122 [CrossRef]
41. Schmekel, B.S. Quasilocal energy of a charged rotating object described by the Kerr-Newman metric. *Phys. Rev. D* **2019**, *100*, 124011. [CrossRef]
42. Burinskii, A. The Dirac electron consistent with proper gravitational and electromagnetic field of the Kerr–Newman solution. *Galaxies* **2021**, *9*, 18. [CrossRef]
43. Letaw, J.R. Vacuum excitation of noninertial detectors on stationary world lines. *Phys. Rev. D* **1981**, *23*, 1709–1714. [CrossRef]
44. Good, M.; Juárez-Aubry, B.A.; Moustos, D.; Temirkhan, M. Unruh-like effects: Effective temperatures along stationary worldlines. *J. High Energy Phys.* **2020**, *06*, 059. [CrossRef]
45. Good, M.R.R.; Temirkhan, M.; Oikonomou, T. Stationary worldline power distributions. *Int. J. Theor. Phys.* **2019**, *58*, 2942–2968. [CrossRef]
46. Kothawala, D.; Padmanabhan, T. Response of Unruh-DeWitt detector with time-dependent acceleration. *Phys. Lett. B* **2010**, *690*, 201–206. [CrossRef]
47. Good, M.R.R.; Zhakenuly, A.; Linder, E.V. Mirror at the edge of the universe: Reflections on an accelerated boundary correspondence with de Sitter cosmology. *Phys. Rev. D* **2020**, *102*, 045020. [CrossRef]
48. Castagnino, M.; Ferraro, R. A toy cosmology: Radiation from moving mirrors, the final equilibrium state and the instantaneous model of particle. *Ann. Phys.* **1985**, *161*, 1–20. [CrossRef]
49. Akhmedov, E.T.; Buividovich, P.V.; Singleton, D.A. De Sitter space and perpetuum mobile. *Phys. Atom. Nucl.* **2012**, *75*, 525–529. [CrossRef]
50. Polyakov, A.M. Decay of vacuum energy. *Nucl. Phys. B* **2010**, *834*, 316–329. [CrossRef]

51. Leonhardt, U. The case for a Casimir cosmology. *Philos. Trans. R. Soc. A* **2020**, *378*, 20190229. [CrossRef] [PubMed]
52. Chen, P. et al. [AnaBHEL Collaboration]. AnaBHEL (Analog Black Hole Evaporation via Lasers) experiment: Concept, design, and status. *Photonics* **2022**, *9*, 1003. [CrossRef]
53. Nico, J.S.; Dewey, M.S.; Gentile, T.R.; Mumm, H.P.; Thompson, A.K.; Fisher, B.M.; Kremsky, I.; Wietfeldt, F.E.; Chupp, T.E.; Cooper, R.L.; et al. Observation of the radioactive decay mode of the free neutron. *Nature* **2006**, *444*, 1059–1062. [CrossRef] [PubMed]
54. Bales, M.J. et al. [RDK II Collaboration]. Precision measurement of the radiative β decay of the free neutron. *Phys. Rev. Lett.* **2016**, *116*, 242501. [CrossRef]
55. Lynch, M.H.; Good, M.R.R. Experimental observation of a moving mirror. *arXiv* **2022**, arXiv:2211.14774. [CrossRef]
56. Chen, P.; Mourou, G. Trajectory of a flying plasma mirror traversing a target with density gradient. *Phys. Plasmas* **2020**, *27*, 123106. [CrossRef]
57. Chen, P.; Mourou, G. Accelerating plasma mirrors to investigate black hole information loss paradox. *Phys. Rev. Lett.* **2017**, *118*, 045001. [CrossRef] [PubMed]

Disclaimer/Publisher's Note: The statements, opinions and data contained in all publications are solely those of the individual author(s) and contributor(s) and not of MDPI and/or the editor(s). MDPI and/or the editor(s) disclaim responsibility for any injury to people or property resulting from any ideas, methods, instructions or products referred to in the content.

Opinion

Physical Mechanisms Underpinning the Vacuum Permittivity

Gerd Leuchs [1,2,*], Margaret Hawton [3] and Luis L. Sánchez-Soto [1,4]

[1] Max-Planck-Institut für die Physik des Lichts, 91058 Erlangen, Germany
[2] Institut für Optik, Information und Photonik, Universität Erlangen-Nürnberg, 91058 Erlangen, Germany
[3] Department of Physics, Lakehead University, Thunder Bay, ON P7B 5E1, Canada
[4] Departamento de Óptica, Facultad de Física, Universidad Complutense, 28040 Madrid, Spain
* Correspondence: gerd.leuchs@mpl.mpg.de

Abstract: The debate about the emptiness of space goes back to the prehistory of science and is epitomized by the Aristotelian 'horror vacui', which can be seen as the precursor of the ether, whose modern version is the dynamical quantum vacuum. In this paper, we suggest to change a common view to 'gaudium vacui' and discuss how the vacuum fluctuations fix the value of the permittivity, ε_0, and permeability, μ_0, by modelling their dynamical response by three-dimensional harmonic oscillators.

Keywords: quantum vacuum; quantum electrodynamics; linear response

1. Introduction

When James Clerk Maxwell introduced the displacement current as a property of empty space and as a source of magnetic field, he struck gold. The result was his famous set of self-consistent equations, which later even turned out to be Lorentz invariant and describe electromagnetism with greatest precision.

Maxwell was visualizing the displacement current as part of the ether [1], an all-pervading medium composed of a subtle substratum. This is a powerful explanatory concept that goes back to the prehistory of science and helped unify our understanding of the physical world for centuries [2]. However, the ether was soon abandoned as a consequence of Albert Einstein's special theory of relativity, which contradicts an absolute reference frame, and the vacuum was considered void (nonetheless, Einstein's relationship with the ether was complex, and changed over time [3]). However, this move was merely an elegant paradigm shift rather than a necessity forced by observation.

Electrodynamics was the new theory of electromagnetic fields interacting with the—at the time—newly discovered elementary particle, the electron. Already then, cumbersome divergences were looming around the corner and we are struggling with them ever since. One was the question whether or not the electron has a finite radius, and if it does not, as hinted by all experiments at higher and higher energies (or, rather, momentum exchange), then the mass diverges and even the charge of the electron diverges on small enough length scales [4]. Another such early divergence was discovered by Max Planck [5]. Previously, he had postulated that the energy had to be quantized in packets of $h\nu$ per mode to derive his famous blackbody radiation formula, where h is the Planck constant and ν is the electromagnetic radiation frequency. However, in 1912, when going to the asymptotic limit for long wavelengths or high temperatures and match it with the experimental observations, Planck noticed that there was an additional contribution $h\nu/2$ to the energy per mode—this is the first time the ground state energy of a quantum harmonic oscillator appeared in the literature.

Later, after Paul Dirac [6] hypothesized the existence of the antielectron, later called positron, and its subsequent experimental discovery, scientists struggled in vain for years to formulate a consistent quantum theory of electrodynamics. The breakthrough came

Citation: Leuchs, G.; Hawton, M.; Sánchez-Soto, L.L. Physical Mechanisms Underpinning the Vacuum Permittivity. *Physics* 2023, 5, 179–192. https://doi.org/10.3390/physics5010014

Received: 3 October 2022
Revised: 30 December 2022
Accepted: 11 January 2023
Published: 8 February 2023

Copyright: © 2023 by the authors. Licensee MDPI, Basel, Switzerland. This article is an open access article distributed under the terms and conditions of the Creative Commons Attribution (CC BY) license (https://creativecommons.org/licenses/by/4.0/).

when Richard Feynman [7] and Julian Schwinger [8] used their approach to first postulate Maxwell's equations and then added the interaction with electrons and positrons [9]: modern quantum electrodynamics (QED) was born. By using the procedure of renormalization, the divergences were modified to much better behaved logarithmic divergences. More elementary particles were discovered, including charged ones, and QED is overall highly successful, but some struggle with divergences remains.

Within the framework of QED, it is understood that the quantum vacuum is not void. By now, there is ample experimental evidence for the nonzero ground state energies of quantum fields populating the vacuum, containing the seeds of multiple virtual processes [10–12]. Wilczek [13] expresses the fundamental characteristics of space and time as properties of the 'grid', the entity one perceives as empty space. The deepest physical theories reveal it to be highly structured; indeed, it appears as the primary ingredient of reality. Several effects manifest themselves when the vacuum is perturbed in specific ways: vacuum fluctuations lead to shifts in the energy level of atoms [14], changes in the boundary conditions produce particles [15], and accelerated motion [16] and gravitation [17] can create thermal radiation. A careful discussion of the nature of these 'vacuum fluctuations' can be found in [18], although we think that, perhaps, the 'vacuum uncertainty' would be a better term.

Since this quantum vacuum is not void any more, it is natural to mull over the prospect of treating it as a medium with electric and magnetic polarizability. This idea can be traced back as far as Furry and Oppenheimer [19], Weisskopf and Pauli [20,21] (see English translation in Ref. [22]), Dicke [23], and Heitler [24].

At this point one might wonder about how the linear response of the quantum vacuum, which one might—but does not have to—relate to a modern Lorentz-invariant ether, is contained in Maxwell's [25]. As this linear response is thought to be already included in Maxwell's equations and since they were axiomatically postulated, the linear response is not explicitly considered anymore in QED. Along this line of thinking, Maxwell's equations already contain the effect of the bare vacuum and only the so-called off-shell contributions will still have to be explicitly considered in QED. Details are given below.

Thus, here, we interpret the response of the bare vacuum as caused in full by the vacuum polarization, i.e., the on-shell contribution. This contribution, however, diverges when attempting to determine it in the frame of QED: when calculating the bare vacuum contribution to Maxwell's equations using the standard QED procedure we do find a closed mathematical expression dependent only on the off-shell momentum value at which the electromagnetic coupling strength diverges.

In an attempt to do a back-of-the-envelope calculation, in this paper, we find a reasonable way to cope with the divergences. This crude derivation uses a relativistic momentum cutoff [23], similar to the one used by Bethe [26] to calculate the Lamb shift in hydrogen. It is surprising how well this works. The numbers come out in the right ballpark.

At face value, Maxwell's equations in vacuum are about electromagnetic fields and the coupling strength between fields and charged particles should not be relevant. However, in the spirit of the discussion above, there is interaction with the vacuum uncertainty, i.e., with virtual electron–positron pairs, and this determines the values of the vacuum permittivity, ε_0, and permeability, μ_0. Thus the coupling strength matters also here. Traditionally, the QED coupling strength is given by Sommerfeld's fine structure constant, $\alpha = e^2/(4\pi\varepsilon_0 \hbar c_{rel})$, in SI units (International System of Units), with e denoting the electron charge. In Maxwell's equations it is somewhat hidden, but it is there [27]. The parameter c_{rel} refers to the limiting speed in special relativity and not necessarily denotes the speed of light, for the purpose of the derivation here. In this paper, we elaborate on the above ideas and show that ε_0 and μ_0 can be estimated from first principles and, thus, also the speed of light.

2. A Dielectric Model of Vacuum Polarization

In textbooks of electromagnetism it is often implicitly assumed that ε_0 and μ_0 are merely measurement system constants. In this vein, they are not considered as fundamental physical properties, but rather artifacts of the SI units, which disappear in Gaussian units. However, this quite simplified viewpoint ignores that, irrespective of the method of allocating a value to ε_0 and μ_0, they just translate into the prediction of Maxwell's equations that, in free space, electromagnetic waves propagate at the speed of light, which has a very specific value and is certainly associated with units. It is therefore more transparent if one includes the susceptibility of the vacuum χ_{vac}, so that in the vacuum Maxwell's displacement reads as $\mathbf{D} = \varepsilon_0 \chi_{\text{vac}} \mathbf{E}$. In the SI system all the dimensions and the numerical value is put into ε_0 such that $\chi_{\text{vac}} = 1$. In comparison, in the Gaussian system of units, $\chi_{\text{vac}} = 1$ is likewise chosen to be 'one' and one might say that the modified definition of the electric field absorbs the properties of the vacuum unit in this case. So, for the vacuum one has $\mathbf{D} = \chi_{\text{vac}} \mathbf{E} = \mathbf{E}$ and the vacuum response is actually hidden in the Gaussian units. In what follows, we use SI units only and the vacuum response is given by the product, $\varepsilon_0 \chi_{\text{vac}}$. Only the product of the two factors has physical significance and writing this as two factors was a result of the historical development.

In a dielectric, it is customary to define the electric displacement $\mathbf{D}$ and the magnetic field $\mathbf{H}$ as

$$\mathbf{D}(\mathbf{r},t) = \varepsilon_0 \mathbf{E}(\mathbf{r},t) + \mathbf{P}(\mathbf{r},t), \qquad (1)$$

$$\mathbf{H}(\mathbf{r},t) = \frac{1}{\mu_0} \mathbf{B}(\mathbf{r},t) - \mathbf{M}(\mathbf{r},t),$$

where $\mathbf{P}$ is the polarization and $\mathbf{M}$ the magnetization induced by the external fields with $\mathbf{r}$ and t denoting the position and time, respectively. In the literature, one can find the observation that $\mathbf{D}(\mathbf{r},t)$ and $\mathbf{H}(\mathbf{r},t)$ are the sum of two completely different physical quantities [28]. However, the authors do not share this view and interpret $\varepsilon_0 \mathbf{E}(\mathbf{r},t)$ and $-\mathbf{B}(\mathbf{r},t)/\mu_0$ as the polarization and magnetization of the vacuum, in this sense we are adding similar quantities. This might appear preposterous in classical electromagnetism, but, as declared in Section 1, the modern view [29] interprets that particle–antiparticle pairs are continually being created in a vacuum filled with the vacuum fluctuations. They live for a brief period of time and then annihilate each other. The lifetime of such a virtual particle pair is governed by its rest energy through the energy–time uncertainty principle [30,31],

$$\Delta \mathcal{E} \, \Delta \tau \gtrsim \hbar, \qquad (2)$$

where $\Delta \mathcal{E}$ is the root-mean-square measure of energy nonconservation and $\Delta \tau$ the time interval, during which this nonconservation is sustained. The creation of this virtual pair requires a surplus energy of at least $2mc_{\text{rel}}^2$, where m is the mass of each partner (we stress again that here c_{rel} is the limiting speed appearing in Lorentz transformations. After all, in this paper, we want to calculate ε_0 and μ_0 based on the properties of the vacuum, and this results then in a value of the speed of light based only on these properties of the vacuum. If the result of the crude model here is found to be close to the known value of the speed of light, this will be an indication of the relevance of the enough simple model).

Therefore, energy conservation must be violated by $\Delta \mathcal{E} \gtrsim 2mc_{\text{rel}}^2$. Equation (2) says that the violation is not detectable in a period shorter than $\hbar/(2mc_{\text{rel}}^2)$ (with $\hbar$ the reduced Planck's constant), so virtual particles can survive about that long. However, nothing can move faster than the relativistic speed limit, so the virtual pair must remain within a distance $d = \hbar/(2mc_{\text{rel}})$; that is, a distance of order of the Compton wavelength,

$$\lambda_C = \frac{\hbar}{mc_{\text{rel}}}. \qquad (3)$$

This also demonstrates that heavy pairs require a larger $\Delta\mathcal{E}$ and thus their effect is concentrated at smaller distances. For that reason, let us so far consider only electron–positron pairs.

In the linear response, one expresses the polarization of matter, $\mathbf{P}_{mat}$, in terms of the corresponding matter susceptibility, χ_{mat}: $\mathbf{P}_{mat}(\omega) = \varepsilon_0 \chi_{mat}(\omega) \mathbf{E}(\omega)$ (and, similarly, for the magnetization) with ω the wave angular frequency. Whenever a medium is dispersive, the linear response is nonlocal in time and integration over past times is required. However, the linear response is local in the frequency domain. Therefore, in order to account for dispersion in the simplest way, let us express the linear response in the frequency domain [32].

As noticed above, in Equation (1), the first term is expected to have an equivalent structure:

$$\mathbf{P}_{vac}(\omega) = \varepsilon_0 \chi_{vac}(\omega) \mathbf{E}(\omega). \tag{4}$$

The vacuum has no resonances and it is homogeneous. The conservation of momentum prohibits the excitation of a virtual pair to a real pair in free space with a plane wave. Far away from resonance, the process is allowed because of the quantum uncertainty of the momentum. In contradistinction, a converging electromagnetic dipole wave may excite real pairs in the vacuum [33]. So, under normal conditions, χ_{vac} has no temporal or spatial frequency dependence and is considered a constant in classical electromagnetism. Historically, as emphasized above, it was chosen to be unity and all the property of the vacuum such as units and numerical value is put into ε_0. Therefore, the familiar expression for $\mathbf{D}$ is

$$\mathbf{D}(\omega) = \varepsilon_0 \chi_{vac} \mathbf{E}(\omega) + \varepsilon_0 \chi_{mat}(\omega) \mathbf{E}(\omega) = \varepsilon_0 [1 + \chi_{mat}(\omega)] \mathbf{E}(\omega). \tag{5}$$

If the value of ε_0 is determined by the structure of the vacuum, it should be possible to calculate it by examining the (polarizing) interaction of photons introduced into the vacuum as test particles [18], as sketched in Figure 1. The possibility that a charged pair can form an atomic bound state (the electron–positron vacuum fluctuation in the lowest energy level at $-2mc_{rel}^2$ that has zero angular momentum is called parapositronium, which is a singlet spin state [34,35]), which can, thus, be well approximated by an oscillator, was discussed by Ruark [36] and further elaborated by Wheeler [37].

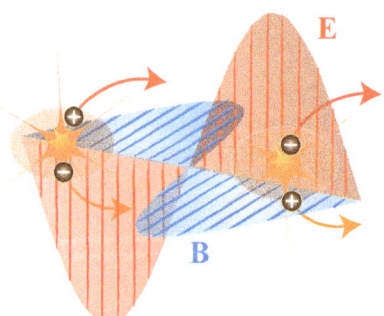

Figure 1. Cartoon view of the particle-antiparticle (denoted by "+" and "−") pairs continually created in the vacuum. The arrows indicate the trajectories of the corresponding particles. See text for more details.

These ideas have recently been readressed [38–41] to calculate ab initio ε_0 by using methods similar to those employed to determine the permittivity in a dielectric. As it is known [42], when interacting with an electric field, an atom in its ground state interacts with the electric field as if it were a harmonic oscillator. Here, we adopt the same strategy to treat the virtual pairs composing the vacuum. This is a reasonable assumption as long as deviations from the equilibrium under the action of an electric or magnetic field are tiny, as they are for the vacuum under normal conditions in a low-energy optics laboratory. In this situation, one can do a Taylor expansion around the point of equilibrium and the harmonic response will dominate. The only parameter needed is the charge of the elementary particles and the effective frequency of the oscillator. The latter is given by the 'spring constant', i.e., the energy gap between the ground state of a virtual pair and first excited level [25,38], where the particles are real. This gap is twice the rest-mass energy, $2mc_{\text{rel}}^2$, of one of the elementary particles of mass m. No other assumptions are required.

The harmonic oscillator assumption allows one to calculate both the induced electric dipoles and magnetic dipoles [43], as sketched in the Appendix A. The only two remaining ambiguities left are (i) whether there are charged elementary particles beyond the ones accounted for in the standard model and (ii) the volume occupied by a single virtual pair in the Dirac sea. According to the position variance of the ground state wave function it should be of the order of the Compton wavelength cubed, but the precise value depends on how dense these virtual pairs are packed.

Let us stress that any radiative or collisional damping is absent in the consideration here as soon as vacuum fluctuations cannot radiate energy or lose energy in collisions with other quanta, because, after these fluctuations vanish, they would permanently leave behind energy, violating the principle of energy conservation [40].

The resulting electric dipole moment is then

$$\wp = \frac{e^2 \hbar^2}{2m^3 c_{\text{rel}}^4} E . \tag{6}$$

This is the time averaged value of a virtual dipole moment which comes and goes, but it is induced by the external field. Consequently, all of these induced dipole moments are in phase with the external field and add up.

Similarly, here, we use the quantum dynamics to calculate the magnetic moment induced by a magnetic field. An external magnetic field applied to the vacuum induces an electric field vortex that accelerates the virtual electron and positron in opposite directions [25]. This yields (see Appendix A):

$$\mathfrak{m} = \frac{e^2 \hbar^2}{2m^3 c_{\text{rel}}^2} B . \tag{7}$$

These are the microscopic dipole moments. Next, let us calculate the macroscopic densities of these dipole moments. We start with the electric case; i.e., the polarization of the vacuum as a dielectric. As mentioned above, the volume occupied by each of these virtual dipoles should be of the order of λ_C^3. As a result, the dipole moment density turns out to be

$$P = \frac{\wp}{\lambda_C^3} \frac{\lambda_C^3}{V} = \frac{e^2}{2\hbar c_{\text{rel}}} \frac{\lambda_C^3}{V} E . \tag{8}$$

The term dividing the Compton wavelength cubed and the volume is of order unity, but no a precise value can be obtained, so, we keep showing this term. The quantity multiplying the field amplitude E plays the role of an effective vacuum permittivity. Interestingly, since the mass drops out, different types of elementary particles having the same electric charge contribute equally to the vacuum polarizability irrespective of their mass. Therefore, one can write:

$$\varepsilon_0 = \frac{1}{2\hbar c_{\text{rel}}} \sum_s q_s^2 \frac{\lambda_{C,s}^3}{V_s} , \tag{9}$$

where the sum is over all elementary particles with charge q_s. Summing over all known elementary particles in the Standard Model and assuming the volume is the Compton wavelength (of particle type s) cubed yields a value for ε_0 which is 2.4 times lower. Considering the simplicity of the approach, it is surprising how close this rough estimate comes to the experimental value of ε_0. Alternatively, we can use the result to determine the volume per virtual particle pair, yielding $V \simeq 0.41\lambda_C^3$. Note, that here we furthermore assumed that the ratio between the Compton wavelength cubed and the volume per pair is the same for all different types of elementary particles, which seems reasonable.

One may ask whether the zero-point energy actually allows heavier particles to dominate [44]. It has been suggested [45] that instead of a single type of particle pairs involved, there is a Gaussian distribution of probabilities of the vacuum energy fluctuations, and consequently a whole range of particle pairs are actually produced, with the center of mass averaged to anywhere in between.

Next, let us estimate the vacuum magnetization. The calculation is straightforward and the final result reads (see Appendix A):

$$\frac{1}{\mu_0} = \frac{c_{\text{rel}}}{2\hbar} \sum_s q_s^2 \frac{\lambda_{C,s}^3}{V_s}, \qquad (10)$$

The vacuum polarization, $\varepsilon_0 \mathbf{E}$, is thus accompanied by vacuum magnetization, $-\mathbf{B}/\mu_0$, and the vacuum is paramagnetic. It is remarkable that in this crude model, the product $\mu_0 \varepsilon_0$ is indeed exactly equal to the inverse square of the limiting speed of Lorentz transformation, c_{rel}, as required by Lorentz covariance. Let us notice that this result is independent of the exact value of the volume per pair and of how many types of elementary particles contribute to the summation over charges in Equations (9) and (10), underlining the general role, played by the speed of light in physics, far beyond the field of optics.

3. Vacuum Polarization in QED

The virtual pairs discussed qualitatively in Section 2, can be well depicted in terms of the time-honored Feynman diagrams. Figure 2 is such a representation of vacuum polarization in the one-loop approximation. In the following we derive an expression for χ_{vac} using the standard technique of QED (those interested in the final result without the derivation, can go straight to Equation (21)).

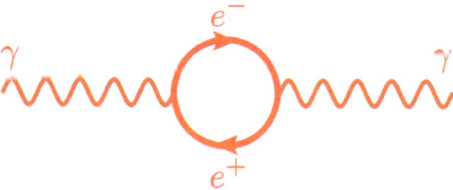

Figure 2. Vacuum polarization in the one-loop approximation. The wavy lines represent an electromagnetic field (γ), while a vertex represents the interaction of the field with the fermions (electron–positron pair, e^-e^+), represented by the loop. The arrows notify the momentum being opposite for particle (electron) and antiparticle (positron). The resulting polarization is maximal for a free electromagnetic field, for which the angular frequency, $\omega = |\mathbf{k}|c$, with $\mathbf{k}$ the wave vector and c denoting the speed of light.

QED typically starts with a Lorentz-invariant Lagrangian density that can be written as

$$\mathcal{L}_{\text{QED}} = \mathcal{L}_{\text{Maxwell}} + \mathcal{L}_{\text{Dirac}} + \mathcal{L}_{\text{int}}. \qquad (11)$$

Here, $\mathcal{L}_{\text{Maxwell}}$ represents the free electromagnetic field, $\mathcal{L}_{\text{Dirac}}$ describes the fermions and the interaction term reads:

$$\mathcal{L}_{\text{int}} = -j^\mu A_\mu, \qquad (12)$$

where j^μ is the external current and $A_\mu = (\Phi, \mathbf{A})$ is the electromagnetic four-potential with Φ the scalar and $\mathbf{A}$ the vector potentials. The Greek letters denote four-dimensional components and take the values 0 (time), 1, 2, and 3 (space). From this Lagrangian density, the wave equations for the fields describing photons and fermions are then derived.

These fields are quantized to permit the creation and annihilation of particles. Charge, linear momentum, and angular momentum are conserved, so annihilation of a photon is accompanied by creation of a particle-antiparticle pair, as illustrated in Figure 2.

An important point for the goal of this study is that the current induced in the vacuum by the four-potential, A^μ, due to virtual pairs can be expressed as [46]

$$j^\mu_{\text{vac}}(k) = c^2_{\text{rel}}\varepsilon_0 \chi_e(k^2) k^2 A^\mu(k), \tag{13}$$

where the Lorentz gauge is used here to simplify the equations.

Note that here, the reciprocal k-space is used. The linear response is represented by the (electron–positron) vacuum susceptibility, $\chi_e(k^2)$. Since this response must be Lorentz invariant, it has to be a function of $k^2 = \omega^2/c^2_{\text{rel}} - \mathbf{k}^2$. The condition $k^2 = 0$, describing a freely propagating photon, is referred to as on-shellness in QED: a real on-shell photon verifies then $\omega^2 = \mathbf{k}^2 c^2_{\text{rel}}$. However, in collisions and other situations where one has nonpropagating fields, such as evanescent waves or near fields, k^2 will typically not be zero. The electron–positron contribution to χ_{vac} in free space will be χ_e.

In position space, still in the Lorentz gauge, Equation (13) becomes

$$j^\mu_{\text{vac}} = \mu_0 \left(-\frac{1}{c^2_{\text{rel}}} \frac{\partial^2}{\partial t^2} + \nabla^2 \right) A^\mu. \tag{14}$$

If, now, one substitutes the fields, $\mathbf{E} = -\partial \mathbf{A}/\partial t - \nabla \Phi$ and $\mathbf{B} = \nabla \times \mathbf{A}$, and takes into account that, for the vacuum, $\mathbf{D} = \varepsilon_0 \mathbf{E}$ and $\mathbf{B} = \mu_0 \mathbf{H}$, one obtains the gauge invariant equation,

$$\mathbf{j}_{\text{vac}} = \frac{\partial \mathbf{D}}{\partial t} - \nabla \times \mathbf{H}. \tag{15}$$

Since it is generally true that in a dielectric, $\mathbf{j} = \partial \mathbf{P}/\partial t + \nabla \times \mathbf{M}$, the $\mathbf{j}_{\text{vac}}$ can be immediately interpreted as the vacuum current of a medium with polarization $\mathbf{D} = \varepsilon_0 \mathbf{E}$ and magnetization $\mathbf{H} = \mathbf{B}/\mu_0$, as was noted above. In Equation (15), the vacuum magnetization current is equal d but is opposite to the polarization current, therefore leading to $\mathbf{j}_{\text{vac}} = 0$.

By making use of the standard technique of Feynmann diagrams, one can show that, at lowest perturbative order, the susceptibility can be expressed as follows [47]:

$$\chi_e(k^2, \Lambda) = 8\pi\alpha \int_0^1 ds\, s(1-s) \int^\Lambda \frac{d^3 p}{(2\pi)^3} \left[p^2 + (mc_{\text{rel}}/\hbar)^2 + s(1-s)k^2 \right]^{-3/2}, \tag{16}$$

where $\alpha = e^2/(4\pi\varepsilon_0 \hbar c_{\text{rel}})$ is the fine structure constant. The integral over the three-momentum $\mathbf{p}$, represents the contribution of a photon of wave vector $\mathbf{k}$ exciting an electron with momentum $\mathbf{p} + s\mathbf{k}$ and a positron with momentum $-\mathbf{p} + (1-s)\mathbf{k}$. This process conserves the three-momentum $\mathbf{p}$, but not the energy. As discussed above, individual pairs with quite high $|\mathbf{p}|$ do not contribute much because they are too ephemeral to polarize much. However, there are so many states with large momentum that their net contribution diverges: the cutoff, Λ, is introduced just to avoid that problem. If one integrates Equation (16) over momenta and expands the result in powers of $1/\Lambda$, one obtains:

$$\chi_e(k^2, \Lambda) = \frac{4\alpha}{\pi} \int_0^1 ds\, s(1-s) \left[\ln\left(\frac{2\hbar\Lambda}{mc_{\text{rel}}}\right) - \frac{1}{2}\ln\left(1 + s(1-s)\frac{\hbar^2 k^2}{m^2 c^2_{\text{rel}}}\right) \right]. \tag{17}$$

One can see that the susceptibility (17) diverges logarithmically in the $\Lambda \to \infty$ limit. This leads to a result that seems to be physically unreasonable: the photon mass is infinite [48], and the contribution of the virtual electron–positron pairs to the vacuum polarization diverges. However, on the other hand, this diverging vacuum susceptibility makes sense because of the screening of a point charge in a dielectric [46]. The observable, or effective charge, positioned in the vacuum is given by $e^2_{\text{eff}} = e^2_{\text{bare}}/\chi_{\text{vac}}(k^2, \Lambda)$. Two arguments can be made to look at the term "squared elementary charge divided by the susceptibility": first, this is the combination in which the quantities appear in the formula for α, and second, what counts is the interaction energy, which is a probe charge e times the potential, $e\Phi(r)$, resulting in the same combination.

In a way, for k close to zero, the infinitely large bare charge, e_{bare}, of the electron and the infinitely large vacuum susceptibility cancel each other and yield a finite effective charge [47]. However, dividing infinities is somehow cumbersome [4]. Alternatively, one can start with the so-called 'regularized' susceptibility and an observable screened charge 'e'. In the following, the regularized quantities are indicated by a caret: $e_{eff}^2 = e^2/\hat{\chi}_{vac}(k^2)$. Far away from the point charge, for $k^2 = 0$, the regularized susceptibility has the finite observable value 'e', and as one moves towards the point charge the susceptibility will approach zero, recovering the infinitely large bare charge. However, in the region of interest one deals with finite values only. The difference between the two approaches is where one hinges the k^2-dependence. In the first approach, one hinges the k^2-dependence at the bare charge, but this makes difficult to carry out any calculation. Therefore, Gottfried and Weisskopf [4] assumed a very large but not infinitely large charge and a very small but not zero diameter of the charge distribution. The disadvantage is that as the calculation moves to an even larger charge and a correspondingly smaller diameter, the resulting susceptibility, $\chi_e(k^2)$, changes drastically in the region far away from the bare charge in order for the increase of the charge at the origin to be compensated. On the other hand, when one hinges calculations to a region in space far enough away from the bare charge, then one deals with finite numbers and functions, and nothing has to be readjusted further away from the bare charge as the bare charge is approaching. So, here, we prefer the second approach.

The standard procedure of dealing with such a divergence is to use the experimentally observed value of the susceptibility at $k^2 = 0$ and use Equation (17) to calculate the k^2-dependence by subtracting two diverging terms to obtain a finite value. Thus, one expresses the susceptibility relative to its regularized on-shell value, $\hat{\chi}_e(0)$, i.e., the value determined experimentally,

$$\hat{\chi}_e(k^2) \equiv \hat{\chi}_e(0) + \lim_{\Lambda \to \infty} [\chi_e(k^2, \Lambda) - \chi_e(0, \Lambda)] \tag{18}$$

$$= \hat{\chi}_e(0) - \frac{2\alpha}{\pi} \int_0^1 ds\, s(1-s)\, \ln\left[1 + s(1-s)\frac{\hbar^2 k^2}{m^2 c_{rel}^2}\right],$$

as the relevant quantity. This is an archetypal example of a regularization of the theory.

The remaining integral can be readily performed, leading to a cumbersome analytical expression [49,50]. However, in the interesting limit $\hbar^2|k^2| \gg m_s^2 c_{rel}^2$, Equation (18) simplifies to

$$\hat{\chi}_e(k^2) = \hat{\chi}_e(0) - \frac{\alpha}{3\pi} \ln\left(\frac{\hbar^2 k^2}{A m^2 c_{rel}^2}\right), \tag{19}$$

where $A = \exp(5/3)$.

As in a standard dielectric, the linear response of the electron–positron vacuum is given by $\hat{\mathbf{P}}_e = \varepsilon_0 \hat{\chi}_e(k^2) \mathbf{E}$. That is, in reciprocal k-space:

$$\mathbf{D}(k) = \varepsilon_0(k^2)\, \mathbf{E}(k), \qquad \mathbf{H}(k) = c_{rel}^2 \varepsilon_0(k^2)\, \mathbf{B}(k). \tag{20}$$

This is quite similar to the classical electromagnetism, where $\mathbf{D} = \varepsilon_0 \mathbf{E}$ and $\mathbf{H} = \mathbf{B}/\mu_0$, but now $\varepsilon_0(k^2) = \varepsilon_0\, \hat{\chi}_e(k^2) \le \varepsilon_0$ and $1/\mu_0(k^2) = 1/[\mu_0 \hat{\chi}_e(k^2)]$. Given the α in the numerator, Equation (19) is a statement about the product $\varepsilon_0 \hat{\chi}_e(k^2)$, not about the separate factors.

Electrons and positrons are not the only types of charged particles. To obtain the susceptibility contributed by other kinds of spin-1/2 particles, one just needs to replace m in the previous expressions and to adjust for the electric charge, q, hidden in α, in case $q^2 \ne e^2$. Charged particles with spin zero also entail replacing the factor of $s(1-s)$ in the integral (17) by $(1-2s^2)/8$ [47]. Summing up over all elementary particle types yields the permittivity of the vacuum: $\varepsilon_0 = \varepsilon_0 \sum_s \hat{\chi}_s(0)$ and $\hat{\chi}_{vac}(0) = \sum_s \hat{\chi}_s(0) = 1$.

In the matter-field coupling constant, α, here, we hold e constant and incorporate the k-dependence into $\varepsilon_0(k^2)$. Since $\varepsilon_0(k^2)^{-1}$ contains all powers of e^2, it incorporates summation over all numbers of pairs. When restricted to an energy scale $\mathcal{E}_{max}$, the sum is over all fermions of mass less than $\mathcal{E}_{max}/c_{rel}^2$ [51–53]. Considering $e_{eff}^2 = e^2/(1 + \hat{\chi}_e(k^2) - \hat{\chi}_e(0))$ is in most ways equivalent to running of the square of effective charge in conventional QED, but the physical interpretation is different.

To obtain $\hat{\chi}_{vac}(k^2)$, one has to sum up over all particles, all of them contribute to the constant value at $k = 0$, summing up to 1, but for small k the running is dominated by the electron–positron vacuum because they have the largest Compton wavelength. So, in the limit of small k, one has: $\hat{\chi}_{vac}(k^2) = 1 + \hat{\chi}_e(k^2) - \hat{\chi}_e(0)$. In a dielectric, it is possible to have a negative induced polarization, ($\chi_{mat} < 0$), when exciting above the resonance of the medium, but $e_{eff}^2 < 0$ makes no physical sense, because in the vacuum there is no such resonance.

The dielectric properties of vacuum differ from those of a material medium in two essential points: $\ln(k^2)$-dependence replaces the usual ω-dependence and Lorentz invariance requires that $\varepsilon_0(k^2)\mu_0(k^2) = 1/c_{rel}^2$. The speed c_{rel} is an universal constant, whereas $\chi_e(k^2)$ and, thus, also the coupling constant, $\alpha(k^2)$, runs. On the photon mass shell, $k^2 = 0$, so a free photon always sees ε_0 and there is no running in this case.

Finally, we argue here that the straightforward back-of-the-envelope calculation sketched in Section 2, is consistent with QED. Actually, the loop in Figure 2 can be thought of as a single polarizable atom with center-of-mass momentum $\hbar k$. If, for simplicity, $k = 0$ is set, the computation of the Feynman diagram involves integrals of the form $\int d^4q \, [q^2 + mc/\hbar]^{-2}$, which entails an exponential decay, $\exp[-(mc/\hbar)|\mathbf{x}|]$, in real space ($\mathbf{x}$). Therefore, the "radius" of such a virtual atom is of order λ_C. Alltogether, the above suggests that virtual pairs can be modelled as oscillating dipoles with frequency $mc_{rel}^2/\hbar$ and volume of order λ_C^3.

Indeed, at large k^2 and to second order in perturbation theory, Equation (19) gives:

$$\varepsilon_0(k^2) \simeq \varepsilon_0 - \frac{1}{12\pi^2 \hbar c} \sum_s q_s^2 \ln\left(\frac{\hbar^2 k^2}{m_s^2 c_{rel}^2}\right) \qquad (21)$$

where a summation over all possible pairs is explicitly included. It is known [54] that at high-momentum (or energy) scale, the coupling constant, $\alpha(k^2)$, in QED becomes infinity. In physical terms, charge screening can make the "renormalized" charge to adopt the finite value observed in the experiment. This is often referred to as 'triviality' [48]. If Λ_L is the value of that momentum, at which $\varepsilon_0(k^2) = 0$ and, equivalently, at which the fine structure constant goes to infinity, usually called the Landau pole [55], then one obtains [27]:

$$\varepsilon_0 = \frac{1}{12\pi^2 \hbar c_{rel}} \sum_s q_s^2 \ln\left(\frac{\hbar^2 \Lambda_L^2}{m_s^2 c_{rel}^2}\right). \qquad (22)$$

While in Equation (21) the dominant term is the one that cannot be calculated, this ambiguity is shifted in Equation (22) to the momentum value, at which the Landau pole is located. This allows us to rewrite Equation (21) as

$$\varepsilon_0(k^2) \simeq \frac{1}{12\pi^2 \hbar c} \sum_s q_s^2 \ln\left(\frac{\Lambda_L^2}{k^2}\right). \qquad (23)$$

Let us note that this equation is only valid for large enough k.

Equation (22) relates ε_0 to Λ_L. Using the experimentally determined value for ε_0, assuming the Standard Model of QCD and summing up over all elementary particles from leptons to the W-boson with their respective charges and masses, one finds $\Lambda_L \hbar/c \simeq$

6×10^{30} GeV/c^2, which is much beyond the Planck mass and is probably an unrealistically high value. Alternatively, one could assume the Minimal Super-symmetric Standard Model (MSSM), which doubles the elementary particles and thus also the sum in Equation (22). This brings Λ_L down significantly to the value $\Lambda_L \hbar / c_{rel} \simeq 1.6 \times 10^{15}$ GeV/c^2, which is close to the momentum range, where the coupling constants are supposed to become equal. Let us note that the QED calculations referred to here are based on the one-loop approximation (see Figure 2) and, at large momenta near the Landau pole, multiple loop contributions will be significant. In that sense the concept of the Landau momentum (or Landau pole) based on the one-loop approximation has limited value. Nevertheless, it well demonstrates the concept.

If one compares the terms in the sum (22) with the ones in Equation (9), one by one, one finds that the two equations, provided that

$$V_s = \frac{\hbar^3}{m_s^3 c_{rel}^3} \frac{6\pi^2}{\ln\left(\frac{\hbar^2 \Lambda_L^2}{m_s^2 c_{rel}^2}\right)}. \tag{24}$$

To fulfill Equation (24) one would have to give up the assumption that the ratio of the Compton wavelength cubed and the volume occupied by a single virtual pair is independent of the type of elementary particle. This may not be reasonable. However, instead of doing the detailed evaluation of the sum in Equation (22), we can perform here some estimations: the masses m_s differ by a factor 10^6, but this factor is diminished by the logarithmic function. As a result, the logarithmic term is almost constant for large enough cutoff Λ_L and, to some approximation, can be taken out of the sum. The term $\ln[\hbar^2 \Lambda_L^2 / (m_s^2 c_{rel}2)]$ varies only little when assuming the Standard Model and its average is 144. If, as an approximation, all logarithmic terms in the sum are replaced by 144, then one obtains the condition that

$$\left\langle \frac{V_s}{\hbar^3} m_s^3 c_{rel}^3 \right\rangle_s = \frac{6\pi^2}{\left\langle \ln\left(\frac{\hbar^2 \Lambda_L^2}{m_s^2 c_{rel}^2}\right) \right\rangle_s} = 0.41. \tag{25}$$

So, one gets the correct value by adjusting the volume per virtual pair in the model discussed here to a reasonable value close to the Compton wavelength cubed, or by choosing the right value for the Landau pole, Λ_L.

Let us stress that in the model here, no divergence appears, the only uncertainty is associated with the volume occupied by each virtual pair. The position variance of the harmonic oscillator ground state wave function gives a crude value for the volume per pair in the right ball park, but it does not give a precise value. There is a different way to estimate this volume in momentum space: in analogy to the derivation of Planck's blackbody radiation formula, one can calculate the number of modes (i. e., standing waves) of the particle's de Broglie wave pattern in a given larger volume, integrate over momentum (inversely proportional to the de Broglie wavelength) and divide the larger volume by the the number of modes obtained. This then determines the volume per mode or per particle-antiparticle pair. However, this integral diverges and one would obtain $V_s = 0$. A crude cure would be to introduce a relativistic cutoff which will also give a volume per pair of the order of the particle Compton wavelength cubed. Staying in configuration space as opposed to momentum space, one seems to avoid this divergence, as suggested by Fried and Gabellini [56], who discuss the advantage of performing QED calculation in configuration space. It remains to find if a more precise value for the volume per pair can be derived using a configuration-space description.

The standard approach in QED is to use plane waves to describe the relative motion of the virtual pairs. In the center of mass reference frame, if the electron has momentum $\hbar \mathbf{k}$, then the positron has momentum $-\hbar \mathbf{k}$. However, the uncertainty principle requires that the lifetime of the pair is quite short, so the distance travelled d is comparably small.

The plane-wave basis seems not suited well enough in order to describe this situation: convergence is quite poor, and the divergences arise. What would be needed is a basis whose ground and first few excited states are of comparable size to d. The merit of the oscillator model is to provide a suitable basis for description of the relative motion in these short lived states.

4. Conclusions

The vacuum permittivity has so far been a purely experimental number. Here, we have worked out a simple enough dielectric model—based just on treating the individual particle–antiparticle pairs as three-dimensional harmonic oscillators, which approximate small deviations from the equilibrium induced by an external electromagnetic field—to point at the intimate relationship between the properties of the quantum vacuum and the constants in Maxwell's equations. From this picture, the vacuum is considered to be understood as an effective medium.

The authors have a hope that, with all the above arguments, the conception that ε_0 and μ_0 are merely measurement system constants, without any physical relevance, will be moderated in physical courses and textbooks.

Author Contributions: Conceptualization, G.L.; formal analysis, G.L., M.H. and L.L.S.-S.; writing, G.L., M.H. and L.L.S.-S. All authors have read and agreed to the published version of the manuscript.

Funding: This research was funded by the Spanish Ministerio de Ciencia e Innovación (Grant PGC2018-099183-B-I00).

Data Availability Statement: Not applicable.

Acknowledgments: Over the years, the ideas in this paper have been further developed and expanded with questions, suggestions, criticism, and advice from many colleagues. Particular thanks for help in various ways goes to Pierre Chavel, Joseph H. Eberly, Berthold-Georg Englert, Michael Fleischhauer, Holger Gies, Uli Katz, Natalia V. Korolkova, Gérard Mourou, Nicolai B. Narozhny, Serge Reynaud, Vsevolod Salakhutdinov, Wolfgang P. Schleich, Anthony E. Siegman, and John Weiner.

Conflicts of Interest: The authors declare no conflict of interest.

Appendix A. Vacuum Fluctuations as Harmonic Oscillators

As discussed in the main text (see Equation (5)), a charged particle-antiparticle pair that results from a fluctuation of the Dirac field will appear in the vacuum as a transient atom. During the time while such an atom exists, it can interact with a photon. For the simplest case, the electron–positron vacuum fluctuation in the lowest energy level (at $-2mc_{\text{rel}}^2$) that has zero angular momentum is called parapositronium, which is a singlet spin state. However, an atom in its ground state interacts with the electric field as if it were a harmonic oscillator with the first two energy levels separated by twice the rest-mass energy, $2mc_{\text{rel}}^2$, of one of the elementary particles of mass m. Here, we use the standard wave functions of the quantized harmonic oscillator. Recall that a harmonic oscillator of two equal moving masses m is like a harmonic oscillator with one moving particle of reduced mass $m/2$.

The virtual pair interacts with an external electric field according to the Hamiltonian $\hat{H}_{\text{int}} = -\wp \cdot \mathbf{E}$. For the full description of the system, three harmonic oscillators are used, one each for the three Cartesian coordinates. In the electric case, we assume a linearly polarized external electric field, which induces a one-dimensional dipole. If one makes the two-level approximation, so that only transitions between the ground (ψ_0) and the first excited (ψ_1) states of the oscillator are relevant, one finds that

$$\wp_{\text{max}} = e \langle \psi_1 | \hat{x} | \psi_0 \rangle, \tag{A1}$$

Using the explicit form of these standard harmonic oscillator wave functions, one immediately obtains the maximum possible electric dipole moment,

$$\wp_{max} = \frac{e\hbar}{\sqrt{2}mc_{rel}}. \tag{A2}$$

According to the standard treatment of a two-level system in an external field [43], the torque vector for this problem is given by $\mathbf{\Omega} = (2\langle\psi_1|\hat{H}_{int}|\psi_0\rangle/\hbar,\ 0,\ \Delta\omega)^\top$, where the subscript $\top$ denotes the transpose. The time-averaged induced dipole moment turns out to be

$$\wp = \frac{\Omega_x}{\Omega_z}\wp_{max} = \frac{2\wp_{max}E}{\hbar\Delta\omega}\wp_{max} = \frac{e^2\hbar^2}{2m^3c_{rel}^4}E, \tag{A3}$$

which is actually Equation (6).

Let us now turn to the induced magnetic dipole, with interaction Hamiltonian, $\hat{H}_{int} = -\mathbf{m}\cdot\mathbf{B}$. The definition of the magnetic moment reads [57]: $\mathbf{m} = \frac{1}{2}e\mathbf{r}\times\dot{\mathbf{r}}$, where the dot denotes time derivative. If, for definiteness, the magnetic field is considered along the z axis, then $\mathbf{m}$ is along the z axis with modulus $\mathfrak{m} = \frac{1}{2}e(x\dot{y}-\dot{x}y) = \frac{1}{2}e\dot{\varphi}(x^2-y^2)$ and φ the polar angle. A magnetic dipole results from a superposition of states of the same parity. Therefore, using now the full three-dimensional harmonic oscillator with its product wave functions, the maximum possible magnetic dipole moment is

$$\mathfrak{m}_{max} = \tfrac{1}{2}e\dot{\varphi}\langle\psi_{000}|\hat{x}^2-\hat{y}^2|\tfrac{1}{\sqrt{2}}(\psi_{200}-\psi_{020})\rangle. \tag{A4}$$

Using again the explicit harmonic oscillator wave functions, one obtains:

$$\mathfrak{m}_{max} = -\tfrac{1}{2}e\dot{\varphi}\frac{\hbar^2}{m^2c_{rel}^2}. \tag{A5}$$

Applying again the two-level atom dynamics, but now between levels 0 and 2, yields the induced magnetic dipole moment:

$$\mathfrak{m} = \frac{\Omega_x}{\Omega_z}\mathfrak{m}_{max} \simeq \frac{2\mathfrak{m}_{max}B}{\hbar\Delta\omega}\mathfrak{m}_{max} = \frac{e^2\hbar^2}{2m^3c_{rel}^2}B, \tag{A6}$$

as found in Equation (7).

Note that the harmonic oscillator approximation is valid for all physical systems in equilibrium, as long as the departure from the equilibrium point is small enough, which is the case here. The point that the excited state spectrum differs greatly is actually not essential under this condition. What is important is the effective 'spring' constant, which is given by the spacing between the ground and the first excited states.

References

1. Bork, A.M. Maxwell, displacement current, and symmetry. *Am. J. Phys.* **1963**, *31*, 854–859. [CrossRef]
2. Whittaker, E.T. *A History of the Theories of Aether and Electricity*; Longmans, Green, and Co.: London, UK, 1910. Available online: https://archive.org/details/historyoftheorie00whitrich/mode/2up (accessed on 5 January 2023).
3. Kostro, L. *Einsten and the Ether*; Apeiron: Montreal, Quebec, Canada, 2000. Available online: https://www.scribd.com/document/145946810/Einstein-and-the-Ether (accessed on 5 January 2023).
4. Gottfried, K.; Weisskopf, V.F. *Concepts of Particle Physics*; Oxford University Press, Inc.: New York, NY, USA, 1986; Volume 2.
5. Planck, M. Über die Begründung des Gesetzes der schwarzen Strahlung. *Ann. Phys.* **1912**, *342*, 642–656. [CrossRef]
6. Dirac, P.A.M. The quantum theory of the electron. *Proc. R. Soc. Lond. A Math. Phys. Engin. Sci.* **1928**, *117*, 610–624. [CrossRef]
7. Feynman, R.P. Space-time approach to quantum electrodynamics. *Phys. Rev.* **1949**, *76*, 769–789. [CrossRef]
8. Schwinger, J. The Theory of quantized fields. III. *Phys. Rev.* **1953**, *91*, 728–740. [CrossRef]
9. Mehra, J.; Milton, K.A. *Climbing the Mountain: The Scientific Biography of Julian Schwinger*; Oxford University Press: Oxford, UK, 2000. [CrossRef]
10. Borchers, H.J.; Haag, R.; Schroer, B. The vacuum state in quantum field theory. *Il Nuovo Cim.* **1963**, *29*, 148–162. [CrossRef]

11. Sciama, D.W. The physical significance of the vacuum state of a quantum field. In *The Philosophy of Vacuum*; Saunders, S., Brown, H.R., Eds.; Clarendon: Oxford, UK, 1991; pp. 137–158.
12. Milonni, P.W. *The Quantum Vacuum*; Academic Press: London, UK, 1994. [CrossRef]
13. Wilczek, F. *The Lightness of Being*; Basic Books: New York, NY, USA, 2008.
14. Lamb, W.E.; Retherford, R.C. Fine structure of the hydrogen atom by a microwave method. *Phys. Rev.* **1947**, *72*, 241–243. [CrossRef]
15. Moore, G.T. Quantum theory of the electromagnetic field in a variable-length one-dimensional cavity. *J. Math. Phys.* **1970**, *11*, 2679–2691. [CrossRef]
16. Unruh, W.G. Notes on black-hole evaporation. *Phys. Rev. D* **1976**, *14*, 870–892. [CrossRef]
17. Hawking, S.W. Particle creation by black holes. *Commun. Math. Phys.* **1975**, *43*, 199–220. [CrossRef]
18. Mainland, G.B.; Mulligan, B. Polarization of vacuum fluctuations: Source of the vacuum permittivity and speed of light. *Found. Phys.* **2020**, *50*, 457–480. [CrossRef]
19. Furry, W.H.; Oppenheimer, J.R. On the theory of the electron and positive. *Phys. Rev.* **1934**, *45*, 245–262. [CrossRef]
20. Pauli, W.; Weisskopf, V. Über die Quantisierung der skalaren relativistischen Wellengleichung. *Helv. Phys. Acta* **1934**, *7*, 709–731. [CrossRef]
21. Weisskopf, V. Über die Elektrodynamik des Vakuums auf Grund des Quanten-Theorie des Elektrons. *Danske Vid. Selsk. Math.-Fys. Medd.* **1936**, *14*, 1–39. Available online: http://gymarkiv.sdu.dk/MFM/kdvs/mfm%2010-19/mfm-14-6.pdf (accessed on 5 January 2023).
22. Weisskopf, V. The electrodynamics of the vacuum based on the quantum theory of the electron; In Miller, A.I. *Early Quantum Electrodynamics*. Cambridge University Press: New York, NY, USA, 1994; pp. 206–226. [CrossRef]
23. Dicke, R.H. Gravitation without a principle of equivalence. *Rev. Mod. Phys.* **1957**, *29*, 363–376. [CrossRef]
24. Heitler, W. *The Quantum Theory of Radiation*; Clarendon Press: Oxford, UK, 1954.
25. Leuchs, G.; Villar, A.S.; Sánchez-Soto, L.L. The quantum vacuum at the foundations of classical electrodynamics. *Appl. Phys. B* **2010**, *100*, 9–13. [CrossRef]
26. Bethe, H.A. The electromagnetic shift of energy levels. *Phys. Rev.* **1947**, *72*, 339–341. [CrossRef]
27. Leuchs, G.; Hawton, M.; Sánchez-Soto, L.L. Quantum field theory and classical optics: Determining the fine structure constant. *J. Phys. Conf. Ser.* **2017**, *793*, 012017. [CrossRef]
28. Tamm, I.E. *Fundamentals of the Theory of Electricity*; Mir Publishers: Moscow, Russia, 1979. Available online: https://ia800200.us.archive.org/0/items/TammElectricity/Tamm-Fundamentals-Of-The-Theory-of-Electricity.pdf (accessed on 5 January 2023).
29. Paraoanu, G.S. The quantum vacuum. In *Romanian Studies in Philosophy of Science*; Pârvu, I., Sandu, G., Toader, I.D., Eds.; Springer International Publishing: Cham, Switzerland, 2015; pp. 181–197. [CrossRef]
30. Hilgevoord, J. The uncertainty principle for energy and time. *Am. J. Phys.* **1996**, *64*, 1451–1456. [CrossRef]
31. Denur, J. The energy-time uncertainty principle and quantum phenomena. *Am. J. Phys.* **2010**, *78*, 1132–1145. [CrossRef]
32. Nussenzveig, H.M. (Ed.) *Causality and Dispersion Relations*; Academic Press: New York, NY, USA, 1972. Available online: https://www.sciencedirect.com/bookseries/mathematics-in-science-and-engineering/vol/95/suppl/C (accessed on 5 Junuary 2023).
33. Narozhny, N.B.; Bulanov, S.S.; Mur, V.D.; Popov, V.S. On e^+e^- pair production by colliding electromagnetic pulses. *JETP Lett.* **2004**, *80*, 382–385. [CrossRef]
34. Deutsch, M.; Brown, S.C. Zeeman effect and hyperfine splitting of positronium. *Phys. Rev.* **1952**, *85*, 1047–1048. [CrossRef]
35. Hughes, V.W.; Marder, S.; Wu, C.S. Hyperfine structure of positronium in its ground state. *Phys. Rev.* **1957**, *106*, 934–947. [CrossRef]
36. Ruark, A.E. Positronium. *Phys. Rev.* **1945**, *68*, 278–278. [CrossRef]
37. Wheeler, J.A. Polyelectrons. *Ann. N. Y. Acad. Sci.* **1946**, *48*, 219–238. [CrossRef]
38. Leuchs, G.; Sánchez-Soto, L.L. A sum rule for charged elementary particles. *Eur. Phys. J. D* **2013**, *67*, 57. [CrossRef]
39. Urban, M.; Couchot, F.; Sarazin, X.; Djannati-Atai, A. The quantum vacuum as the origin of the speed of light. *Eur. Phys. J. D* **2013**, *67*, 58. [CrossRef]
40. Mainland, G.B.; Mulligan, B. Theoretical calculation of the fine-structure constant and the permittivity of the vacuum. *arXiv* **2017**, arXiv:1705.11068. [CrossRef]
41. Mainland, G.B.; Mulligan, B. How vacuum fluctuations determine the properties of the vacuum. *J. Phys. Conf. Ser.* **2019**, *1239*, 012016. [CrossRef]
42. Feynman, R.P.; Leighton, R.B.; Sands, M.L. *The Feynman Lectures on Physics*; Basic Books: New York, NY, USA, 2011; Volume III. Available online: https://www.feynmanlectures.caltech.edu/III_toc.html (accessed on 5 January 2023).
43. Allen, L.; Eberly, J.H. *Optical Resonance and Two-Level Atoms*; Dover Publications: New York, NY, USA, 1987.
44. Hajdukovic, D.S. On the relation between mass of a pion, fundamental physical constants and cosmological parameters. *EPL (Europhys. Lett.)* **2010**, *89*, 49001. [CrossRef]
45. Margan, E. *Some Intriguing Consequences of the Quantum Vacuum Fluctuations in the Semi-Classical Formalism*; Technical Report; Jozef Stefan Institute: Ljubljana, Slovenia, 2011. Available online: https://www-f9.ijs.si/~margan/Articles/SomeConsequences.pdf (accessed on 5 January 2023).
46. Leuchs, G.; Hawton, M.; Sánchez-Soto, L.L. QED response of the vacuum. *Physics* **2020**, *2*, 14–21. [CrossRef]
47. Prokopec, T.; Woodard, R. Vacuum polarization and photon mass in inflation. *Am. J. Phys.* **2004**, *72*, 60–72. [CrossRef]

48. Bogoliubov, N.N.; Shirkov, D.V. *The Theory of Quantized Fields*; John Wiley & Sons., Inc.: New York, NY, USA, 1980. Available online: https://archive.org/details/IntroductionToTheoryOfQuantizedFields/page/n1/mode/2up (accessed on 5 January 2023).
49. Itzykson, C.; Zuber, J.-B. *Quantum Field Theory*; Mc Graw Hill Publishing Company, Inc.: New York, NY, USA, 1980. Available online: https://archive.org/details/quantumfieldtheo0000itzy (accessed on 5 January 2023).
50. Peskin, M.E.; Schroeder, D.V. *An Introduction to Quantum Field Theory*; CRC Press: Boca Raton, FL, USA, 2016. [CrossRef]
51. Eidelman, S.; Jegerlehner, F. Hadronic contributions to $(g-2)$ of the leptons and to the effective fine structure constant $\alpha(M_Z^2)$. *Z. Phys. C* **1995**, *67*, 585–601. [CrossRef]
52. Hogan, C.J. Why the universe is just so? *Rev. Mod. Phys.* **2000**, *72*, 1149–1161. [CrossRef]
53. Hoecker, A. The hadronic contribution to the muon anomalous magnetic moment and to the running electromagnetic fine structure constant at M_Z—Overview and latest results. *Nucl. Phys. B Proc. Suppl.* **2011**, *218*, 189–200. [CrossRef]
54. Gell-Mann, M.; Low, F.E. Quantum electrodynamics at small distances. *Phys. Rev.* **1954**, *95*, 1300–1312. [CrossRef]
55. Landau, L.D.; Abrikosov, A.A.; Khalatnikov, I.M. On the removal of infinities in quantum electrodynamics. *Dokl. Akad. Nauk SSSR* **1954**, *95*, 497–502; English translation: *Collected Papers of L.D. Landau*; Ter Harr, D., Ed.; Pergamon Press Ltd.; Gordon and Breach, Science Publishers, Inc.: Oxford, UK, 1965; pp. 607–610. [CrossRef]
56. Fried, H.; Gabellini, Y. On the summation of Feynman graphs. *Ann. Phys.* **2012**, *327*, 1645–1667. [CrossRef]
57. Cohen-Tannoudji, C.; Dupont-Roc, J.; Grynberg, G. *Photons and Atoms: Introduction to Quantum Electrodynamics*; Wiley-VCH Verlag GmbH & Co. KGaA: Weinheim, Germany, 2004. [CrossRef]

Disclaimer/Publisher's Note: The statements, opinions and data contained in all publications are solely those of the individual author(s) and contributor(s) and not of MDPI and/or the editor(s). MDPI and/or the editor(s) disclaim responsibility for any injury to people or property resulting from any ideas, methods, instructions or products referred to in the content.

Article

Two New Methods in Stochastic Electrodynamics for Analyzing the Simple Harmonic Oscillator and Possible Extension to Hydrogen

Daniel C. Cole

Department of Mechanical Engineering, Boston University, Boston, MA 02215, USA; dccole@bu.edu

Abstract: The position probability density function is calculated for a classical electric dipole harmonic oscillator bathed in zero-point plus Planckian electromagnetic fields, as considered in the physical theory of stochastic electrodynamics (SED). The calculations are carried out via two new methods. They start from a general probability density expression involving the formal integration over all probabilistic values of the Fourier coefficients describing the stochastic radiation fields. The first approach explicitly carries out all these integrations; the second approach shows that this general probability density expression satisfies a partial differential equation that is readily solved. After carrying out these two fairly long analyses and contrasting them, some examples are provided for extending this approach to quantities other than position, such as the joint probability density distribution for positions at different times, and for position and momentum. This article concludes by discussing the application of this general probability density expression to a system of great interest in SED, namely, the classical model of hydrogen.

Keywords: stochastic electrodynamics; classical physical dynamics; hydrogen; harmonic oscillator; nonlinear dynamics

Citation: Cole, D.C. Two New Methods in Stochastic Electrodynamics for Analyzing the Simple Harmonic Oscillator and Possible Extension to Hydrogen. *Physics* **2023**, *5*, 229–246. https://doi.org/10.3390/physics5010018

Received: 19 November 2022
Revised: 6 January 2023
Accepted: 17 January 2023
Published: 21 February 2023

Copyright: © 2023 by the author. Licensee MDPI, Basel, Switzerland. This article is an open access article distributed under the terms and conditions of the Creative Commons Attribution (CC BY) license (https://creativecommons.org/licenses/by/4.0/).

1. Introduction

This paper involves the physics of stochastic electrodynamics (SED) and the exploration of a new approach for analyzing probabilities associated with charged particle motion due to the interaction with stochastic electromagnetic radiation. SED involves the movement of classical charged point particles while interacting with a specific form of fluctuating classical electromagnetic radiation. Despite SED being completely classical, agreement has been shown between SED and quantum mechanics (QM) and even quantum electrodynamics (QED), under an interesting range of conditions. The "classical physics" aspects of SED consist of electromagnetic radiation that obeys Maxwell's classical, microscopic electromagnetic equations, while the classical charged particles obey the relativistic Lorentz–Dirac classical equation of motion [1,2]. The physical predictions of SED that agree with QM and QED hold for classical systems with a linear differential equation of motion. A very good demonstration of this point is Ref. [3] for the simple harmonic oscillator (SHO). Even the complicated fully retarded van der Waals forces between atoms modelled by electric dipole oscillators fulfill this agreement, as do Casimir forces between continuum materials; this agreement holds in both cases for all temperature conditions. Interestingly enough, at one point many who have studied and explored SED thought that SED might form the basis for QM and QED. However, complications have since been found to persuade most researchers that this is not the case; for more details, see Refs. [3–8].

The disagreement between SED and QM for all arbitrary "atomic systems" has some bearing for motivating the investigation in this paper. An interesting point to explain why all "atomic systems" covered by QM do not also hold for SED was first made and analyzed by Boyer [9] and subsequently followed up in a different way by the present author [10]. The point was this: the binding force for all atomic systems in nature is due to the Coulombic force. Hence, not just any binding force inserted into the SED description, other than the Coulombic-based one, should be expected to share agreement with real atomic systems in nature; moreover, SED should not be expected to agree with nonphysical "atomic systems" containing arbitrary binding forces in QM. The difficulty here is that such Coulombic-based systems are inherently nonlinear for the equation of motion for the classical electrons interacting with the classical nucleus; hence, such systems are far more complicated to analyze in SED than when the differential equation of motion is linear. More about this point is discussed in the concluding section of Section 4, but this forms much of the motivation in this study for examining a new way of calculating probabilities in SED.

Key aspects of SED concern classical charged particles interacting with classical electromagnetic radiation at some temperature T, with the recognition that at $T = 0$, the radiation is nonzero, with particular properties that enable a statistical equilibrium between charged particles and radiation. Some of the specific properties of classical electromagnetic zero-point (ZP) radiation at $T = 0$ include that the radiation frequency distribution must be Lorentz invariant [11,12] and the fundamental definition of $T = 0$ must be obeyed by ZP radiation [13–15]. In SED, the $T = 0$ stochastic radiation is referred to as ZP radiation. The stochastic radiation for $T \geq 0$ is referred here as "ZP plus Planckian" (ZPP) radiation.

This study involves exploring the use of the following expression:

$$P_{3x}(\mathbf{x}) = \int_{-\infty}^{\infty} dA_1 \cdots \int_{-\infty}^{\infty} dA_N \cdots \int_{-\infty}^{\infty} dB_1 \cdots \int_{-\infty}^{\infty} dB_N$$
$$\times P_{F,A\text{-}B}(A_1, \cdots, A_N, B_1, \cdots, B_N) \delta^3\left[\mathbf{x} - \mathbf{x}_{A_1, \cdots, B_N}(t)\right], \quad (1)$$

for the probability density of finding a point charged particle at position $\mathbf{x}$, in the steady state condition. The restriction to "steady state" can be removed, but that results in a time dependence $P_{3x}(\mathbf{x}, t)$, which is not treated here. Also, beside the above expression for $P_{3x}(\mathbf{x})$, such as those for the energy and momentum, are discussed in Section 2.2.

The "3x" notation in Equation (1) indicates that the function $P_{3x}(\mathbf{x})$ refers to the position vector point in 3D space, while the "F,A-B" notation refers to the probability density function for the Fourier coefficients of the electromagnetic radiation described next. Specifically, $A_1, \cdots, A_N, B_1, \cdots, B_N$ in Equation (1) represent the coefficients in the Fourier expression for the electric and magnetic fields that the particle is "immersed" within. For more details about these Fourier coefficients, see Refs. [4,16]; however, these coefficients are also explained in Section 2.1 when expressions for the radiation electromagnetic fields are introduced. At the end of the calculations, $N \to \infty$ is imposed. Finally, $\mathbf{x}_{A_1, \cdots, B_N}(t)$ in Equation (1) is the steady state trajectory of the particle and is a function of $A_1, \cdots, A_N, B_1, \cdots, B_N$, while $\delta^3\left[\mathbf{x} - \mathbf{x}_{A_1, \cdots, B_N}(t)\right]$ is the Dirac delta function in 3D.

All "A" and "B" coefficients are real quantities and are integrated from $-\infty$ to $+\infty$. Their values control the steady state solution for the particle's trajectory of $\mathbf{x}_{A_1, \cdots, B_N}(t)$. The probability density of these fields, $P_{F,A\text{-}B}(A_1, \cdots, A_N, \cdots, B_N)$ dictates their contribution to how $\delta^3\left[\mathbf{x} - \mathbf{x}_{A_1, \cdots, B_N}(t)\right]$ "selects" the contribution to the final particle probability density $P_{3x}(\mathbf{x})$.

Two solution methods are examined in this paper. In Section 2.1, a slight variation to Equation (1) is used to fully evaluate the analytic probability density $P_{1x}(x)$ for the position of a one-dimensional (1D) SHO. Here, each Fourier coefficient is explicitly integrated over. In Section 2.2, a few other examples using expressions similar to Equation (1) are also discussed for the 1D and 3D SHOs, including the particle's joint probability densities of $P_{x,x}(x_1, t_1; x_2, t_2)$ and $P_{3x,3p}(\mathbf{x}, \mathbf{p})$, where $\mathbf{p}$ is the particle's momentum, as well as the probability density for the kinetic plus potential energy of the oscillator.

The remainder of this paper has two more Sections. In Section 3, a second method, different from the direct integration method in Section 2.1, is described for deducing the analytic expression for $P_{1x}(x)$. This second method uses a partial differential equation (PDE) approach. Finally, Section 4 provides some concluding remarks, including a discussion of generalizing this study to more complicated systems, in particular the classical hydrogen atom.

2. Calculating Analytic SHO Probability Density Functions from Initial General Expression

2.1. Direct Integration Method for Analytic Expression of SHO Probability Density, $P_{1x}(x)$

The use of probability density expressions like Equation (1) has been explored in Refs. [17,18], but mainly for the stochastic electric field values of the $T \geq 0$ radiation fields. In this paper, the following is analyzed: the stochastic properties of a classical charged point electron, bound by an SHO potential and bathed in stochastic classical electromagnetic ZPP radiation.

To start, let us describe the radiation fields by considering a large region of space, where "large" means as compared to the size of the space that any charged particles, "bound" by a classical potential, occupy via traversing within the confines of this classical potential. Let us consider a rectangular parallelepiped region that the radiation fields are confined within, where the rectangular parallelepiped has dimensions in the space of L_x, L_y, and L_z, along the x, y, and z axes. Although other shapes can be considered (see, e.g., Ref. [19]), the rectangular parallelepiped offers mathematical simplicity, especially since at the end of the calculations, L_x, L_y, and L_z are typically taken in the limit of infinite in size.

In what follows, the ZP or ZPP radiation fields are represented as an infinite sum of plane waves, with periodic boundary conditions (bcs) imposed. This imposition enables the use of Fourier series to describe the fields. For a large region of space, this periodicity does not effect the physical analysis, but it does simplify the subsequent mathematical analysis. Hence, the following sum of plane waves is used for the "free" electric $\mathbf{E}(\mathbf{x},t)$ and magnetic $\mathbf{B}(\mathbf{x},t)$ radiation fields in this large parallelepiped volume [4] (see p. 76, Equations (3.65) and (3.66) in Ref. [4]):

$$\mathbf{E}_{rad}(\mathbf{x},t) = \frac{1}{(L_x L_y L_z)^{1/2}} \sum_{n_x,n_y,n_z=-\infty}^{\infty} \sum_{\lambda=1,2} \hat{\varepsilon}_{\mathbf{k}_n,\lambda} \left[\begin{array}{c} A_{\mathbf{k}_n,\lambda} \cos(\mathbf{k}_n \cdot \mathbf{x} - \omega_n t) \\ +B_{\mathbf{k}_n,\lambda} \sin(\mathbf{k}_n \cdot \mathbf{x} - \omega_n t) \end{array} \right], \quad (2)$$

$$\mathbf{B}_{rad}(\mathbf{x},t) = \frac{1}{(L_x L_y L_z)^{1/2}} \sum_{n_x,n_y,n_z=-\infty}^{\infty} \sum_{\lambda=1,2} \left(\hat{\mathbf{k}}_n \times \hat{\varepsilon}_{\mathbf{k}_n,\lambda} \right) \left[\begin{array}{c} A_{\mathbf{k}_n,\lambda} \cos(\mathbf{k}_n \cdot \mathbf{x} - \omega_n t) \\ +B_{\mathbf{k}_n,\lambda} \sin(\mathbf{k}_n \cdot \mathbf{x} - \omega_n t) \end{array} \right]. \quad (3)$$

Here,

$$\mathbf{k}_n = \frac{2\pi n_x}{L_x}\hat{\mathbf{x}} + \frac{2\pi n_y}{L_y}\hat{\mathbf{y}} + \frac{2\pi n_z}{L_z}\hat{\mathbf{z}}, \quad (4)$$

and n_x, n_y, and n_z are integers. Moreover, $\omega_n = c|\mathbf{k}_n|$, $\mathbf{k}_n \cdot \hat{\varepsilon}_{\mathbf{k}_n,\lambda} = \mathbf{k}_n \cdot \hat{\varepsilon}_{\mathbf{k}_n,\lambda'} = 0$, and $\hat{\varepsilon}_{\mathbf{k}_n,\lambda} \cdot \hat{\varepsilon}_{\mathbf{k}_n,\lambda'} = 0$ for $\lambda \neq \lambda'$, where λ and λ' indicate the linear polarization direction, and c denotes the speed of light. Here, λ and λ' are essentially indices that take on only two values, so each might be represented by the values 1 or 2. Moreover, $\hat{\mathbf{k}}_n = \mathbf{k}_n/|\mathbf{k}_n|$; similarly, all other vectors with a hat are meant to be unit vectors, and all quantities in bold are vectors.

It should be noted that one can show, using the free space Maxwell's equations, that Equations (2) and (3) satisfy the wave equations of $\nabla^2 \mathbf{E}(\mathbf{x},t) = \frac{1}{c^2}\frac{\partial^2}{\partial t^2}\mathbf{E}(\mathbf{x},t)$ and $\nabla^2 \mathbf{B}(\mathbf{x},t) = \frac{1}{c^2}\frac{\partial^2}{\partial t^2}\mathbf{B}(\mathbf{x},t)$. Moreover, the presence of $\hat{\varepsilon}_{\mathbf{k}_n,\lambda}$ and $\left(\hat{\mathbf{k}}_n \times \hat{\varepsilon}_{\mathbf{k}_n,\lambda}\right)$ in Equations (2) and (3), plus the cited relationships of $\mathbf{k}_n \cdot \hat{\varepsilon}_{\mathbf{k}_n,\lambda} = \mathbf{k}_n \cdot \hat{\varepsilon}_{\mathbf{k}_n,\lambda'} = 0$, and $\hat{\varepsilon}_{\mathbf{k}_n,\lambda} \cdot \hat{\varepsilon}_{\mathbf{k}_n,\lambda'} = 0$ for $\lambda \neq \lambda'$, enable all four free space Maxwell's equations to be satisfied.

In SED, the stochastic nature of the radiation fields in Equations (2) and (3) arises from the probability distribution of the Fourier coefficients $A_1, \cdots, A_N, B_1, \cdots, B_N$ over a large ensemble of equally sized space regions, which in the case considered here has dimensions L_x, L_y, L_z. The fields within the ensemble of space regions are characterized by the temperature $T \geq 0$; consequently, the Fourier coefficients will have a probabilistic distribution over the ensemble. For each member of the ensemble, the Fourier coefficients are fixed; only when examining each cavity in the ensemble will the Fourier coefficients be different and follow a probabilistic distribution in values.

Although this description is followed in SED, this basic behavior goes back to Planck in the first half of Planck's major treatise [20] and later Einstein and Hopf [21,22] (see English translations in Refs. [23,24], respectively).The main difference with this much older studies and SED is that in SED the assumption is not made that the radiation fields fall to zero at $T = 0$.

Here, a classical charged point particle, with charge, q, is considered oscillating in one dimension, namely, the $\hat{x}$ direction, constrained along the $\hat{x}$ direction by a SHO potential, $\frac{1}{2}m\omega_0^2 x^2$, giving rise to a binding force, $-m\omega_0^2 x(t)$, along the $\hat{x}$ direction. If one imagines a sphere of uniform charge density and net charge $-q$ that the $+q$ point charge oscillates within, then the SHO force acting on the $+q$ charge can be pictured as originating in this way. This neutral system will look like an electric dipole oscillator at distances far from the oscillator system.

As for the oscillating $+q$ charge, one can describe its motion using the nonrelativistic approximation of the Lorentz–Dirac equation:

$$m\ddot{x}(t) = -m\omega_0^2 x(t) + m\Gamma \dddot{x}(t) + qE_{\text{rad},x}(\mathbf{x}=\mathbf{0},t) , \qquad (5)$$

where $\Gamma = \frac{2}{3}\frac{q^2}{mc^3}$, and $m\Gamma\dddot{x}$ is the nonrelativistic expression for the radiation reaction for a charged point particle of mass m and charge q. $E_{\text{rad},x}$ is the net electric field in the $\hat{x}$ direction due to the sum of the radiation fields, assuming them to be either ZP or ZPP. The reason for $E_{\text{rad},x}(\mathbf{x}=\mathbf{0},t)$ is that the dipole approximation is being made when evaluating the electric field component of the Lorentz force. The magnetic field component of the Lorentz force is assumed to be much smaller in magnitude than $qE_{\text{rad},x}(\mathbf{x}=\mathbf{0},t)$ and is ignored here.

Finally, a common approximation to the weak "force" of $m\Gamma\dddot{x}(t)$, due to the small magnitude of Γ, is first that $m\ddot{x} \approx -m\omega_0^2 x$, or $\ddot{x} \approx -\omega_0^2 x$, so that

$$\dddot{x} = \frac{d}{dt}\ddot{x} \approx \frac{d}{dt}\left(-\omega_0^2 x\right) = -\omega_0^2 \dot{x} . \qquad (6)$$

After dividing through by m, Equation (5) becomes:

$$\ddot{x}(t) = -\omega_0^2 x(t) - \Gamma\omega_0^2 \dot{x}(t) + \frac{q}{m}E_{\text{rad},x}(\mathbf{x}=\mathbf{0},t) . \qquad (7)$$

Using Equation (7) and the properties of the ZP and ZPP radiation fields, Boyer showed in Ref. [3] that a detailed agreement exists between SED versus QM and QED, for the stochastic properties of this SHO system, for $T \geq 0$.

Rewriting Equation (2) in the dipole approximation,

$$E_{\text{rad},x}(\mathbf{x}=\mathbf{0},t) = \frac{1}{(L_x L_y L_z)^{1/2}} \sum_p (\hat{x} \cdot \hat{\varepsilon}_p)\left[A_p \cos(\omega_p t) - B_p \sin(\omega_p t)\right]$$

$$= \frac{1}{(L_x L_y L_z)^{1/2}} \sum_p (\hat{x} \cdot \hat{\varepsilon}_p) \text{Re}\left[(A_p + iB_p)e^{i\omega_p t}\right] , \qquad (8)$$

where the sum over the index p means the full sum over n_x, n_y, n_z and λ in Equation (2), and the second line in Equation (8) arises due to the A and B coefficients being real.

The steady state particular solution to Equation (7) can be shown to be

$$x_{\text{ss}}(t) = \frac{(q/m)}{\sqrt{L_x L_y L_z}} \sum_p (\hat{x} \cdot \hat{\varepsilon}_p) \text{Re}\left[\frac{e^{i\omega_p t}(A_p + iB_p)}{\left(-\omega_p^2 + \omega_0^2 + i\Gamma\omega_p\omega_0^2\right)}\right] , \qquad (9)$$

as can be checked by substituting Equation (9) into Equation (5) and by looking at times large enough that the homogeneous solution dies out.

Let us relabel $x_{ss}(t)$ as $x_{A_1,\cdots,B_N}(t)$, to emphasize the dependence on the Fourier coefficients, as in Equation (1). Moreover, to help make better use of Equation (9), let us make the following definitions:

$$x_{A,p}(t) \equiv \frac{(q/m)}{\sqrt{L_x L_y L_z}} (\hat{x} \cdot \hat{\epsilon}_p) A_p \, \text{Re} \left[\frac{e^{i\omega_p t}}{-\omega_p^2 + \omega_0^2 + i\Gamma \omega_p \omega_0^2} \right] \quad (10)$$

and

$$x_{B,p}(t) \equiv \frac{(q/m)}{\sqrt{L_x L_y L_z}} (\hat{x} \cdot \hat{\epsilon}_p) B_p \, \text{Re} \left[\frac{i e^{i\omega_p t}}{-\omega_p^2 + \omega_0^2 + i\Gamma \omega_p \omega_0^2} \right]$$

$$= -\frac{(q/m)}{\sqrt{L_x L_y L_z}} (\hat{x} \cdot \hat{\epsilon}_p) B_p \, \text{Im} \left[\frac{e^{i\omega_p t}}{-\omega_p^2 + \omega_0^2 + i\Gamma \omega_p \omega_0^2} \right] , \quad (11)$$

Simplifying further, let us define $x'_{A,p}(t)$ and $x'_{B,p}(t)$ via:

$$x_{A,p}(t) \equiv A_p x'_{A,p}(t) \quad (12)$$

and

$$x_{B,p}(t) \equiv B_p x'_{B,p}(t) . \quad (13)$$

In SED, the Fourier coefficients $A_1, \cdots, A_N, B_1, \cdots, B_N$ are assumed to be independent random variables with Gaussian probability density distribution,

$$P_F(A_p) = \frac{1}{\sqrt{2\pi \left(\sigma_p^A\right)^2}} \exp\left(-\frac{A_p^2}{2\left(\sigma_p\right)^2}\right) . \quad (14)$$

The same distribution holds for the Fourier coefficient B_p. The label "F" is added to specify that this probability density function P_F refers to the Fourier coefficients. Moreover, σ_p depends on $\mathbf{k_n}$ as in Equation (4), as well as the temperature T. This dependence has been studied considerably in SED. In particular, the functional form of σ_p at $T = 0$ is a cornerstone for SED and is expressed by $[\sigma(\omega_n, T = 0)]^2 = 2\pi\hbar\omega_n$ where $\hbar$ is the Planck's reduced constant. It is this functional form at $T = 0$ that was referred to in Section 1 and that is deduced first via the imposition of Lorentz invariance by Marshall [11] and Boyer [12], and much later by the author by imposing the thermodynamic definition of $T = 0$ [13–15].

In general, for $T \geq 0$,

$$\sigma_p \to [\sigma(\omega_\mathbf{n}, T)]^2 = 2\pi\hbar\omega_\mathbf{n} + \frac{4\pi\hbar\omega_\mathbf{n}}{\exp\left(\frac{\hbar\omega_\mathbf{n}}{k_B T}\right) - 1} = 2\pi\hbar\omega_\mathbf{n} \coth\left(\frac{\hbar\omega_\mathbf{n}}{2k_B T}\right) , \quad (15)$$

where $\omega_\mathbf{n} = c|\mathbf{k_n}|$.

Replacing $\delta^3[\mathbf{x} - \mathbf{x}_{A_1,\cdots,B_N}(t)]$ in Equation (1) with the 1D Dirac delta function of

$$\frac{1}{2\pi} \int_{-\infty}^{\infty} ds \, e^{is(x - x_{A_1,\cdots,B_N}(t))} ,$$

and realizing that $P_{F,A\text{-}B}(A_1,\cdots,A_N,B_1,\cdots,B_N)$ equals $P_F(A_1)\cdots P_F(A_N)P_F(B_1)\cdots P_F(B_N)$ due to the random variable independence of these Fourier coefficients, then the 1D position probability density function for this 1D SHO is

$$P_{1x}(x)$$
$$= \int_{-\infty}^{\infty} dA_1 \cdots \int_{-\infty}^{\infty} dA_N \cdots \int_{-\infty}^{\infty} dB_1 \cdots \int_{-\infty}^{\infty} dB_N P_{F,A\text{-}B}(A_1,\cdots,A_N,B_1,\cdots,B_N)\delta\left[x - x_{A_1,\cdots,B_N}(t)\right]$$
$$= \int_{-\infty}^{\infty} ds \int_{-\infty}^{\infty} dA_1 \cdots \int_{-\infty}^{\infty} dB_N \frac{1}{2\pi} e^{is\left(x - x_{A_1,\cdots,B_N}(t)\right)}$$
$$\times \frac{1}{\sqrt{2\pi(\sigma_1)^2}} \exp\left(-\frac{A_1^2}{2\sigma_1^2}\right) \cdots \frac{1}{\sqrt{2\pi(\sigma_N)^2}} \exp\left(-\frac{B_N^2}{2\sigma_N^2}\right)$$
$$= \frac{1}{2\pi}\frac{1}{2\pi\sigma_1^2} \cdots \frac{1}{2\pi\sigma_N^2} \int_{-\infty}^{\infty} ds\, e^{isx}$$
$$\times \int_{-\infty}^{\infty} dA_1 \exp\left(-\frac{A_1^2}{2\sigma_1^2} - isA_1 x'_{A,1}\right) \cdots \int_{-\infty}^{\infty} dB_N \exp\left(-\frac{B_N^2}{2\sigma_N^2} - isB_N x'_{B,N}\right) . \quad (16)$$

The evaluation of Equation (16) can be conducted by completing squares. Specifically:

$$\int_{-\infty}^{\infty} dA_p \exp\left(-\frac{A_p^2}{2\sigma_p^2} - isA_p x'_{A,p}\right)$$
$$= \int_{-\infty}^{\infty} dA_p \exp\left[-\frac{\left(A_p^2 + 2isA_p x'_{A,p}\sigma_p^2\right)}{2\sigma_p^2}\right]$$
$$= \int_{-\infty}^{\infty} dA_p \exp\left\{-\frac{\left[A_p^2 + 2isA_p x'_{A,p}\sigma_p^2 - s^2\left(x'_{A,p}\right)^2\sigma_p^4\right]}{2\sigma_p^2} - \frac{s^2\left(x'_{A,p}\right)^2\sigma_p^4}{2\sigma_p^2}\right\}$$
$$= \exp\left[-\frac{s^2\left(x'_{A,p}\right)^2\sigma_p^2}{2}\right] \int_{-\infty}^{\infty} dA_p \exp\left\{-\frac{\left(A_p + isx'_{A,p}\sigma_p^2\right)^2}{2\sigma_p^2}\right\}$$
$$= \exp\left[-\frac{s^2\left(x'_{A,p}\right)^2\sigma_p^2}{2}\right] \sqrt{2\pi\sigma_p^2} . \quad (17)$$

The same holds true for the B terms.
Hence, from Equations (16) and (17):

$$P_{1x}(x) = \frac{1}{2\pi} \int_{-\infty}^{\infty} ds\, e^{isx} \left\{\exp\left[-\frac{s^2\left(x'_{A,1}\right)^2\sigma_1^2}{2}\right] \cdots \exp\left[-\frac{s^2\left(x'_{A,N}\right)^2\sigma_N^2}{2}\right]\right\}$$
$$\times \left\{\exp\left[-\frac{s^2\left(x'_{B,1}\right)^2\sigma_1^2}{2}\right] \cdots \exp\left[-\frac{s^2\left(x'_{B,N}\right)^2\sigma_N^2}{2}\right]\right\}$$
$$= \frac{1}{2\pi} \int_{-\infty}^{\infty} ds \exp\left[isx - \frac{s^2}{2}\sum_p\left[\left(x'_{A,p}\sigma_p\right)^2 + \left(x'_{B,p}\sigma_p\right)^2\right]\right] . \quad (18)$$

To better identify the significance of this expression, let us calculate

$$\left\langle [x_{A_1,\cdots,B_N}(t)]^2 \right\rangle = \left\langle [x_{\text{ss}}(t)]^2 \right\rangle = \left\langle \left\{ \frac{(q/m)}{\sqrt{L_x L_y L_z}} \sum_p (\hat{x} \cdot \hat{\varepsilon}_p) \operatorname{Re}\left[\frac{e^{i\omega_p t}(A_p + iB_p)}{\left(-\omega_p^2 + \omega_0^2 + i\Gamma\omega_p\omega_0^2\right)} \right] \right\}^2 \right\rangle , \qquad (19)$$

where the angle brackets mean that the ensemble average is to be taken. Let us impose that

$$\langle A_{\mathbf{k_n},\lambda} \rangle = \langle B_{\mathbf{k_n},\lambda} \rangle = 0 , \qquad (20)$$

from Equation (14), which also holds for the $B_{\mathbf{k_n},\lambda}$ variables, and also impose the assumption of independent random variables,

$$\left\langle A_{\mathbf{k_n},\lambda} B_{\mathbf{k_{n'}},\lambda'} \right\rangle = 0 , \qquad (21)$$

and:

$$\left\langle A_{\mathbf{k_n},\lambda} A_{\mathbf{k_{n'}},\lambda'} \right\rangle = \left\langle B_{\mathbf{k_n},\lambda} B_{\mathbf{k_{n'}},\lambda'} \right\rangle = 0 , \text{ if } \mathbf{n} \neq \mathbf{n'} \text{ or } \lambda \neq \lambda' , \qquad (22)$$

while

$$\left\langle A_{\mathbf{k_n},\lambda} A_{\mathbf{k_n},\lambda} \right\rangle = \left\langle B_{\mathbf{k_n},\lambda} B_{\mathbf{k_n},\lambda} \right\rangle = [\sigma(\omega_{\mathbf{n}}, T)]^2 , \qquad (23)$$

are assumed to be functions of the frequency of the radiation and of the temperature T, which were previously labelled as σ_p^2.

Continuing with the evaluation of Equation (19), let us first note that:

$$x_{\text{ss}}(t) = \frac{(q/m)}{\sqrt{L_x L_y L_z}} \sum_p (\hat{x} \cdot \hat{\varepsilon}_p) \left\{ \begin{array}{l} A_p \operatorname{Re}\left[\frac{e^{i\omega_p t}}{(-\omega_p^2 + \omega_0^2 + i\Gamma\omega_p\omega_0^2)} \right] \\ -B_p \operatorname{Im}\left[\frac{e^{i\omega_p t}}{(-\omega_p^2 + \omega_0^2 + i\Gamma\omega_p\omega_0^2)} \right] \end{array} \right\} . \qquad (24)$$

Taking the statistical properties into account of Equations (21)–(23), we obtain:

$$\left\langle [x_{\text{ss}}(t)]^2 \right\rangle = \frac{(q/m)^2}{L_x L_y L_z} \sum_p (\hat{x} \cdot \hat{\varepsilon}_p)^2 \sigma_p^2 \left(\left\{ \operatorname{Re}\left[\frac{e^{i\omega_p t}}{(-\omega_p^2 + \omega_0^2 + i\Gamma\omega_p\omega_0^2)} \right] \right\}^2 + \left\{ \operatorname{Im}\left[\frac{e^{i\omega_p t}}{(-\omega_p^2 + \omega_0^2 + i\Gamma\omega_p\omega_0^2)} \right] \right\}^2 \right)$$

$$= \frac{(q/m)^2}{L_x L_y L_z} \sum_p (\hat{x} \cdot \hat{\varepsilon}_p)^2 \left| \frac{1}{\left(-\omega_p^2 + \omega_0^2 + i\Gamma\omega_p\omega_0^2\right)} \right|^2 \sigma_p^2 . \qquad (25)$$

To simplify later expressions, let

$$\sigma_x^2 \equiv \left\langle [x_{\text{ss}}(t)]^2 \right\rangle . \qquad (26)$$

Now one just needs to relate the terms $\sum_p \sigma_p^2 \left[\left(x'_{A_p}\right)^2 + \left(x'_{B_p}\right)^2 \right]$ in Equation (18) to $\left\langle [x_{\text{ss}}(t)]^2 \right\rangle$. As shown below, they are exactly equal. From Equations (10)–(13):

$$\sum_p \sigma_p^2 \left[\left(x'_{A_p}\right)^2 + \left(x'_{B_p}\right)^2 \right]$$

$$= \sum_p \sigma_p^2 \frac{(q/m)^2}{L_x L_y L_z} (\hat{x} \cdot \hat{\varepsilon}_p)^2 \left(\left\{ \operatorname{Re}\left[\frac{e^{i\omega_p t}}{-\omega_p^2 + \omega_0^2 + i\Gamma\omega_p\omega_0^2} \right] \right\}^2 + \left\{ \operatorname{Im}\left[\frac{e^{i\omega_p t}}{-\omega_p^2 + \omega_0^2 + i\Gamma\omega_p\omega_0^2} \right] \right\}^2 \right)$$

$$= \sum_p \frac{(q/m)^2}{L_x L_y L_z} (\hat{x} \cdot \hat{\varepsilon}_p)^2 \frac{1}{\left|-\omega_p^2 + \omega_0^2 + i\Gamma\omega_p\omega_0^2\right|^2} \sigma_p^2 = \left\langle [x_{\text{ss}}(t)]^2 \right\rangle . \qquad (27)$$

Equation (18) then becomes, after completing the square of s and then integrating over s:

$$
\begin{aligned}
P_{1x}(x) &= \frac{1}{2\pi} \int_{-\infty}^{\infty} ds \, \exp\left[isx - \frac{s^2}{2}\left\langle [x_{ss}(t)]^2 \right\rangle\right] \\
&= \frac{1}{2\pi} \int_{-\infty}^{\infty} ds \, \exp\left[-\frac{\sigma_x^2}{2}\left(s^2 - \frac{2isx}{\sigma_x^2} - \frac{x^2}{(\sigma_x^2)^2}\right) - \frac{x^2}{2\sigma_x^2}\right] \\
&= \frac{1}{2\pi} \sqrt{\frac{2\pi}{\sigma_x^2}} \exp\left(-\frac{x^2}{2\sigma_x^2}\right) \\
&= \frac{1}{\sqrt{2\pi\sigma_x^2}} \exp\left(-\frac{x^2}{2\sigma_x^2}\right) .
\end{aligned}
\qquad (28)
$$

This Gaussian result has been deduced in SED previously by researchers in SED, but not as far as this author knows, starting via the general probability expression in Equation (1). Although the above is a much longer derivation than deductions published earlier, it is still illuminating, as discussed in Sections 2.2, 3 and 4 below.

Moreover, while the expression (28) is connected with QED [3], one can relate it to the relevant expression in QM as calculated from Schrödinger's equation when taking into account the probability density at temperature T, and summing over all probability density functions $|\psi_n(x)|^2$, each times $\frac{1}{Z}\exp(-E_n/kT)$, where $Z = \sum_n \exp(-E_n/kT)$. To obtain this agreement, and essentially dropping the QED effects, in SED one would make what some call the continuum and resonant approximations, where first the sum over $\mathbf{n}$ is approximated as a 3D integral, and then later the charge q is assumed to be small.

More specifically, the continuum approximation consists of the following approximations, from Equation (4):

$$
dk_x = 2\pi \frac{dn_x}{L_x} \;,\; dk_y = 2\pi \frac{dn_y}{L_y} \;,\; dk_z = 2\pi \frac{dn_z}{L_z} \;,
\qquad (29)
$$

so that

$$
\sum_{\mathbf{n}} \cdots = \sum_{n_1,n_2,n_3} \cdots \approx \int dn_x \int dn_y \int dn_z \cdots \approx \frac{L_x L_y L_z}{(2\pi)^3} \int dk_x \int dk_y \int dk_z \cdots .
\qquad (30)
$$

Consequently,

$$
\begin{aligned}
\left\langle [x_{ss}(t)]^2 \right\rangle &= \sum_p \sigma_p^2 \frac{(q/m)^2}{L_x L_y L_z} (\hat{\mathbf{x}} \cdot \hat{\varepsilon}_p)^2 \frac{1}{\left|-\omega_p^2 + \omega_0^2 + i\Gamma\omega_p\omega_0^2\right|^2} \\
&= \frac{(q/m)^2}{L_x L_y L_z} \sum_{n_x,n_y,n_z=-\infty}^{\infty} \sigma_{\mathbf{k_n}}^2 \sum_{\lambda=1,2} (\hat{\mathbf{x}} \cdot \hat{\varepsilon}_{\mathbf{k_n},\lambda})^2 \frac{1}{\left|-\omega_{\mathbf{k_n}}^2 + \omega_0^2 + i\Gamma\omega_{\mathbf{k_n}}\omega_0^2\right|^2} \\
&\approx \frac{(q/m)^2}{(2\pi)^3} \int_{-\infty}^{\infty} dk_x \int_{-\infty}^{\infty} dk_y \int_{-\infty}^{\infty} dk_z \, \sigma_{\mathbf{k}}^2 \sum_{\lambda=1,2} (\hat{\mathbf{x}} \cdot \hat{\varepsilon}_{\mathbf{k},\lambda})^2 \frac{1}{\left|-\omega_{\mathbf{k}}^2 + \omega_0^2 + i\Gamma\omega_{\mathbf{k}}\omega_0^2\right|^2} .
\end{aligned}
\qquad (31)
$$

One can show that

$$
\sum_{\lambda=1,2} (\hat{\mathbf{x}} \cdot \hat{\varepsilon}_{\mathbf{k},\lambda})^2 = \sum_{\lambda=1,2} (\hat{\varepsilon}_{\mathbf{k},\lambda,x})^2 = \left[1 - \left(\frac{k_x^2}{\mathbf{k}}\right)^2\right] .
\qquad (32)
$$

Now, using spherical coordinates in Equation (31) and noting that $\sigma_\mathbf{k}$ actually depends only on the magnitude of $\mathbf{k}$, or $|\mathbf{k}| = k$, and likewise, $\omega_\mathbf{k} = \omega_k = ck$, one finds:

$$\left\langle [x_{ss}(t)]^2 \right\rangle \approx \frac{(q/m)^2}{(2\pi)^3} \int_0^\infty dk \int_0^\pi d\theta \int_0^{2\pi} d\phi\, k(\sin\theta) \sigma_k^2 \frac{\left[1 - \left(\frac{k_x^2}{k}\right)^2\right]}{\left[(-\omega_k^2 + \omega_0^2)^2 + (\Gamma \omega_k \omega_0^2)^2\right]} \,. \tag{33}$$

Integrating over θ and ϕ and noting by symmetry that

$$\int_0^\pi d\theta \sin\theta \int_0^{2\pi} d\phi \left[1 - \left(\frac{k_x^2}{k}\right)^2\right] = 4\pi - \frac{4\pi}{3} = \frac{8\pi}{3} \,, \tag{34}$$

then:

$$\left\langle [x_{ss}(t)]^2 \right\rangle \approx \frac{(q/m)^2}{(2\pi)^3} \frac{8\pi}{3} \frac{1}{c^3} \int_0^\infty d\omega \frac{\omega^2 \sigma^2(\omega, T)}{\left[(-\omega^2 + \omega_0^2)^2 + (\Gamma \omega \omega_0^2)^2\right]} \,. \tag{35}$$

For small values of q in $\Gamma = \frac{2}{3} \frac{q^2}{mc^3}$ (for an electron, Γ is very small, about 6.27×10^{-24} s), $\left[(-\omega^2 + \omega_0^2)^2 + (\Gamma \omega \omega_0^2)^2\right]^{-1}$ becomes strongly peaked when $\omega \approx \omega_0$. The above integral can then be well approximated by

$$\int_0^\infty d\omega \frac{\omega_0^2 \sigma^2(\omega_0, T)}{(\omega - \omega_0)^2 (2\omega_0)^2 + (\Gamma \omega_0^3)^2} \approx \int_{-\infty}^\infty d\omega \frac{\omega_0^2 \sigma^2(\omega_0, T)}{(\omega - \omega_0)^2 (2\omega_0)^2 + (\Gamma \omega_0^3)^2} \,. \tag{36}$$

Making use of the integral,

$$\int_{-\infty}^\infty d\omega \frac{1}{\omega^2 A^2 + B^2} = \frac{\pi}{AB} \,. \tag{37}$$

Equation (35) becomes:

$$\left\langle [x_{ss}(t)]^2 \right\rangle \approx \frac{(q/m)^2}{(2\pi)^3} \frac{8\pi}{3} \frac{1}{c^3} \omega_0^2 \sigma^2(\omega_0, T) \frac{\pi}{(2\omega_0)\left(\frac{2}{3}\frac{q^2}{mc^3}\omega_0^3\right)}$$

$$= \frac{\hbar}{2m\omega_0} \coth\left(\frac{\hbar \omega_0}{2kT}\right) \,. \tag{38}$$

This is more recognizable for QM, and even more so for $T \to 0$, becoming

$$\left\langle [x_{ss}(t)]^2 \right\rangle_{T=0} = \frac{\hbar}{2m\omega_0} \,. \tag{39}$$

2.2. Examples of Other Analytic SHO Probability Density Functions That Can Similarly Be Deduced

The method of Section 2.1 can be used to obtain many other types of probability density functions. Using

$$m\ddot{\mathbf{x}}(t) = -m\omega_0^2 \mathbf{x}(t) + m\Gamma \dddot{\mathbf{x}}(t) + q\mathbf{E}_{\text{rad}}(\mathbf{x} = \mathbf{0}, t) \,, \tag{40}$$

one can certainly generalize the previous 1D SHO to a 3D SHO with the probability density function of $P_{3x}(\mathbf{x})$ in Equation (1).

Moreover, the position and momentum joint probability density function for this 3D SHO can be expressed by

$$P_{3x,3p}(\mathbf{x},\mathbf{p}) = \int_{-\infty}^{\infty} dA_1 \cdots \int_{-\infty}^{\infty} dB_N P_{F,A\text{-}B}(A_1,\cdots,A_N,B_1,\cdots,B_N)\delta^3[\mathbf{x}-\mathbf{z}_{A_1\ldots B_N}(t)]\delta^3\left[\mathbf{p}-\mathbf{p}_{A_1\ldots B_N}(t)\right] , \quad (41)$$

which can be used to find an analytic expression for $P_{3x,3p}(\mathbf{x},\mathbf{p})$ in a similar manner as carried out in Section 2.1 for the 1D SHO and $P_{1x}(x)$.

In addition, one can express the probability density for the nonrelativistic energy of a 3D SHO via

$$P_E(\mathcal{E}) = \int dA_1 \cdots \int dB_N P_{F,A\text{-}B}(A_1,\cdots,A_N,B_1,\cdots,B_N)\delta^3\{\mathcal{E}-[\text{KE}(t)+\text{PE}(t)]\} , \quad (42)$$

where

$$\text{KE}(t) + \text{PE}(t) = \frac{m}{2}[\dot{x}_{ss}(t)]^2 + \frac{m\omega_0^2}{2}[x_{ss}(t)]^2 . \quad (43)$$

As another example, using the same method, one can calculate the following 1D SHO joint probability position density distribution $P_{1x,1x}(x_1,t_1;x_2,t_2)$. The final result reads:

$$P_{1x,1x}(x_1,t_1;x_2,t_2)$$
$$= \int_{-\infty}^{\infty} dA_1 \cdots \int_{-\infty}^{\infty} dB_N P_{F,A\text{-}B}(A_1,\cdots,A_N,B_1,\cdots,B_N)\delta[x_1-x_{A_1\ldots B_N}(t_1)]\delta[x_2-x_{A_1\ldots B_N}(t_2)]$$
$$= \int_{-\infty}^{\infty} dA_1 \cdots \int_{-\infty}^{\infty} dB_N P_{F,A\text{-}B}(A_1,\cdots,A_N,B_1,\cdots,B_N)$$
$$\times \frac{1}{2\pi}\int_{-\infty}^{\infty} ds_1 e^{is_1(x_1-x_{A_1,\cdots,B_N}(t_1))} \frac{1}{2\pi}\int_{-\infty}^{\infty} ds_2 e^{is_2(x_2-x_{A_1,\cdots,B_N}(t_2))}$$
$$= \frac{\exp\left(-\frac{[x_1^2\langle x_{ss}^2(t)\rangle + x_2^2\langle x_{ss}^2(t)\rangle - 2x_1 x_2 \langle x_{ss}(t_1)x_{ss}(t_2)\rangle]}{2[\langle x_{ss}^2(t)\rangle^2 - \langle x(t_1)x(t_2)\rangle^2]}\right)}{2\pi\sqrt{\langle x_{ss}^2(t)\rangle^2 - \langle x_{ss}(t_1)x_{ss}(t_2)\rangle^2}} . \quad (44)$$

Here, the ensemble average of the square of the steady state solution $x_{ss}(t)$, or $\langle x_{ss}^2(t)\rangle$, is independent of t, as shown in Equation (31), and as given in the more familiar continuum and resonant approximation in Equation (38), as $\frac{\hbar}{2m\omega_0}\coth\left(\frac{\hbar\omega_0}{2kT}\right)$. However, the ensemble average of the product $x_{ss}(t_1)x_{ss}(t_2)$, or $\langle x_{ss}(t_1)x_{ss}(t_2)\rangle$, depends on the time difference $t_1 - t_2$, and is given by

$$\langle x_{ss}(t_1)x_{ss}(t_2)\rangle = \frac{q^2}{m^2(L_xL_yL_z)}\sum_{\mathbf{n},\lambda}(\hat{\mathbf{x}}\cdot\hat{\boldsymbol{\varepsilon}}_{\mathbf{n},\lambda})^2(\sigma_{\mathbf{n},\lambda})^2\frac{\cos[\omega_{\mathbf{n}}(t_2-t_1)]}{\left(-\omega_{\mathbf{n}}^2+\omega_0^2\right)^2+\left(\Gamma\omega_0^2\omega_{\mathbf{n}}\right)^2} . \quad (45)$$

In the continuum and resonant approximation, this simplifies to

$$\langle x_{ss}(t_1)x_{ss}(t_2)\rangle \approx \frac{\hbar}{2m\omega_0}\coth\left(\frac{\hbar\omega_0}{2kT}\right)\cos[\omega_0(t_2-t_1)] . \quad (46)$$

Thus, the general expressions are quite straightforwad to formulate, as in Equations (1), (16), (41), (42) and (44), although carrying out all the integrations to arrive at an analytic expression, as in Equations (28) and (44), can be quite nontrivial.

3. A PDE Approach for Deducing the SHO Probability Density Function $P_{1x}(x)$

Rather than directly integrating over all the A_p and B_p radiation Fourier coefficients in Equation (16) to obtain the analytic expression for $P_{1x}(x)$ in Equation (28), here it is shown that Equation (16) satisfies a PDE that enables $P_{1x}(x)$ in Equation (28) to be deduced. In some ways, this approach is less complicated than the direct integration method of Section 2.1 and might provide insight for more complicated systems than the SHO, such as the classical hydrogen case.

Without integrating over all $A_1, \cdots, B_N$ variables as in Section 2.1, nor by only showing that Equation (28) solves Equation (47) below, here, let us directly take the 1D version of Equation (1) and illustrate how this expression satisfies

$$\frac{\partial}{\partial(\sigma_x^2)} P_{1x}(x) = \frac{1}{2} \frac{\partial^2}{\partial x^2} P_{1x}(x) \ , \tag{47}$$

where σ_x^2 is given by Equations (25) and (26), or in the continuum resonant approximations by Equation (38). More specifically, as is shown below, in order for Equation (47) to hold with $P_{1x}(x)$ given by the 1D version of Equation (1) (i.e., the top line of Equation (16)), then σ_x^2 must be given by Equation (25). Moreover, upon imposing that $P_{1x}(x) \to 0$ as $|x| \to \infty$, with symmetry about $x = 0$, and that the function monotonically decreases as $|x|$ increases, one obtains the solution Equation (28) for Equation (47).

To show that $P_{1x}(x)$ in the 1D version of Equation (1) satisfies Equation (47), let us start with

$$\frac{\partial}{\partial(\sigma_x^2)} P_{1x}(x)$$

$$= \frac{\partial}{\partial(\sigma_x^2)} \int_{-\infty}^{\infty} dA_1, \cdots, \int_{-\infty}^{\infty} dA_N \int_{-\infty}^{\infty} dB_1, \cdots, \int_{-\infty}^{\infty} dB_N$$

$$P_F(A_1), \cdots, P_F(A_N), P_F(B_1), \cdots, P_F(B_N) \delta[x - x_{A_1, \cdots, B_N}(t)]$$

$$= \int_{-\infty}^{\infty} dA_1, \cdots, \int_{-\infty}^{\infty} dA_N \int_{-\infty}^{\infty} dB_1, \cdots, \int_{-\infty}^{\infty} dB_N \delta[x - x_{A_1, \cdots, B_N}(t)]$$

$$\times \frac{\partial}{\partial(\sigma_x^2)} [P_F(A_1) \cdots P_F(A_N) P_F(B_1) \cdots P_F(B_N)] \ . \tag{48}$$

Clearly,

$$\frac{\partial}{\partial(\sigma_x^2)} [P_F(A_1) \cdots P_F(B_N)]$$

$$= \left[\frac{\partial}{\partial(\sigma_x^2)} P_F(A_1) \right] P_F(A_2) \cdots P_F(A_N) P_F(B_1) P_F(B_2) \cdots P_F(B_N)$$

$$+ P_F(A_1) \left[\frac{\partial}{\partial(\sigma_x^2)} P_F(A_2) \right] \cdots P_F(A_N) P_F(B_1) P_F(B_2) \cdots P_F(B_N)$$

$$+ \cdots + P_F(A_1) P_F(A_2) \cdots P_F(A_N) P_F(B_1) P_F(B_2) \cdots \left[\frac{\partial}{\partial(\sigma_x^2)} P_F(B_N) \right] \ . \tag{49}$$

Moreover, since one can readily show that

$$\frac{\partial}{\partial \left(\sigma_p^2 \right)} \left\{ \frac{1}{\sqrt{2\pi\sigma_p^2}} \exp\left[-\frac{(A_p)^2}{2\sigma_p^2} \right] \right\} = \frac{1}{2} \frac{\partial^2}{\partial A_p^2} \left\{ \frac{1}{\sqrt{2\pi\sigma_p^2}} \exp\left[-\frac{(A_p)^2}{2\sigma_p^2} \right] \right\} \ , \tag{50}$$

then,

$$\frac{\partial}{\partial(\sigma_x^2)}P_F(A_p) = \frac{\partial(\sigma_p^2)}{\partial(\sigma_x^2)}\frac{\partial}{\partial(\sigma_p^2)}P_F(A_p)$$

$$= \frac{\partial(\sigma_p^2)}{\partial(\sigma_x^2)}\frac{1}{2}\frac{\partial^2}{\partial A_p^2}\left\{\frac{1}{\sqrt{2\pi\sigma_p^2}}\exp\left[-\frac{(A_p)^2}{2\sigma_p^2}\right]\right\}, \quad (51)$$

and likewise for $P_F(B_p)$.

From Equations (48)–(51):

$$\frac{\partial}{\partial\sigma_x^2}P_{1x}(x)$$

$$= \int_{-\infty}^{\infty}dA_1\cdots\int_{-\infty}^{\infty}dB_N\left\{\begin{array}{l}\frac{\partial(\sigma_1^2)}{\partial(\sigma_x^2)}\frac{1}{2}\frac{d^2P_F(A_1)}{dA_1^2}P_F(A_2)\cdots P_F(B_N)\\ +P_F(A_1)\frac{\partial(\sigma_2^2)}{\partial(\sigma_x^2)}\frac{1}{2}\frac{d^2P_F(A_2)}{dA_2^2}P_F(A_3)\cdots P_F(B_N)\\ +\cdots+[P_F(A_1)\cdots P_F(B_{N-1})]\frac{\partial(\sigma_N^2)}{\partial(\sigma_x^2)}\frac{1}{2}\frac{d^2P_F(B_N)}{dB_N^2}\end{array}\right\}\delta[x-x_{A_1,\cdots,B_N}(t)]. \quad (52)$$

Each of the 2N terms in the sum within the curly brackets above can be integrated by parts twice. Considering the pth A_p term,

$$\frac{1}{2}\int_{-\infty}^{\infty}dA_1 P_F(A_1)\cdots\int_{-\infty}^{\infty}dA_{p-1}P_F(A_{p-1})\left\{\int_{-\infty}^{\infty}dA_p\frac{\partial(\sigma_p^2)}{\partial(\sigma_x^2)}\frac{d^2P_F(A_p)}{dA_p^2}\right\}$$

$$\times\int_{-\infty}^{\infty}dA_{p+1}P_F(A_{p+1})\cdots\int_{-\infty}^{\infty}dA_N P_F(A_N)\int_{-\infty}^{\infty}dB_1 P_F(B_1)\cdots\int_{-\infty}^{\infty}dB_N P_F(B_N)\delta[x-x_{A_1,\cdots,B_N}(t)], \quad (53)$$

and integrating by parts for this pth term yields the following. Note that neither σ_x^2 from Equation (25) nor σ_p^2 from Equation (15) depend on A_p or B_p, so $\partial(\sigma_p^2)/\partial(\sigma_x^2)$ is not involved with this integration by parts, for either A_p or B_p.

Hence:

$$\int_{-\infty}^{\infty}dA_p\frac{\partial(\sigma_p^2)}{\partial(\sigma_x^2)}\frac{d^2P_F(A_p)}{dA_p^2}\delta[x-x_{A_1,\cdots,B_N}(t)]$$

$$= \frac{\partial(\sigma_p^2)}{\partial(\sigma_x^2)}\left(\begin{array}{l}\int_{-\infty}^{\infty}dA_p\frac{d}{dA_p}\left\{\frac{dP_F(A_p)}{dA_p}\delta[x-x_{A_1,\cdots,B_N}(t)]\right\}\\ -\int_{-\infty}^{\infty}dA_p\left\{\frac{dP_F(A_p)}{dA_p}\frac{d}{dA_p}\delta[x-x_{A_1,\cdots,B_N}(t)]\right\}\end{array}\right)$$

$$= \frac{\partial(\sigma_p^2)}{\partial(\sigma_x^2)}\left[\begin{array}{l}\left\{\frac{dP_F(A_p)}{dA_p}\delta[x-x_{A_1,\cdots,B_N}(t)]\right\}\Big|_{A_p\to-\infty}^{A_p\to\infty}\\ -\int_{-\infty}^{\infty}dA_p\left(\begin{array}{l}\frac{d}{dA_p}\left\{P_F(A_p)\frac{d}{dA_p}\delta[x-x_{A_1,\cdots,B_N}(t)]\right\}\\ -P_F(A_p)\frac{d^2}{dA_p^2}\delta[x-x_{A_1,\cdots,B_N}(t)]\end{array}\right)\end{array}\right]. \quad (54)$$

From Equation (14), $P_F(A_p)$ and $\frac{d}{dA_p}P_F(A_p)$ both go to zero exponentially as $A_p \to \pm\infty$, so the first term in the last line above is zero. Continuing:

$$\frac{\partial(\sigma_p^2)}{\partial(\sigma_x^2)} \int_{-\infty}^{\infty} dA_p \frac{d^2 P_F(A_p)}{dA_p^2} \delta[x - x_{A_1,\cdots,B_N}(t)]$$

$$= \frac{\partial(\sigma_p^2)}{\partial(\sigma_x^2)} \left[\begin{array}{l} -P_F(A_p)\frac{d}{dA_p}\delta[x - x_{A_1,\cdots,B_N}(t)]\Big|_{A_p \to -\infty}^{A_p \to \infty} \\ + \int_{-\infty}^{\infty} dA_p P_F(A_p) \frac{d^2}{dA_p^2} \delta[x - x_{A_1,\cdots,B_N}(t)] \end{array} \right]$$

$$= \frac{\partial(\sigma_p^2)}{\partial(\sigma_x^2)} \int_{-\infty}^{\infty} dA_p P_F(A_p) \frac{d^2}{dA_p^2} \delta[x - x_{A_1,\cdots,B_N}(t)]. \tag{55}$$

Thus:

$$\frac{\partial}{\partial \sigma_x^2} P_{1x}(x)$$

$$= \frac{1}{2} \int_{-\infty}^{\infty} dA_1 \cdots \int_{-\infty}^{\infty} dB_N P_F(A_1) \cdots P_F(A_N) P_F(B_1) \cdots P_F(B_N)$$

$$\times \sum_{p=1}^{N} \frac{\partial(\sigma_p^2)}{\partial(\sigma_x^2)} \left(\frac{d^2}{dA_p^2} + \frac{d^2}{dB_p^2} \right) \delta[x - x_{A_1,\cdots,B_N}(t)]. \tag{56}$$

Since

$$\frac{d^2 \delta[x - x_{A_1\ldots B_N}(t)]}{dA_p^2} = \frac{d}{dA_p} \left\{ \left(\frac{\partial x_{A_1\ldots B_N}}{\partial A_p} \right) \frac{\partial \delta[x - x_{A_1\ldots B_N}(t)]}{\partial x_{A_1\ldots B_N}} \right\}, \tag{57}$$

then with $x_{A_1,\cdots,B_N}(t) = x_{ss}(t)$ in Equation (9),

$$\left(\frac{\partial x_{A_1\ldots B_N}}{\partial A_p} \right) = \frac{(q/m)}{\sqrt{L_x L_y L_z}} \varepsilon_{p,x} \operatorname{Re}\left(\frac{e^{i\omega_p t}}{-\omega_p^2 + \omega_0^2 + i\Gamma \omega_p \omega_0^2} \right), \tag{58}$$

and

$$\left(\frac{\partial x_{A_1\ldots B_N}}{\partial B_p} \right) = -\frac{(q/m)}{\sqrt{L_x L_y L_z}} \varepsilon_{p,x} \operatorname{Im}\left(\frac{e^{i\omega_p t}}{-\omega_p^2 + \omega_0^2 + i\Gamma \omega_p \omega_0^2} \right). \tag{59}$$

Both $\left(\frac{\partial x_{A_1\ldots B_N}}{\partial A_p}\right)$ and $\left(\frac{\partial x_{A_1\ldots B_N}}{\partial B_p}\right)$ are independent of A_p and B_p for all p. This follows from the linearity of $x_{A_1\ldots B_N}$ with the A's and B's, which in turn is due to the special case of the SHO obeying the linear ordinary dufferential equation (ODE) in Equation (5).

Hence, from Equation (57):

$$\frac{d^2 \delta[x - x_{A_1\ldots B_N}(t)]}{dA_p^2} = \left(\frac{\partial x_{A_1\ldots B_N}}{\partial A_p} \right)^2 \frac{\partial^2 \delta[x - x_{A_1\ldots B_N}(t)]}{\partial x_{A_1\ldots B_N}^2}, \tag{60}$$

and, likewise,

$$\frac{d^2 \delta[x - x_{A_1\ldots B_N}(t)]}{dB_p^2} = \left(\frac{\partial x_{A_1\ldots B_N}}{\partial B_p} \right)^2 \frac{\partial^2 \delta[x - x_{A_1\ldots B_N}(t)]}{\partial x_{A_1\ldots B_N}^2}. \tag{61}$$

Consequently, from Equations (56), (60) and (61):

$$\frac{\partial}{\partial \sigma_x^2} P_{1x}(x) = \frac{1}{2} \int_{-\infty}^{\infty} dA_1 \cdots \int_{-\infty}^{\infty} dB_N P_F(A_1) \cdots P_F(A_N) P_F(B_1) \cdots P_F(B_N)$$

$$\times \sum_{p=1}^{N} \frac{\partial(\sigma_p^2)}{\partial(\sigma_x^2)} \left[\left(\frac{\partial x_{A_1\ldots B_N}}{\partial A_p}\right)^2 + \left(\frac{\partial x_{A_1\ldots B_N}}{\partial B_p}\right)^2 \right] \frac{\partial^2 \delta[x - x_{A_1\ldots B_N}(t)]}{\partial x_{A_1\ldots B_N}^2}. \quad (62)$$

Using Equations (58) and (59),

$$\left[\left(\frac{\partial x_{A_1\ldots B_N}}{\partial A_p}\right)^2 + \left(\frac{\partial x_{A_1\ldots B_N}}{\partial B_p}\right)^2 \right]$$

$$= \left[\frac{(q/m)}{\sqrt{L_x L_y L_z}} \varepsilon_{p,x} \operatorname{Re}\left(\frac{e^{i\omega_p t}}{-\omega_p^2 + \omega_0^2 + i\Gamma \omega_p \omega_0^2}\right) \right]^2 + \left[-\frac{(q/m)}{\sqrt{L_x L_y L_z}} \varepsilon_{p,x} \operatorname{Im}\left(\frac{e^{i\omega_p t}}{-\omega_p^2 + \omega_0^2 + i\Gamma \omega_p \omega_0^2}\right) \right]^2$$

$$= \frac{(q/m)^2 (\varepsilon_{p,x})^2}{L_x L_y L_z} \left[\frac{1}{\left(-\omega_p^2 + \omega_0^2\right)^2 + \left(\Gamma \omega_p \omega_0^2\right)^2} \right]. \quad (63)$$

Hence:

$$\frac{\partial}{\partial \sigma_x^2} P_{1x}(x)$$

$$= \frac{1}{2} \int_{-\infty}^{\infty} dA_1 \cdots \int_{-\infty}^{\infty} dB_N P_F(A_1) \cdots P_F(A_N) P_F(B_1) \cdots P_F(B_N)$$

$$\times \left\{ \sum_{p=1}^{N} \frac{\partial(\sigma_p^2)}{\partial(\sigma_x^2)} \frac{(q/m)^2 (\varepsilon_{p,x})^2}{L_x L_y L_z} \left[\frac{1}{\left(-\omega_p^2 + \omega_0^2\right)^2 + \left(\Gamma \omega_p \omega_0^2\right)^2} \right] \right\} \frac{\partial^2 \delta[x - x_{A_1\ldots B_N}(t)]}{\partial x_{A_1\ldots B_N}^2}. \quad (64)$$

From Equation (25),

$$\frac{\partial \sigma_x^2}{\partial \sigma_x^2} = 1 = \frac{(q/m)^2}{(L_x L_y L_z)} \sum_p (\varepsilon_{p,x})^2 \frac{1}{\left(-\omega_n^2 + \omega_0^2\right)^2 + \left(\Gamma \omega_0^2 \omega_n\right)^2} \frac{\partial(\sigma_p^2)}{\partial(\sigma_x^2)}. \quad (65)$$

This is where the functional form of σ_x^2 in Equation (25) enters in to enable Equation (17) to be solved by the 1D version of Equation (1).

From Equation (65), the quantity in curly brackets in Equation (64) equals unity, and Equation (64) becomes

$$\frac{\partial}{\partial \sigma_x^2} P_{1x}(x,t)$$

$$= \frac{1}{2} \int_{-\infty}^{\infty} dA_1 \cdots \int_{-\infty}^{\infty} dB_N P_F(A_1) \cdots P_F(A_N) P_F(B_1) \cdots P_F(B_N) \frac{\partial^2 \delta[x - x_{A_1\ldots B_N}(t)]}{\partial x_{A_1\ldots B_N}^2}$$

$$= \frac{1}{2} \frac{\partial^2}{\partial x^2} \int_{-\infty}^{\infty} dA_1 \cdots \int_{-\infty}^{\infty} dB_N P_F(A_1) \cdots P_F(A_N) P_F(B_1) \cdots P_F(B_N) \delta[x - x_{A_1\ldots B_N}(t)]$$

$$= \frac{1}{2} \frac{\partial^2}{\partial x^2} P_{1x}(x,t). \quad (66)$$

This completes the proof of Equation (47), carried out without the full integration of $P_F(A_1) \cdots P_F(A_N) P_F(B_1) \cdots P_F(B_N)$, as in Section 2.1. In a sense, certainly integrations were carried out over A_p and B_p via the double integration by parts in Equations (53)–(55).

However, this was a far simpler task than the more complicated operations in Section 2.1 of completing squares for all A_p and B_p variables and then integrating them. Here, the double integration by parts only required that $P_F(A_p) \to 0$ and $\frac{d}{dA_p}P_F(A_p) \to 0$ as $|A_p| \to \infty$ in Equations (54) and (55), and similarly for B_p. This behavior for $P_F(A_p)$ and $\frac{d}{dA_p}P_F(A_p)$ was assured by the Gaussian functional form of $P_F(A_p)$ in Equation (50). Moreover, this Gaussian form was important in the steps of Equations (50) and (51). In Section 2.1, the Gaussian form of $P_F(A_p)$ was important for a different reason, namely, to enable the squares to be carried out when integrating.

4. Concluding Remarks

Two methods were shown in this paper for finding the analytic expression (i.e., Equation (28)) for the probability density of the position of a classical charged point particle in an SHO potential, where the charge is bathed in classical electromagnetic random radiation at a temperature T. Both of these methods used the 1D form of Equation (1) as the starting point. Section 2.1 obtained the analytic expression by explicitly integrating over each of the Fourier coefficients; the calculation was fairly lengthy. In contrast, Section 3 showed that the 1D version of Equation (1) satisfied a PDE, which in turn enabled Equation (28) to be deduced.

Some other relevant points of this study are the following. First, Equation (1) should hold for dynamic systems other than the SHO. Moreover, it should also hold for the SHO in more generality than considered here, namely, not just for the steady state part of the oscillatory motion, but also including the initial transitory motion. Including the probability density of the initial conditions in Equation (1) would enable this to be accomplished. The probability density then changes from $P_{1x}(x)$ to $P_{1x}(x,t)$, as it will now depend on time.

A dynamic system of considerable interest to be considered here is the classical hydrogen, with a $-e$ classical charged point particle as the classical electron, and a much more massive nucleus with charge $+e$ for the proton as the nucleus. In SED, this would again be "bathed" in classical electromagnetic random radiation at temperature T. This system is of interest as it represents a real atomic system, as opposed to the solvable, but hypothetical SHO. Hydrogen is the simplest of atomic systems, so it is a suitable system to be analyzed in detail. Many researchers in SED have tackled this problem, but it still remains an open problem.

Equation (1) should hold for this classical hydrogen system. The quantity $\mathbf{x}_{A_1,\cdots,B_N}(t)$ in $\delta^3[\mathbf{x} - \mathbf{x}_{A_1,\cdots,B_N}(t)]$ would become the trajectory of the classical electron given its initial conditions and how the radiation fields influence the electron's trajectory, just as occurs for the SHO treated here. The equation of motion for the classical electron should be the Lorentz–Dirac equation, where the relativistic version is used, although it would be interesting to obtain and compare the resulting $P_{3x}(\mathbf{x})$ when the nonrelativistic approximation of the Lorentz–Dirac equation is used. It should be noted that the "easiest" and clearest situation to be considered for this problem would be the $T = 0$ case, since we expect that the electron would be bound and would not move off to $|\mathbf{x}| = \infty$ in space. For $T > 0$, there is a nonzero probability that the electron will "ionize" and $|\mathbf{x}(t)| \to \infty$, so this situation is more difficult to analyze with the present scheme.

Despite that Equation (1) should be valid for this classical hydrogen atom, there is a significant difference in using this $P_{3x}(\mathbf{x})$ formulation for the classical hydrogen atom versus the SHO model analyzed here. The "success" of Sections 2.1 and 3 came about because of the following: a nonrelativistic equation of motion was used (Equation (5)), the dipole approximation was made for the radiation's electric field, and the radiation magnetic field effects were ignored. These approximations enabled an analytic result for the motion, $x_{ss}(t) = x_{A_1,\cdots,B_N}(t)$, to be obtained: Equation (9). This expression for $x_{ss}(t)$ is linear in the Fourier coefficients of the radiation's electric field. This linear analytic expression for the particle's motion was used in each of the methods in Sections 2.2 and 3, to arrive at analytic expressions for $P_{1x}(x)$ and σ_x^2: Equations (25), (26) and (28). Without the linear analytic expression of Equation (9), the methods in Sections 2.1 and 3 could not have been carried out in the manners described.

In contrast, an analytic expression is not known for calculating the classical electron's probability distribution for the mentioned classical hydrogen atom, whether one treats the electron's trajectory relativistically, or even with a simpler nonrelativistic approximation. Although the expression for $P_{3x}(\mathbf{x})$ in Equation (1) should be correct for this classical hydrogen atom, other mathematical or numerical methods would need to be developed to carry out similar approaches in Sections 2.1 and 3 of direct integrations or showing the expression satisfies an appropriate PDE, from which $P_{3x}(\mathbf{x})$ can be deduced.

To date, no researcher has found an analytic means within SED for deducing $P_{3x}(\mathbf{x})$ for the classical hydrogen atom. Consequently, the author along with Y. Zou carried out simulation methods in 2003 for deducing $P_{3x}(\mathbf{x})$ for hydrogen, with some degree of success [25]. More recently, extensive simulations have been carried out by Nieuwenhuizen and Liska [26,27] with interesting results, but always with some fraction of the ensemble of hydrogen systems resulting in electrons leaving, or "ionizing," away from the classical nucleus. Despite this problem, all three simulation efforts [25–27] do not result in the classical electron "falling," or spiraling, into the nucleus due to energy radiating off from the classical electron's orbital motion. Thus, these simulations in SED "solve" the old atomic collapse problem of the simple classical atomic model by Rutherford.

Although these simulations are insightful, there are reasons for concern. Simulations in Refs. [25,26] were not relativistic, while Ref. [27] was certainly more relativistic than the others, but still not completely so, and each have various physical approximations. Perhaps of even more concern is that all of these studies deal with a chaotic system, where small errors in electron trajectories cannot of course be avoided numerically, but that build up to large errors quickly. Could these account for the apparent ionizations of some classical electrons in the ensemble of systems investigated? No matter how much the numerical resolutions of the simulations are reduced, this effect cannot go away, as known from chaos theory.

An analytic solution is indeed ideal for overcoming such problems, but may not be possible to obtain, whether nonrelativistically or relativistically. Nevertheless, this goal of exploring Equation (1) for obtaining analytical results was part of the motivation for this study. What is interesting to note is that using Feynman's path integral method in QM [28] was certainly exceptionally successful for a range of systems, but for a long time, starting from about 1948 [29] until Duru's and Kleinert's paper in 1979 [30], the hydrogen atom was not solved via this path integral method. Moreover, it should be mentioned that Equation (1) has some resemblance to a "path integral" formulation, although the Fourier coefficients of the stochastic radiation field are integrated over instead of the possible paths of the particle.

Recapitulating, in this paper, the general expression of Equation (1) was used to obtain an analytic expression of $P_{1x}(x)$ for the 1D electric dipole SHO, one of the first systems analyzed in SED. The calculations were fairly long for each of the two methods discussed, but certainly tractable. Despite Equation (1) being correct for the classical hydrogen atom, evaluating Equation (1) for this system is far more complicated task for reasons discussed.

Funding: This research received no external funding.

Data Availability Statement: Not applicable.

Conflicts of Interest: The authors declare no conflict of interest.

Abbreviations

The following abbreviations are used in this paper:

ODE   ordinary differential equation
PDE   partial differential equation
QED   quantum electrodynamics

QM	quantum mechanics
SED	stochastic electrodynamics
SHO	simple harmonic oscillator
1D	one-dimensional
3D	three-dimensional

References

1. Teitelboim, C. Splitting of the maxwell tensor: Radiation reaction without advanced fields. *Phys. Rev. D* **1970**, *1*, 1572–1582. [CrossRef]
2. Teitelboim, C.; Villarroel, D.; van Weert, C.G. Classical electrodynamics of retarded fields and point particles. *Riv. Nuovo C* **1980**, *3*, 1. [CrossRef]
3. Boyer, T.H. General connection between random electrodynamics and quantum electrodynamics for free electromagnetic fields and for dipole oscillator systems. *Phys. Rev. D* **1975**, *11*, 809–830. [CrossRef]
4. de la Peña, L.; Cetto, A.M. *The Quantum Dice. An Introduction to Stochastic Electrodynamics*; Kluwer Academic Publishers/Springer Science+Business Media B.V.: Dordrecht, The Netherlands, 1996. [CrossRef]
5. Boyer, T.H. Stochastic electrodynamics: The closest classical approximation to quantum theory. *Atoms* **2019**, *7*, 29. [CrossRef]
6. Boyer, T.H. Random electrodynamics: The theory of classical electrodynamics with classical electromagnetic zero–point radiation. *Phys. Rev. D* **1975**, *11*, 790–808. [CrossRef]
7. Cole, D.C. Reviewing and extending some recent work on stochastic electrodynamics. In *Essays on Formal Aspects of Electromagnetic Theory*; Lakhtakia, A., Ed.; World Scientific: Singapore, 1993; pp. 501–532. [CrossRef]
8. Boyer, T.H. The classical vacuum. *Sci. Am.* **1985**, *253*, 70–78. [CrossRef]
9. Boyer, T.H. Scaling symmetry and thermodynamic equilibrium for classical electromagnetic radiation. *Found. Phys.* **1989**, *19*, 1371–1383. [CrossRef]
10. Cole, D.C. Classical electrodynamic systems interacting with classical electromagnetic random radiation. *Found. Phys.* **1990**, *20*, 225–240. [CrossRef]
11. Marshall, T.W. Statistical electrodynamics. *Proc. Camb. Philos. Soc.* **1965**, *61*, 537–546. [CrossRef]
12. Boyer, T.H. Derivation of the blackbody radiation spectrum without quantum assumptions. *Phys. Rev.* **1969**, *182*, 1374–1383. [CrossRef]
13. Cole, D.C. Derivation of the classical electromagnetic zero–point radiation spectrum via a classical thermodynamic operation involving van der Waals forces. *Phys. Rev. A* **1990**, *42*, 1847–1862. [CrossRef]
14. Cole, D.C. Entropy and other thermodynamic properties of classical electromagnetic thermal radiation. *Phys. Rev. A* **1990**, *42*, 7006–7024. [CrossRef]
15. Cole, D.C. Reinvestigation of the thermodynamics of blackbody radiation via classical physics. *Phys. Rev. A* **1992**, *45*, 8471–8489. [CrossRef]
16. Bohm, D. *Quantum Theory*; Prentice–Hall, Inc.: Englewood Cliffs, NJ, USA, 1951.
17. Cole, D.C. Energy considerations of classical electromagnetic zero-point radiation and a specific probability calculation in stochastic electrodynamics. *Atoms* **2019**, *7*, 50. [CrossRef]
18. Cole, D.C. Probability calculations within stochastic electrodynamics. *Front. Phys.* **2020**, *8*, 127. [CrossRef]
19. Cole, D.C. Thermodynamics of blackbody radiation via classical physics for arbitrarily shaped cavities with perfectly conducting walls. *Found. Phys.* **2000**, *30*, 1849–1867. [CrossRef]
20. Planck, M. *The Theory of Heat Radiation*; P. Blakiston's Son & Co.: Philadelphia, PA, USA, 1814. Available online: https://www.gutenberg.org/files/40030/40030-pdf.pdf (accessed on 12 January 2023).
21. Einstein, A.; Hopf, L. Über einen Satz der Wahrscheinlichkeitsrechnung und seine Anwendung in der Strahlungstheorie. *Ann. Phys.* **1910**, *338*, 1096–1104. [CrossRef]
22. Einstein, A.; Hopf, L. Statistische Untersuchung der Bewegung eines Resonators in einem Strahlungsfeld. *Ann. Phys.* **1910**, *338*, 1105–1115. [CrossRef]
23. Einstein, A.; Hopf, L. On a Theorem of the Probability Calculus and its Application in the Theory of Radiation. In *The Collected Papers of Albert Einstein. Volume 3: The Swiss Years: Writings 1909–1911*; Klein, M.J., Kox, A.J., Renn, J., Schulman., R., Eds.; Princeton University Press: Princeton, NJ, USA, 1994; pp. 211–219. Available online: https://einsteinpapers.press.princeton.edu/vol3-trans/225 (accessed on 12 January 2023).
24. Einstein, A.; Hopf, L. Statistical Investigation of Resonators's Motion in a Radiation Field. In *The Collected Papers of Albert Einstein. Volume 3: The Swiss Years: Writings 1909–1911*; Klein, M.J., Kox, A.J., Renn, J., Schulman., R., Eds.; Princeton University Press: Princeton, NJ, USA, 1994; pp. 220–230. Available online: https://einsteinpapers.press.princeton.edu/vol3-trans/234 (accessed on 12 January 2023).
25. Cole, D.C.; Zou, Y. Quantum mechanical ground state of hydrogen obtained from classical electrodynamics. *Phys. Lett. A* **2003**, *317*, 14–20. [CrossRef]
26. Nieuwenhuizen, T.M.; Liska, M.T.P. Simulation of the hydrogen ground state in stochastic electrodynamics. *Phys. Scr.* **2015**, *2015*, 014006. [CrossRef]

27. Nieuwenhuizen, T.M.; Liska, M.T.P. Simulation of the hydrogen ground state in stochastic electrodynamics-2: Inclusion of relativistic corrections. *Found. Phys.* **2015**, *45*, 1190–1202. [CrossRef]
28. Feynman, R.P.; Hibbs, A.R. *Quantum Mechanics and Path Integrals*; McGraw–Hill: New York, NY, USA, 1965.
29. Feynman, R.P. Space-time qpproach to Non-relativistic quantum mechanics. *Rev. Mod. Phys.* **1948**, *20*, 367–387. [CrossRef]
30. Duru, I.H.; Kleinert, H. Solution of the path integral for the H atom. *Phys. Lett. B* **1979**, *84*, 185–188. [CrossRef]

Disclaimer/Publisher's Note: The statements, opinions and data contained in all publications are solely those of the individual author(s) and contributor(s) and not of MDPI and/or the editor(s). MDPI and/or the editor(s) disclaim responsibility for any injury to people or property resulting from any ideas, methods, instructions or products referred to in the content.

Article

van der Waals Dispersion Potential between Excited Chiral Molecules via the Coupling of Induced Dipoles

A. Salam

Department of Chemistry, Wake Forest University, Winston-Salem, NC 27109-7486, USA; salama@wfu.edu; Tel.: +1-336-758-3713

Abstract: The retarded van der Waals dispersion potential between two excited chiral molecules was calculated using an approach, in which electric and magnetic dipole moments are induced in each particle by fluctuations in the vacuum electromagnetic field. An expectation value of the coupling of the moments at different centres to the dipolar interaction tensors was taken over excited matter states and the ground state radiation field, the former yielding excited molecular polarisabilities and susceptibilities, and the latter field–field spatial correlation functions. The dispersion potential term proportional to the mixed dipolar polarisability is discriminatory, dependent upon molecular handedness, and contains additional terms due to transitions that de-excite each species as well as the usual u-integral term over imaginary frequency, which applies to both upward and downward transitions. Excited state dispersion potentials of a comparable order of magnitude involving paramagnetic and diamagnetic couplings were also computed. Pros and cons of the method adopted are compared to other commonly used approaches.

Keywords: dispersion forces; excited states; vacuum fluctuations; molecular chirality; quantum electrodynamics

1. Introduction

Classic examples of phenomena that are attributed to vacuum fluctuations of the electromagnetic field [1] include spontaneous emission [2] and the Lamb shift [3,4]. In the case of inter-particle interactions, a fundamental coupling that arises from the zero-point energy associated with the ground state of the radiation field is the well-known Casimir-van der Waals dispersion force between two or more particles [5–8]. For atoms and non-polar molecules, this is the only interaction contributing to the inter-particle energy shift, and is responsible for the manifestation of solid and liquid phases of such forms of matter at low temperature.

The $\frac{1}{2}\hbar\omega$ of energy per mode possessed by the vacuum field, where ω is the circular frequency, albeit infinite in magnitude since there are an infinite number of oscillatory modes, is a direct consequence of quantising electromagnetic radiation, and is a quintessential feature of quantum electrodynamics (QED) theory [9–15], rigorously accounting for the photon. Here $\hbar$ is the reduced Planck constant. The exchange of such gauge bosons between electrons, whether free or bound, mediates the interaction between particles of matter. For instance, the propagation of a single virtual photon, originating due to spontaneous emission by an excited entity undergoing decay, and whose excitation energy is captured on absorption of the photon by an acceptor species in close proximity, describes a multitude of processes involving the migration of energy between various types of chromophoric units, processes commonly collected under the umbrella term "excitation energy transfer" [16–19]. This is another example of a vacuum field-induced effect since there are no photons prior to or after the coupling between particles.

By way of contrast, in a perturbative calculation of the dispersion interaction, in which both species are in the ground electronic state and no photons are present, the energy shift

Citation: Salam, A. van der Waals Dispersion Potential between Excited Chiral Molecules via the Coupling of Induced Dipoles. *Physics* **2023**, *5*, 247–260. https://doi.org/10.3390/physics5010019

Received: 1 January 2023
Revised: 8 February 2023
Accepted: 13 February 2023
Published: 24 February 2023

Copyright: © 2023 by the author. Licensee MDPI, Basel, Switzerland. This article is an open access article distributed under the terms and conditions of the Creative Commons Attribution (CC BY) license (https://creativecommons.org/licenses/by/4.0/).

is viewed as arising from the exchange of two virtual photons between the pair of atoms or molecules [12,13]. Additional computational difficulties arise when using perturbation theory if one or both of the particles are in excited electronic states [20]. These centre around the proper identification of terms associated with real photon emission and whether excitation energy is localised or is in fact exchanged between centres in a reversible manner. Quite a few recent publications have dealt with these aspects in an attempt to arrive at the correct functional form for the potential [21–26].

Other calculational techniques have been adopted to further understand the nature of the dispersion force when not all of the interacting particles are in their ground state. One approach is to evaluate the quantum electrodynamical radiation fields in the neighbourhood of a charged source and then calculate the response of the second particle, via its polarisability, to the electromagnetic field emanating from the first particle, with appropriate expectation values taken over ground and/or excited states of matter, yielding the pertinent multipolar contribution to the dispersion energy shift [27,28]. An advantage of employing this physical picture and calculational method is its straightforward extension to obtain the result for the N-body electric dipole polarisable [29] or N-body arbitrarily electric multipole polarisable [30] dispersion potentials.

An especially physically intuitive approach to calculate the Casimir–Polder interaction energy, in that it highlights the key role played by the vacuum electromagnetic field, is to consider the energy shift as arising from the coupling of electric dipole moments induced at each atom or molecule by the ground state of the field to the retarded electric dipole–dipole interaction tensor [31]. Because this last quantity features in the amplitude for resonance energy transfer, the method may be employed to readily evaluate the additional contribution to the dispersion potential that is due to real photon emission when one or both species are electronically excited. This approach will be employed below to calculate the dispersion energy shift between two excited chiral molecules, and two other potentials of a similar order of magnitude. For these cases, the usual electric dipole approximation needs to be relaxed, and higher multipole moment terms such as the magnetic dipole have to be included in the treatment. This is because optically active molecules possess fewer or no elements of symmetry relative to achiral compounds and consequently have less restrictive spectroscopic selection rules apply to them.

One interesting aspect of the dispersion interaction energy between chiral molecules is that it is discriminatory, depending on the handedness of the interacting pair. The chirality dependent ground state dispersion potential has been previously evaluated using the three methods described above, namely, the perturbation and response theories, and the induced moment method [32–35], with excited state energies only evaluated using response theory [36]. Recently, perturbation theory has been employed within the framework of macroscopic QED theory [6] to calculate the dispersion interactions between one or two chiral molecules in the presence of a chiral plate [37], or when situated in a magnetodielectric medium [38], with novel features emerging as a result of placing bodies in complex environments such as altering the sign of the dispersion force as the relative separation distances are varied. The results obtained will enable an assessment to be made of the feasibility of applying the fluctuating moment method to systems in excited electronic states that are characterised by multipoles higher than the electric dipole relative to more conventional approaches. These results will also complement other studies dealing with interactions amongst enantiomers.

The paper is organised as follows. A brief overview of the induced moment method applicable to chiral molecules is presented in Section 2. The calculation of the excited state dispersion potential between two chiral molecules is detailed in Section 3. An energy shift of a similar order of magnitude is then obtained in Section 4, that between an electric dipole polarisable molecule and a paramagnetically susceptible one. The diamagnetic counterpart to this last contribution, also of an identical order of magnitude to the two previous potentials, is evaluated in Section 5. A brief summary is given in Section 6.

2. Moments Induced in a Chiral Molecule by Electromagnetic Radiation

Consider a chiral molecule, ξ, located at the position $\vec{R}_\xi$. Lacking an improper axis of rotation, such species may belong to one of the following molecular point groups: C_1, C_n, D_n, T, and O, with $n \geq 2$. In the first point group listed, spectroscopic selection rules permit transitions to all orders of multipole moment distributions between electronic states. Often, it is sufficient to invoke the dipole approximation to describe chiral molecules since the vector dot product of an electric dipole moment ($\vec{\mu}$) and a magnetic dipole moment ($\vec{m}$) vector yields a pseudoscalar quantity, which changes sign when substituting one enantiomer by its mirror-image structure. In what follows, only these two multipole moments are retained and the electric quadrupole moment, $\vec{Q}$, is neglected. Although $\vec{Q}$ is of a comparable order of magnitude to $\vec{m}$, with both a factor of the fine structure constant smaller than $\vec{\mu}$, mixed electric dipole-quadrupole dependent contributions to the dispersion interaction vanish for freely tumbling systems, and are not considered henceforth.

Thus, within the dipolar approximation, the electric and magnetic dipole moments induced in a chiral molecule by electromagnetic radiation of mode $\vec{k}, \lambda$, where $\vec{k}$ is the wave vector and λ is the index of polarisation of the propagating radiation fields, are:

$$\mu_i^{ind}(\xi; \vec{k}, \lambda) = \varepsilon_0^{-1}\alpha_{ij}(\xi; k)d_j^\perp(\vec{k}, \lambda; \vec{R}_\xi) + G_{ij}(\xi; k)b_j(\vec{k}, \lambda; \vec{R}_\xi), \tag{1}$$

and

$$m_j^{ind}(\xi; \vec{k}, \lambda) = \varepsilon_0^{-1}G_{ij}(\xi; k)d_i^\perp(\vec{k}, \lambda; \vec{R}_\xi) + \chi_{ij}(\xi; k)b_i(\vec{k}, \lambda; \vec{R}_\xi), \tag{2}$$

where the Latin letter subscripts denote Cartesian tensor components in the space-fixed frame of reference, and the Einstein summation rule is in effect for indices that repeat. Here, ε_0 denotes the permittivity of free space. In the relations (1) and (2), $\alpha_{ij}(\xi; k)$ is the dynamic electric dipole polarisability tensor, $G_{ij}(\xi; k)$ is the mixed electric-magnetic dipole analogue, and $\chi_{ij}(\xi; k)$ is the magnetic dipole polarisability tensor or the paramagnetic susceptibility tensor. Their explicit forms are given by

$$\alpha_{ij}(\xi; k) = \sum_t \left\{ \frac{\mu_i^{st}(\xi)\mu_j^{ts}(\xi)}{E_{ts} - \hbar ck} + \frac{\mu_j^{st}(\xi)\mu_i^{ts}(\xi)}{E_{ts} + \hbar ck} \right\}, \tag{3}$$

$$G_{ij}(\xi; k) = \sum_t \left\{ \frac{\mu_i^{st}(\xi)m_j^{ts}(\xi)}{E_{ts} - \hbar ck} + \frac{m_j^{st}(\xi)\mu_i^{ts}(\xi)}{E_{ts} + \hbar ck} \right\}, \tag{4}$$

and

$$\chi_{ij}(\xi; k) = \sum_t \left\{ \frac{m_i^{st}(\xi)m_j^{ts}(\xi)}{E_{ts} - \hbar ck} + \frac{m_j^{st}(\xi)m_i^{ts}(\xi)}{E_{ts} + \hbar ck} \right\}, \tag{5}$$

where $\mu_i^{st}(\xi) = <s|\mu_i(\xi)|t>$ and $m_i^{st}(\xi) = <s|m_i(\xi)|t>$ are the transition electric and magnetic dipole moment matrix elements between electronic states $|s>$ and $|t>$, with energies E_s and E_t, respectively, and $E_{ts} = E_t - E_s$ symbolizing the energy differences between these states. Here c stands for the speed of light.

For a specific mode $\vec{k}, \lambda$, the second quantised microscopic Maxwell field operators appearing in Equations (1) and (2) are the familiar Fourier series mode expansions for the transverse electric displacement field, $\vec{d}^\perp(\vec{r})$, and the magnetic field, $\vec{b}(\vec{r})$,

$$\vec{d}^\perp(\vec{k}, \lambda; \vec{r}) = i\left(\frac{\hbar ck\varepsilon_0}{2V}\right)^{1/2} [\vec{e}^{(\lambda)}(\vec{k})a^{(\lambda)}(\vec{k})e^{i\vec{k}\cdot\vec{r}} - \vec{\overline{e}}^{(\lambda)}(\vec{k})a^{\dagger(\lambda)}(\vec{k})e^{-i\vec{k}\cdot\vec{r}}], \tag{6}$$

and

$$\vec{b}(\vec{k},\lambda;\vec{r}) = i\left(\frac{\hbar k}{2\varepsilon_0 cV}\right)^{1/2} [\vec{b}^{(\lambda)}(\vec{k})a^{(\lambda)}(\vec{k})e^{i\vec{k}\cdot\vec{r}} - \vec{\bar{b}}^{(\lambda)}(\vec{k})a^{\dagger(\lambda)}(\vec{k})e^{-i\vec{k}\cdot\vec{r}}]. \tag{7}$$

These radiation fields are linear functions of the bosonic annihilation and creation operators for a $\vec{k},\lambda$-mode photon, $a^{(\lambda)}(\vec{k})$ and $a^{\dagger(\lambda)}(\vec{k})$, respectively, with $\vec{e}^{(\lambda)}(\vec{k})$ and $\vec{b}^{(\lambda)}(\vec{k})$ as the complex unit electric and magnetic polarisation vectors, and V is the quantisation volume of the box.

The leading contribution to the interaction energy, ΔE, between two molecules, A and B, arises from the coupling of the induced electric dipole moments at each centre. Keeping only the first term of Equation (1),

$$\Delta E = \sum_{\vec{k},\lambda} \mu_i^{ind}(A;\vec{k},\lambda)\mu_j^{ind}(B;\vec{k},\lambda) Re V_{ij}(k,\vec{R}), \tag{8}$$

where the sum is executed over all radiation field modes, the inter-nuclear displacement, $R = |\vec{R}_B - \vec{R}_A|$, and $V_{ij}(k,\vec{R})$ is the retarded electric dipole–dipole tensor that couples the two induced dipoles. It is given by the familiar expression [12,13]

$$V_{ij}(k,\vec{R}) = \frac{1}{4\pi\varepsilon_0 R^3}[(\delta_{ij} - 3\hat{R}_i\hat{R}_j)(1 - ikR) - (\delta_{ij} - \hat{R}_i\hat{R}_j)k^2R^2]e^{ikR}, \tag{9}$$

where δ_{ij} is the Kronecker delta.

Power and Thirunamachandran showed [31] how Equation (8) led to the Casimir–Polder potential between two ground state atoms or molecules as well as to the energy shift when one of the pair is in an excited electronic state. In this paper, their method is extended to chiral molecules and other magnetic systems.

3. Dispersion Potential between Two Excited Chiral Molecules

Let us start with employing the fluctuating moment method to calculate the dispersion interaction energy between two optically active molecules. Species A is initially in the excited electronic state $|p\rangle$ and may undergo upward or downward virtual transitions to level $|n\rangle$, with B undergoing similar transitions from $|r\rangle \leftarrow |q\rangle$. In addition to the coupling between two induced electric dipoles as given in Equation (8), there is an analogous term involving the coupling of two magnetic dipoles, which also interact via the retarded interaction tensor (9). Furthermore, an electric dipole induced at one site may interact with an induced magnetic dipole of the second particle. This time coupling occurs through the interaction tensor [12,13]

$$U_{ij}(k,\vec{R}) = -\frac{ik}{4\pi\varepsilon_0 c}\varepsilon_{ijk}\nabla_k \frac{e^{ikR}}{R} = \frac{1}{4\pi\varepsilon_0 cR^3}\varepsilon_{ijk}\hat{R}_k[ikR + k^2R^2]e^{ikR}, \tag{10}$$

where ε_{ijk} is the Levi–Civita tensor. Hence, the energy shift may be expressed as

$$\Delta E = \sum_{\vec{k},\lambda} \{[\mu_i^{ind}(A)\mu_j^{ind}(B) + c^{-2}m_i^{ind}(A)m_j^{ind}(B)]Re V_{ij}(k,\vec{R}) \\ + [\mu_i^{ind}(A)m_j^{ind}(B) + m_i^{ind}(A)\mu_j^{ind}(B)]Im U_{ij}(k,\vec{R})\}. \tag{11}$$

Equation (11) is the starting point in the evaluation of energy shifts dependent upon magnetic dipole coupling including chiral molecules to leading order. For terms proportional to the handedness of the two molecules, manifested by the mixed electric-magnetic dipole polarisability (4), substituting for the second term of Equation (1) and the first term of Equation (2), yields:

$$\Delta E = \sum_{\vec{k},\lambda} \{[G_{ik}(A;k)G_{jl}(B;k)b_k(\vec{k},\lambda;\vec{R}_A)b_l(\vec{k},\lambda;\vec{R}_B)$$
$$+\varepsilon_0^{-2}c^{-2}G_{ki}(A;k)G_{lj}(B;k)d_k^\perp(\vec{k},\lambda;\vec{R}_A)d_l^\perp(\vec{k},\lambda;\vec{R}_B)]ReV_{ij}(k,\vec{R}) \quad (12)$$
$$+\varepsilon_0^{-1}[G_{ik}(A;k)G_{lj}(B;k)b_k(\vec{k},\lambda;\vec{R}_A)d_l^\perp(\vec{k},\lambda;\vec{R}_B)$$
$$+G_{ki}(A;k)G_{jl}(B;k)d_k^\perp(\vec{k},\lambda;\vec{R}_A)b_l(\vec{k},\lambda;\vec{R}_B)]ImU_{ij}(k,\vec{R})\}.$$

To evaluate the dispersion potential between two electronically excited chiral molecules, with the electromagnetic field in the vacuum state, the expectation value of Equation (12) is taken over the state $|p^A, q^B; 0(\vec{k},\lambda)>$. The molecular factors yielded excited state mixed electric-magnetic dipole polarisabilities of the form given by Equation (4). From Equation (12), it can be seen that for the radiation field part, four separate field-field spatial correlation functions need to be evaluated over the ground state of the electromagnetic field. These are straightforwardly obtained from the Maxwell field operators (6) and (7), and have earlier been given [39] for an N-photon state of the radiation field, $|N(\vec{k},\lambda)>$, in the calculation of the modification of the ground state dispersion force between two chiral molecules due to an intense radiation field. For the vacuum electromagnetic field, these correlation functions are:

$$<0(\vec{k},\lambda)|d_k^\perp(\vec{R}_A)d_l^\perp(\vec{R}_B)|0(\vec{k},\lambda)> = \left(\frac{\hbar ck\varepsilon_0}{2V}\right)e_k^{(\lambda)}(\vec{k})\bar{e}_l^{(\lambda)}(\vec{k})e^{-i\vec{k}\cdot\vec{R}}, \quad (13)$$

$$<0(\vec{k},\lambda)|b_k(\vec{R}_A)b_l(\vec{R}_B)|0(\vec{k},\lambda)> = \left(\frac{\hbar k}{2\varepsilon_0 cV}\right)b_k^{(\lambda)}(\vec{k})\bar{b}_l^{(\lambda)}(\vec{k})e^{-i\vec{k}\cdot\vec{R}}, \quad (14)$$

$$<0(\vec{k},\lambda)|d_k^\perp(\vec{R}_A)b_l(\vec{R}_B)|0(\vec{k},\lambda)> = \left(\frac{\hbar k}{2V}\right)e_k^{(\lambda)}(\vec{k})\bar{b}_l^{(\lambda)}(\vec{k})e^{-i\vec{k}\cdot\vec{R}}, \quad (15)$$

$$<0(\vec{k},\lambda)|b_k(\vec{R}_A)d_l^\perp(\vec{R}_B)|0(\vec{k},\lambda)> = \left(\frac{\hbar k}{2V}\right)b_k^{(\lambda)}(\vec{k})\bar{e}_l^{(\lambda)}(\vec{k})e^{-i\vec{k}\cdot\vec{R}}. \quad (16)$$

3.1. Contribution from Upward and Downward Transitions

For ease of presentation, contributions from both upward and downward transitions that have identical functional form were distinguished from contributions that solely arise from downward transitions. We considered the former type of term first. Examining the first term of Equation (12), substituting Equation (14) produces

$$\sum_{\vec{k},\lambda}\left(\frac{\hbar k}{2\varepsilon_0 cV}\right)G_{ik}(A;k)G_{jl}(B;k)b_k^{(\lambda)}(\vec{k})\bar{b}_l^{(\lambda)}(k)e^{-i\vec{k}\cdot\vec{R}}ReV_{ij}(k,\vec{R}). \quad (17)$$

It is worth pointing out that the radiation field part is similar to that featured in the evaluation of the Casimir–Polder potential [31], with magnetic rather than electric polarisation vectors appearing in Equation (17). To proceed further, the sum over photon modes must be performed. For the polarization index sum, the following identities may be employed:

$$\sum_\lambda e_i^{(\lambda)}(\vec{k})\bar{e}_j^{(\lambda)}(\vec{k}) = \sum_\lambda b_i^{(\lambda)}(\vec{k})\bar{b}_j^{(\lambda)}(\vec{k}) = \delta_{ij} - \hat{k}_i\hat{k}_j, \quad (18)$$

while the wave vector sum is converted to an integral via

$$\frac{1}{V}\sum_{\vec{k}} \to \frac{1}{(2\pi)^3}\int d^3\vec{k}. \quad (19)$$

In spherical polar coordinates, $d^3\vec{k} = k^2 dk d\Omega$, with $d\Omega$ an element of the solid angle. Thus, Equation (17) reads:

$$\frac{\hbar}{16\pi^3\varepsilon_0 c}\int dk d\Omega k^3 G_{ik}(A;k)G_{jl}(B;k)(\delta_{kl}-\hat{k}_k\hat{k}_l)e^{-i\vec{k}\cdot\vec{R}}\mathrm{Re}V_{ij}(k,\vec{R}). \qquad (20)$$

The angular average is carried out using

$$\frac{1}{4\pi}\int d\Omega(\delta_{ij}-\hat{k}_i\hat{k}_j)e^{\pm i\vec{k}\cdot\vec{R}} = (\delta_{ij}-\hat{R}_i\hat{R}_j)\frac{\sin kR}{kR} + (\delta_{ij}-3\hat{R}_i\hat{R}_j)\left(\frac{\cos kR}{k^2R^2} - \frac{\sin kR}{k^3R^3}\right). \qquad (21)$$

After substituting $\mathrm{Re}V_{ij}(k,\vec{R})$ (9), Equation (20) reads:

$$\frac{\hbar}{16\pi^3\varepsilon_0^2 c^3}\int_0^\infty dk k^3 G_{ik}(A;k)G_{jl}(B;k)\left[(\delta_{kl}-\hat{R}_k\hat{R}_l)\frac{\sin kR}{kR} + (\delta_{kl}-3\hat{R}_k\hat{R}_l)\left(\frac{\cos kR}{k^2R^2}-\frac{\sin kR}{k^3R^3}\right)\right]$$
$$\times \left[(\delta_{ij}-3\hat{R}_i\hat{R}_j)(\cos kR + kR\sin kR) - (\delta_{ij}-\hat{R}_i\hat{R}_j)k^2R^2\cos kR\right], \qquad (22)$$

which holds for A and B in fixed mutual orientation. The isotropic contribution to the potential may be obtained by applying the result for the random orientational averaging of the tensor Equation (4) [40]

$$<G_{ij}(\xi;k)> = \frac{1}{3}\delta_{ij}\delta_{\lambda\mu}G_{\lambda\mu}(\xi;k) = \delta_{ij}G(\xi;k), \qquad (23)$$

where $\frac{1}{3}\delta_{\lambda\mu}G_{\lambda\mu}(\xi;k) = G(\xi;k)$ is the isotropic polarisability, and Greek letter subscripts denote the Cartesian tensor components in the body-fixed frame of reference. Contracting the geometric factors in Equation (22) gives:

$$-\frac{\hbar}{16\pi^3\varepsilon_0^2 cR^2}\int_0^\infty dk k^4 G(A;k)G(B;k)\left\{\sin 2kR\left(1-\frac{5}{k^2R^2}+\frac{3}{k^4R^4}\right) + \cos 2kR\left(\frac{2}{kR}-\frac{6}{k^3R^3}\right)\right\}$$
$$= -\frac{\hbar}{16\pi^3\varepsilon_0^2 R^2}\int_0^\infty dk k^4 G(A;k)G(B;k)\,\mathrm{Im}\left[1+\frac{2i}{kR}-\frac{5}{k^2R^2}-\frac{6i}{k^3R^3}+\frac{3}{k^4R^4}\right]e^{2ikR}. \qquad (24)$$

Finally, transforming to the complex variable $k = iu$, taking the integral into the complex plane by rotating the line of integration by $\pi/2$, yields:

$$-\frac{\hbar}{16\pi^3\varepsilon_0^2 cR^2}\int_0^\infty du u^4 e^{-2uR}G(A;iu)G(B;iu)\left[1+\frac{2}{uR}+\frac{5}{u^2R^2}+\frac{6}{u^3R^3}+\frac{3}{u^4R^4}\right]. \qquad (25)$$

where $G(\xi;iu)$ is the isotropic dynamic mixed electric-magnetic dipole polarisability evaluated at the imaginary frequency, $\omega = icu$. The second term of Equation (12) produces a contribution identical to Equation (25), which therefore doubles up.

Considering the third term of Equation (12), substituting Equation (16) gives:

$$\sum_{\vec{k},\lambda}\left(\frac{\hbar k}{2\varepsilon_0 V}\right)G_{ik}(A;k)G_{lj}(B;k)b_k^{(\lambda)}(\vec{k})\overline{e}_l^{(\lambda)}(\vec{k})e^{-i\vec{k}\cdot\vec{R}}\mathrm{Im}U_{ij}(k,\vec{R}). \qquad (26)$$

Use is now made of the polarisation sum,

$$\sum_\lambda e_i^{(\lambda)}(\vec{k})\overline{b}_j^{(\lambda)}(\vec{k}) = \varepsilon_{ijk}\hat{k}_k, \qquad (27)$$

along with the continuum approximation to the wave vector sum (19). The required angular integration is given by

$$\frac{1}{4\pi}\int \hat{k}_k e^{\pm i\vec{k}\cdot\vec{R}}d\Omega = \mp i\left(\frac{\cos kR}{kR} - \frac{\sin kR}{k^2R^2}\right)\hat{R}_k, \qquad (28)$$

so that Equation (26) becomes, after inserting $\mathrm{Im}U_{ij}(k,\vec{R})$ from Equation (10):

$$-\frac{i\hbar}{16\pi^3\varepsilon_0^2 cR^3}\varepsilon_{ijm}\varepsilon_{kln}\hat{R}_m\hat{R}_n \int_0^\infty dk k^3 G_{ik}(A;k)G_{lj}(B;k)\left[\frac{\cos kR}{kR} - \frac{\sin kR}{k^2R^2}\right][kR\cos kR + k^2R^2\sin kR]. \quad (29)$$

Performing a tumbling average using Equation (23) and contracting tensors, Equation (29) becomes:

$$-\frac{i\hbar}{16\pi^3\varepsilon_0^2 cR^2} \int_0^\infty dk k^4 G(A;k)G(B;k)\left\{\sin 2kR + \frac{2}{kR}\cos 2kR - \frac{1}{k^2R^2}\sin 2kR\right\}$$
$$= -\frac{i\hbar}{16\pi^3\varepsilon_0^2 cR^2} \int_0^\infty dk k^4 G(A;k)G(B;k)\text{Im}\left[1 + \frac{2i}{kR} - \frac{1}{k^2R^2}\right]e^{2ikR}. \quad (30)$$

Substituting the complex variable $k = iu$ and rotating the line of integration as above yields:

$$\frac{\hbar}{16\pi^3\varepsilon_0^2 cR^2} \int_0^\infty du u^4 e^{-2uR} G(A;iu)G(B;iu)\left[1 + \frac{2}{uR} + \frac{1}{u^2R^2}\right]. \quad (31)$$

Recognising that the fourth term of Equation (12) produces an identical contribution to Equation (31), the u-integral contribution to the excited state dispersion potential between two chiral (c) molecules is given by twice the sum of Equations (25) and (31),

$$\Delta E^u_{c-c} = -\frac{\hbar}{8\pi^3\varepsilon_0^2 cR^2} \int_0^\infty du u^4 e^{-2uR} G(A;iu)G(B;iu)\left[\frac{4}{u^2R^2} + \frac{6}{u^3R^3} + \frac{3}{u^4R^4}\right], \quad (32)$$

which is applicable to both upward and downward transitions in A and B since there are no limitations on the intermediate state sums over levels $|n\rangle$ and $|r\rangle$ in the excited state polarisabilities, $G(\xi;iu)$, $\xi = A, B$.

3.2. Additional Contribution from Downward Transitions

Let us now examine the terms contributing to the energy shift due to the emission of a real photon from the excited electronic states of A and B. These are in addition to the upward transitions captured in the u-integral term (32). The total contribution arising from de-excitation in each molecule can be written as

$$\Delta E^{RES} = \Delta E^{A-RES} + \Delta E^{B-RES}. \quad (33)$$

The starting expressions for each of the two terms are easily obtained from Equation (12). For species A, one has:

$$\Delta E^{A-RES} = \sum_{k,\lambda} \{[G_{ik}(A;k)G_{jl}(B;k)b_k(\vec{R}_A)b_l(\vec{R}_B)$$
$$+\varepsilon_0^{-2}c^{-2}G_{ki}(A;k)G_{lj}(B;k)d_k^\perp(\vec{R}_A)d_l^\perp(\vec{R}_B)]V_{ij}^{RES}(k_{pn},\vec{R}) \quad (34)$$
$$+\varepsilon_0^{-1}[G_{ik}(A;k)G_{lj}(B;k)b_k(\vec{R}_A)d_l^\perp(\vec{R}_B)$$
$$+G_{ki}(A;k)G_{jl}(B;k)d_k^\perp(\vec{R}_A)b_l(\vec{R}_B)]U_{ij}^{RES}(k_{pn},\vec{R})\},$$

where $V_{ij}^{RES}(k_{pn},\vec{R})$ and $U_{ij}^{RES}(k_{pn},\vec{R})$ are the resonant contributions of the coupling tensors (9) and (10), respectively, evaluated at the wave vector of the downward transition occurring in A, $k_{pn} = \omega_{pn}/c$. An expression similar to Equation (34) may be written for ΔE^{B-RES} with $V_{ij}^{RES}(k_{qr},\vec{R})$ and $U_{ij}^{RES}(k_{qr},\vec{R})$ appearing instead, reflecting de-excitation in B at the resonant frequency $\omega_{qr} = ck_{qr}$ for the transition $|r\rangle \leftarrow |q\rangle$. Similar to the evaluation of the u-integral term, an expectation value was taken over Equation (33) with the state $|p^A, q^B; 0(\vec{k},\lambda)\rangle$. Use was made of the vacuum field–field spatial correlation functions (13)–(16) for the radiation field part. For species A and B, excited state molecular polarisabilities $G_{ij}(\xi;k)$ featured. Examining the first term of Equation (34), and following similar steps that led to Equation (20), yields:

$$\frac{1}{8\pi^3\varepsilon_0 c^2} \sum_n \int dk d\Omega k^4 \frac{\mu_i^{pn}(A) m_k^{np}(A)}{k_{np}^2 - k^2} G_{jl}(B;k)(\delta_{kl} - \hat{k}_k\hat{k}_l)e^{-i\vec{k}\cdot\vec{R}} V_{ij}^{RES}(k_{pn}, \vec{R}), \quad (35)$$

after substituting $G_{ik}(A;k)$ (4). Instead of Equation (21), it is convenient to use the following form for the integration over the solid angle $d\Omega$:

$$\frac{1}{4\pi}\int d\Omega (\delta_{ij} - \hat{k}_i\hat{k}_j)e^{\pm i\vec{k}\cdot\vec{R}} = \frac{1}{2ik^3}(-\nabla^2\delta_{ij} + \nabla_i\nabla_j)\frac{1}{R}(e^{ikR} - e^{-ikR}). \quad (36)$$

Equation (35) then becomes, after carrying out the integral over k,

$$\frac{1}{4\pi\varepsilon_0 c^2} \sum_{\substack{n \\ E_p > E_n}} \mu_i^{pn}(A) m_k^{np}(A) G_{jl}(B;k_{pn})(-\nabla^2\delta_{kl} + \nabla_k\nabla_l)\frac{e^{-ik_{pn}R}}{R} V_{ij}^{RES}(k_{pn}, \vec{R})$$

$$= -\frac{1}{16\pi^2\varepsilon_0^2 c^2} \sum_{\substack{n \\ E_p > E_n}} \mu_i^{pn}(A) m_k^{np}(A) G_{jl}(B;k_{pn})[(-\nabla^2\delta_{ij} + \nabla_i\nabla_j)\frac{e^{ik_{pn}R}}{R}][(-\nabla^2\delta_{kl} + \nabla_k\nabla_l)\frac{e^{-ik_{pn}R}}{R}], \quad (37)$$

where in the last line, $V_{ij}^{RES}(k_{pn}, \vec{R})$ (9) is inserted. Interestingly, B responds through its polarisability, $G_{jl}(B;k_{pn})$, to the excitation energy of the downward transition in A at the frequency $\omega_{pn} = ck_{pn}$. The energy shift term (37) applies for A and B in fixed relative orientation and the sum is restricted to states for which $E_{pn} > 0$. Performing an orientational average, absorbing the constant $1/3$ into each molecular factor, and contracting, produces

$$-\frac{1}{8\pi^2\varepsilon_0^2 c^2 R^6} \sum_{\substack{n \\ E_p > E_n}} [\vec{\mu}^{pn}(A)\cdot\vec{m}^{np}(A)]G(B;k_{pn})[3 + k_{pn}^2 R^2 + k_{pn}^4 R^4]. \quad (38)$$

Let us now evaluate the fourth term of Equation (34), this time with the help of Equation (15), which yields

$$\frac{1}{8\pi^3\varepsilon_0 c}\varepsilon_{klm}\sum_n \int dk d\Omega k^4 \frac{\mu_k^{pn}(A) m_i^{np}(A)}{k_{np}^2 - k^2} G_{jl}(B;k)\hat{k}_m e^{-i\vec{k}\cdot\vec{R}} U_{ij}^{RES}(k_{pn}, \vec{R}). \quad (39)$$

Employing an alternative form for the angular average (28),

$$\frac{1}{4\pi}\int d\Omega \hat{k}_k e^{\pm i\vec{k}\cdot\vec{R}} = \mp\frac{1}{2k^2}\nabla_k\frac{1}{R}(e^{ikR} - e^{-ikR}), \quad (40)$$

and performing the k-integration, Equation (39) becomes:

$$-\frac{i}{4\pi\varepsilon_0 c}\sum_{\substack{n \\ E_p > E_n}}\mu_k^{pn}(A) m_i^{np}(A)G_{jl}(B;k_{pn})k_{pn}\varepsilon_{klm}\nabla_m \frac{e^{-ik_{pn}R}}{R} U_{ij}^{RES}(k_{pn}, \vec{R})$$

$$= -\frac{1}{(4\pi\varepsilon_0 c)^2}\sum_{\substack{n \\ E_p > E_n}}\mu_k^{pn}(A) m_i^{np}(A)G_{jl}(B;k_{pn})k_{pn}^2\varepsilon_{ijn}\varepsilon_{klm}\nabla_m \frac{e^{-ik_{pn}R}}{R}\nabla_n \frac{e^{ik_{pn}R}}{R}, \quad (41)$$

where in the last line $U_{ij}^{RES}(k_{pn}, \vec{R})$ is substituted. After random averaging, accounting for the factor $(1/3)^2$ that arises, Equation (41) becomes:

$$-\frac{1}{8\pi^2\varepsilon_0^2 c^2 R^6}\sum_{\substack{n \\ E_p > E_n}}[\vec{\mu}^{pn}(A)\cdot\vec{m}^{np}(A)]G(B;k_{pn})[k_{pn}^2 R^2 + k_{pn}^4 R^4]. \quad (42)$$

Noting that the second and third terms of Equation (34) produce contributions identical to Equations (38) and (42), respectively, adding these two terms and doubling, results in the additional contribution arising from downward transitions having the form,

$$\Delta E_{c-c}^{A-RES} = -\frac{1}{4\pi^2 \varepsilon_0^2 c^2 R^6} \sum_{\substack{n \\ E_p > E_n}} [\vec{\mu}^{pn}(A) \cdot \vec{m}^{np}(A)] G(B; k_{pn})[3 + 2k_{pn}^2 R^2 + 2k_{pn}^4 R^4]. \quad (43)$$

From Equation (43), the additional contribution arising from downward transitions in excited B, the second term of Equation (33), can be written down immediately as

$$\Delta E_{c-c}^{B-RES} = -\frac{1}{4\pi^2 \varepsilon_0^2 c^2 R^6} \sum_{\substack{r \\ E_q > E_r}} G(A; k_{qr})[\vec{\mu}^{qr}(B) \cdot \vec{m}^{rq}(B)][3 + 2k_{qr}^2 R^2 + 2k_{qr}^4 R^4], \quad (44)$$

where now species A responds via its excited state polarisability, $G(A; k_{qr})$, to the decay occurring due to the emission in B at the frequency $\omega_{qr} = ck_{qr}$. Just like the u-integral (32), the two terms of ΔE^{RES} are discriminatory, dependent upon the chirality of A and B, changing sign when a mirror-image counterpart replaces one enantiomer. The two contributions (43) and (44), are added to Equation (32) to give the total dispersion potential between two excited chiral molecules, namely,

$$\Delta E_{c-c} = \Delta E_{c-c}^u + \Delta E_{c-c}^{A-RES} + \Delta E_{c-c}^{B-RES}. \quad (45)$$

The result of Equation (45) agrees with an earlier evaluation using response theory [36].

4. Dispersion Energy Shift between an Electric Dipole Polarisable Molecule and a Paramagnetically Susceptible One

For the leading order, chirality in a molecule is characterised by the presence of electric and magnetic dipole moments. A contribution to the dispersion potential of an identical order of magnitude to that between two chiral molecules considered in the previous section is that between an electric dipole polarisable molecule and a magnetic dipole susceptible one. Both of these interaction energies contain a total of two electric and two magnetic dipole moments across the two sites. On letting A be an excited electrically polarisable species and B an excited paramagnetically susceptible entity, the relevant induced dipoles from Equations (1) and (2) to be used in the method deployed arise from the first term of Equation (1) and the second term of Equation (2), which couple to the interaction tensor (10), producing an energy shift:

$$\Delta E = \sum_{\vec{k},\lambda} \mu_i^{ind}(A; \vec{k}, \lambda) m_j^{ind}(B; \vec{k}, \lambda) \mathrm{Im} U_{ij}(k, \vec{R})$$
$$= \sum_{\vec{k},\lambda} \varepsilon_0^{-1} \alpha_{ik}(A; k) \chi_{jl}(B; k) d_k^\perp(\vec{k}, \lambda; \vec{R}_A) b_l(\vec{k}, \lambda; \vec{R}_B) \mathrm{Im} U_{ij}(k, \vec{R}). \quad (46)$$

Taking the expectation value of Equation (46) over the state $|p^A, q^B; 0(\vec{k}, \lambda)>$, and making use of the field-field correlation function (15), one obtains:

$$\Delta E = \sum_{\vec{k},\lambda} \left(\frac{\hbar k}{2\varepsilon_0 V}\right) \alpha_{ik}(A; k) \chi_{jl}(B; k) e_k^{(\lambda)}(\vec{k}) \overline{b}_l^{(\lambda)}(\vec{k}) e^{-i\vec{k} \cdot \vec{R}} \mathrm{Im} U_{ij}(k, \vec{R}). \quad (47)$$

Performing the polarisation sum using Equation (27), converting the $\vec{k}$-sum to an integral using Equation (19), and carrying out the angular integral using Equation (28), one arrives at

$$\Delta E = \frac{i\hbar}{4\pi^2\varepsilon_0}\varepsilon_{klm}\hat{R}_m \int_0^\infty dk k^3 \alpha_{ik}(A;k)\chi_{jl}(B;k)\left[\frac{\cos kR}{kR} - \frac{\sin kR}{k^2R^2}\right]\mathrm{Im}U_{ij}(k,\vec{R})$$
$$= \frac{i\hbar}{16\pi^3\varepsilon_0^2 cR^3}\varepsilon_{ijn}\varepsilon_{klm}\hat{R}_m\hat{R}_n \int_0^\infty dk k^3 \alpha_{ik}(A;k)\chi_{jl}(B;k)\left[\frac{\cos kR}{kR} - \frac{\sin kR}{k^2R^2}\right][kR\cos kR + k^2R^2\sin kR], \tag{48}$$

by inserting Equation (10) for $\mathrm{Im}U_{ij}(k,\vec{R})$. This result holds for a pair of anisotropic molecules. To obtain the isotropic potential, we make use of the orientational averages for the second rank tensors,

$$<X_{ij}(\xi;k)> = \frac{1}{3}\delta_{ij}\delta_{\lambda\mu}X_{\lambda\mu}(\xi;k) = \delta_{ij}X(\xi;k), \tag{49}$$

for $X(\xi;k) = \alpha(\xi;k)$ or $\chi(\xi;k)$. Contracting the tensors and simplification using trigonometric identities relating single and double angle arguments yields:

$$\Delta E = \frac{i\hbar}{16\pi^3\varepsilon_0^2 R^2}\int_0^\infty dk k^4 \alpha(A;k)\chi(B;k)\left\{\sin 2kR + \frac{2}{kR}\cos 2kR - \frac{1}{k^2R^2}\sin 2kR\right\}$$
$$= \frac{i\hbar}{16\pi^3\varepsilon_0^2 cR^2}\int_0^\infty dk k^4 \alpha(A;k)\chi(B;k)\,\mathrm{Im}\left[1 + \frac{2i}{kR} - \frac{1}{k^2R^2}\right]e^{2ikR}. \tag{50}$$

Rotating the integral from the real to the imaginary axis and substituting the complex variable $k = iu$, Equation (50) results in the u-integral contribution to the dispersion potential between an excited electric dipole polarisable molecule and an excited magnetic dipole susceptible molecule being given by

$$\Delta E^u_{e-m} = \frac{\hbar}{16\pi^3\varepsilon_0^2 cR^2}\int_0^\infty du u^4 e^{-2uR}\alpha(A;iu)\chi(B;iu)\left[1 + \frac{2}{uR} + \frac{1}{u^2R^2}\right], \tag{51}$$

where $\alpha(A;iu)$ and $\chi(B;iu)$ are the excited state polarisabilities evaluated at the imaginary frequency $\omega = icu$ and is straightforwardly obtained from expressions (3) and (5), respectively. The result (51) is identical in form to the expression obtained previously for the ground state interaction energy using perturbation and response theories [34,41,42], and is likewise repulsive.

The resonant terms are evaluated from

$$\Delta E^{RES}_{e-m} = \Delta E^{A-RES}_{e-m} + \Delta E^{B-RES}_{e-m}$$
$$= \sum_{\vec{k},\lambda}\varepsilon_0^{-1}\alpha_{ik}(A;k)\chi_{jl}(B;k)d_k^\perp(\vec{k},\lambda;\vec{R}_A)b_l(\vec{k},\lambda;\vec{R}_B)[U^{RES}_{ij}(k_{pn},\vec{R}) + U^{RES}_{ij}(k_{qr},\vec{R})]. \tag{52}$$

The evaluation of Equation (52) in a manner similar to the chiral–chiral example, using Equation (40) for the angular average, leads to the isotropic contributions,

$$\Delta E^{RES}_{e-m} = -\frac{1}{8\pi^2\varepsilon_0^2 c^2 R^6}\sum_{\substack{n \\ E_p > E_n}}|\vec{\mu}^{pn}(A)|^2\chi(B;k_{pn})[k^2_{pn}R^2 + k^4_{pn}R^4]$$
$$-\frac{1}{8\pi^2\varepsilon_0^2 c^2 R^6}\sum_{\substack{r \\ E_q > E_r}}|\vec{m}^{qr}(B)|^2\alpha(A;k_{qr})[k^2_{qr}R^2 + k^4_{qr}R^4], \tag{53}$$

neither of which, like the u-integral, are discriminatory. The two terms in Equation (53) apply only to downward transitions from the excited state. Each susceptibility responds to the emission frequency of the other particle, with the energy shift exhibiting inverse square dependent far-zone behaviour. The total excited state dispersion potential between electric and magnetic dipole polarisable systems is given by the sum of Equations (51) and (53),

$$\Delta E_{e-m} = \Delta E^u_{e-m} + \Delta E^{RES}_{e-m}, \tag{54}$$

agreeing with an earlier study [36].

5. Contribution from Diamagnetic Coupling

In Section 4, the paramagnetic contribution to the excited state dispersion interaction between an electric dipole polarisable molecule and a magnetic dipole susceptible molecule was studied. No account, however, was taken of the diamagnetic coupling term, which produces a contribution of identical order of magnitude to both the potentials evaluated thus far. This contribution is now considered. As previously stated, particle A is let to be an excited electric dipole polarisable species. However, here, B is selected to be an excited diamagnetic molecule. In the multipolar coupling scheme of non-relativistic QED [12–15,43,44], the diamagnetic coupling term of particle ξ is

$$H^d(\xi) = \frac{e^2}{8m} \sum_a \{(\vec{q}_a(\xi) - \vec{R}_\xi) \times \vec{b}(\vec{R}_\xi)\}^2, \tag{55}$$

where $\vec{q}_a(\xi)$ is the position of electron a relative to the centre of molecule ξ, $\vec{R}_\xi$. For an isotropic diamagnetic source located at the origin, Equation (55) reduces to $\frac{e^2}{12m} \sum_a q_a^2(\xi) \vec{b}^2(0)$. Therefore, in the presence of a magnetic field, the induced electronic coordinate of molecule B is given by

$$q_j^{ind}(B) = \chi_{jl}^d(B;0) b_l(0), \tag{56}$$

where the isotropic frequency independent excited state diamagnetic susceptibility is defined as

$$\chi^d(\xi;0) = -\frac{e^2}{12m} \sum_a < q_a^2(\xi) >^{qq}, \tag{57}$$

where the excited state matrix element of $q_a^2(\xi)$ over the state $|q>$ is $< q| < q_a^2(\xi) > |q> = < q_a^2(\xi) >^{qq}$. The induced electric dipole moment arising from the first term of Equation (1) and Equation (56) couple, yielding an energy shift,

$$\begin{aligned}\Delta E &= \sum_{\vec{k},\lambda} \mu_i^{ind}(A; \vec{k}, \lambda) q_j^{ind}(B) \mathrm{Im} U_{ij}(k, \vec{R}) \\ &= \sum_{\vec{k},\lambda} \varepsilon_0^{-1} \alpha_{ik}(A;k) \chi_{jl}^d(B;0) d_k^\perp(\vec{k},\lambda;\vec{R}_A) b_l(\vec{k},\lambda;\vec{R}_B) \mathrm{Im} U_{ij}(k,\vec{R}). \end{aligned} \tag{58}$$

Next, the expectation value of Equation (58) was taken over the state $|p^A, q^B; 0(\vec{k}, \lambda)>$, giving the p-th and q-th excited state electric dipole polarisability and diamagnetic susceptibility, respectively, and the electric-magnetic field–field spatial correlation function (15):

$$\Delta E = \sum_{\vec{k},\lambda} \left(\frac{\hbar k}{2\varepsilon_0 V}\right) \alpha_{ik}(A;k) \chi_{jl}^d(B;0) e_k^{(\lambda)}(\vec{k}) \bar{b}_l^{(\lambda)}(\vec{k}) e^{-i\vec{k}\cdot\vec{R}} \mathrm{Im} U_{ij}(k, \vec{R}), \tag{59}$$

after omitting the molecular state labels. The remainder of the calculation follows that given in Section 4 when evaluating the paramagnetic contribution to the dispersion energy. For A and B in fixed relative orientation,

$$\Delta E = \frac{i\hbar}{16\pi^3 \varepsilon_0^2 cR^3} \varepsilon_{ijn}\varepsilon_{klm} \hat{R}_m \hat{R}_n \int_0^\infty dk k^3 \alpha_{ik}(A;k) \chi_{jl}^d(B;0) \left[\frac{\cos kR}{kR} - \frac{\sin kR}{k^2 R^2}\right] [kR\cos kR + k^2 R^2 \sin kR]. \tag{60}$$

Utilising the result for the rotationally averaged diamagnetic susceptibility, $< \chi_{jl}^d(B;0) > = \delta_{jl} \chi^d(B;0)$, the isotropic diamagnetic contribution to the dispersion potential is

$$\Delta E_{e-d}^u = \frac{\hbar}{16\pi^3 \varepsilon_0^2 cR^2} \int_0^\infty du u^4 e^{-2uR} \alpha(A;iu) \chi^d(B;0) \left[1 + \frac{2}{uR} + \frac{1}{u^2 R^2}\right], \tag{61}$$

when expressed in terms of the complex variable $k = iu$.

Combining the diamagnetic u-integral term (61) with the corresponding paramagnetic term (51), gives:

$$\Delta E^u_{p+d} = \frac{\hbar}{16\pi^3\varepsilon_0^2 c R^2} \int_0^\infty du\, u^4 e^{-2uR} \alpha(A; iu) \chi^m(B; iu) \left[1 + \frac{2}{uR} + \frac{1}{u^2 R^2}\right], \tag{62}$$

where the magnetic susceptibility tensor, χ^m, is a sum of paramagnetic (p) and diamagnetic (d) components, recalling that the latter is frequency independent,

$$\chi^m(B; iu) = \chi^p(B; iu) + \chi^d(B; 0), \tag{63}$$

with the isotropic paramagnetic susceptibility given by

$$\chi^p(B; iu) = \frac{2}{3} \sum_r \frac{|\vec{m}^{qr}(B)|^2 E_{rq}}{E_{rq}^2 + (\hbar c u)^2}, \tag{64}$$

and with the diamagnetic susceptibility $\chi^d(B; 0)$ given by (57). A functional form similar to Equation (62) was obtained using perturbation and response theories for the ground state dispersion potential [13,34,41,45].

For the additional contributions to the u-integral term arising solely from de-excitation, diamagnetic B does not respond to the downward transitions in A, $|n\rangle \leftarrow |p\rangle$, so that from the first term of Equation (53),

$$\Delta E^{A-RES}_{e-d} = -\frac{1}{8\pi^2\varepsilon_0^2 c^2 R^6} \sum_{\substack{n \\ E_p > E_n}} |\vec{\mu}^{pn}(A)|^2 \chi^d(B;0)[k_{pn}^2 R^2 + k_{pn}^4 R^4]. \tag{65}$$

Particle A, on the other hand, responds to the downward transitions occurring in B, $|r\rangle \leftarrow |q\rangle$, modifying the second term of Equation (53) to give for diamagnetic B the contribution,

$$\Delta E^{B-RES}_{e-d} = -\frac{e^2}{48\pi^2\varepsilon_0^2 c^2 m R^6} \sum_{\substack{r \\ E_q > E_r}} <|q^{qr}(B)|^2> \alpha(A; k_{qr})[k_{qr}^2 R^2 + k_{qr}^4 R^4], \tag{66}$$

with

$$\Delta E^{RES}_{e-d} = \Delta E^{A-RES}_{e-d} + \Delta E^{B-RES}_{e-d}, \tag{67}$$

a sum of Equations (65) and (66). Hence, the total dispersion potential between an excited electric dipole polarisable molecule and an excited diamagnetic one is given by

$$\Delta E_{e-d} = \Delta E^u_{e-d} + \Delta E^{RES}_{e-d} \tag{68}$$

with ΔE^u_{e-d} given by Equation (61).

6. Summary

Within a quantum field framework, retarded van der Waals dispersion potentials between atoms or molecules in the ground state are commonly evaluated using diagrammatic time-dependent perturbation theory. This method, however, gives rise to computational difficulties when one or both of the pair are electronically excited, since resonant terms have to be accounted for. An alternative treatment of this problem involved employing response theory, in which each particle responds, through its electric or magnetic susceptibility, to the source Maxwell fields of the other entity. However, this method requires first calculating the second quantised electric and magnetic radiation fields in the vicinity of a source multipole moment before the energy shift can be evaluated, with the additional burden that fields second order in the moments are needed at the very least [27,34,46].

To overcome some of these problems, in this paper, the excited state dispersion potential between optically active molecules is examined from a different physical point of view, one that was previously considered for the calculation of the Casimir–Polder potential. Fluctuations in the vacuum electromagnetic field induce multipole moments in atoms or molecules, which in turn couple through pertinent retarded interaction tensors. Expectation values are taken over excited matter states and the ground state of the radiation field. The latter yielded known expressions for the field–field spatial correlation functions and serve to highlight the prominent role played by the electromagnetic vacuum when calculating dispersion forces.

For two excited chiral molecules characterised by electric and magnetic dipole moments, all three terms contributing to the dispersion interaction energy—the familiar u-integral term involving excited state mixed electric-magnetic dipole polarisabilities of each particle, and two extra contributions arising from downward only transitions in each species—are found to be discriminatory, depending on the handedness of molecules A and B. Two other dispersion potentials involving magnetic interactions that were of a similar order of magnitude to the chiral–chiral energy shift are also computed. These included the potential between an electric dipole polarisable molecule and a second that is either paramagnetically or diamagnetically susceptible. Neither energy shift contribution changes sign on interchanging enantiomers, with the second coupling term being independent of frequency, but which may be combined with the paramagnetic part to produce an overall magnetically susceptible contribution, as found earlier for the ground state [34,38,42].

Interest in dispersion energies between chiral molecules lies in their discriminatory behaviour as well as in outlining the role played by magnetic and diamagnetic coupling. It complements other research areas that involve optically active species such as chiral light–matter interactions, analytical based methods for achieving enantiomer excess and separation, and the synthesis of chiral compounds and drugs in organic chemistry and the pharmaceutical industry [35–38].

Funding: This research received no external funding.

Data Availability Statement: Not applicable.

Conflicts of Interest: The author declares no conflict of interest.

References

1. Milonni, P.W. *The Quantum Vacuum*; Academic Press: San Diego, CA, USA, 1994.
2. Dirac, P.A.M. The Quantum Theory of Emission and Absorption of Radiation. *Proc. R. Soc. A Math. Phys. Eng. Sci.* **1927**, *114*, 243–265. Available online: https://www.jstor.org/stable/94746 (accessed on 8 February 2022).
3. Lamb, W.E., Jr.; Retherford, R.C. Fine Structure of the Hydrogen Atom. *Phys. Rev.* **1947**, *72*, 241–243. [CrossRef]
4. Bethe, H.A. The Electromagnetic Shift of Energy Levels. *Phys. Rev.* **1947**, *72*, 339–341. [CrossRef]
5. Casimir, H.B.G.; Polder, D. The Influence of Retardation on the London van der Waals Forces. *Phys. Rev.* **1948**, *73*, 360–372. [CrossRef]
6. Buhmann, S.Y. *Dispersion Forces I*; Springer: Berlin/Heidelberg, Germany, 2012. [CrossRef]
7. Salam, A. *Non-Relativistic QED Theory of the van der Waals Dispersion Interaction*; Springer International Publishing AG: Cham, Switzerland, 2016. [CrossRef]
8. Passante, R. Dispersion Interactions between Neutral Atoms and the Quantum Electrodynamical Vacuum. *Symmetry* **2018**, *10*, 735. [CrossRef]
9. Schwinger, J.S. (Ed.) *Selected Papers on Quantum Electrodynamics*; Dover Publications, Inc.: Mineola, NY, USA, 1958.
10. Power, E.A. *Introductory Quantum Electrodynamics*; Longmans, Green & Co. Ltd.: London, UK, 1964.
11. Healy, W.P. *Non-Relativistic Quantum Electrodynamics*; Academic Press, Inc.: London, UK, 1982.
12. Craig, D.P.; Thirunamachandran, T. *Molecular Quantum Electrodynamics*; Dover Publications, Inc.: Mineola, NY, USA, 1998.
13. Salam, A. *Molecular Quantum Electrodynamics*; John Wiley & Sons, Inc.: Hoboken, NJ, USA, 2010. [CrossRef]
14. Andrews, D.L.; Bradshaw, D.S.; Forbes, K.A.; Salam, A. Quantum Electrodynamics in Modern Optics and Photonics: Tutorial. *J. Opt. Soc. Am. B* **2020**, *37*, 1153–1172. [CrossRef]
15. Woolley, R.G. *Foundations of Molecular Quantum Electrodynamics*; Cambridge University Press: Cambridge, UK, 2022. [CrossRef]
16. Scholes, G.D. Long-Range Energy Transfer in Molecular Systems. *Ann. Rev. Phys. Chem.* **2003**, *54*, 57–87. [CrossRef]
17. Andrews, D.L.; Demidov, A.A. (Eds.) *Resonance Energy Transfer*; Wiley: Chichester, UK, 1999.

18. Salam, A. The Unified Theory of Resonance Energy Transfer According to Molecular Quantum Electrodynamics. *Atoms* **2018**, *6*, 56. [CrossRef]
19. Jones, G.A.; Bradshaw, D.S. Resonance Energy Transfer: From Fundamental Theory to Recent Applications. *Front. Phys.* **2019**, *7*, 100. [CrossRef]
20. Power, E.A.; Thirunamachandran, T. Dispersion Forces between Molecules with One or Both Molecules Excited. *Phys. Rev. A* **1995**, *51*, 3660–3666. [CrossRef]
21. Safari, H.; Karimpour, M.R. Body-Assisted van der Waals Interaction between Excited Atoms. *Phys. Rev. Lett.* **2015**, *114*, 013201. [CrossRef] [PubMed]
22. Berman, P.R. Interaction Energy of Non-Identical Atoms. *Phys. Rev. A* **2015**, *91*, 042127. [CrossRef]
23. Donaire, M.; Guérout, R.; Lambrecht, A. Quasiresonant van der Waals Interaction between Nonidentical Atoms. *Phys. Rev. Lett.* **2015**, *115*, 033201. [CrossRef] [PubMed]
24. Milonni, P.W.; Rafsanjani, S.M.H. Distance Dependence of Two-Atom Dipole Interactions with One Atom in an Excited State. *Phys. Rev. A* **2015**, *92*, 062711. [CrossRef]
25. Donaire, M. Two-Atom Interaction Energies with One Atom in an Excited State: van der Waals Potentials Versus Level Shifts. *Phys. Rev. A* **2016**, *93*, 052706. [CrossRef]
26. Barcellona, P.; Passante, R.; Rizzuto, L.; Buhmann, S.Y. van der Waals Interactions between Excited Atoms in Generic Environments. *Phys. Rev. A* **2016**, *94*, 012705. [CrossRef]
27. Power, E.A.; Thirunamachandran, T. Quantum Electrodynamics with Nonrelativistic Sources. V. Electromagnetic Field Correlations and Intermolecular Interactions between Molecules in Either Ground or Excited States. *Phys. Rev. A* **1993**, *47*, 2539–2551. [CrossRef]
28. Passante, R.; Rizzuto, L. Effective Hamiltonians in Nonrelativistic Quantum Electrodynamics. *Symmetry* **2021**, *13*, 2375. [CrossRef]
29. Power, E.A.; Thirunamachandran, T. The Non-Additive Dispersion Energies for N Molecules: A Quantum Electrodynamical Theory. *Proc. R. Soc. A Math. Phys. Eng. Sci.* **1985**, *401*, 267–279.
30. Aldegunde, J.; Salam, A. Dispersion Energy Shifts Among N Bodies with Arbitrary Electric Multipole Polarisability: Molecular QED Theory. *Mol. Phys.* **2015**, *113*, 226–231. [CrossRef]
31. Power, E.A.; Thirunamachandran, T. Casimir-Polder Potential as an Interaction between Induced Dipoles. *Phys. Rev. A* **1993**, *48*, 4761–4763. [CrossRef] [PubMed]
32. Mavroyannis, C.; Stephen, M.J. Dispersion Forces. *Mol. Phys.* **1962**, *5*, 629–638. [CrossRef]
33. Jenkins, J.K.; Salam, A.; Thirunamachandran, T. Discriminatory Dispersion Interaction Energies between Chiral Molecules. *Mol. Phys.* **1994**, *82*, 835–840. [CrossRef]
34. Jenkins, J.K.; Salam, A.; Thirunamachandran, T. Retarded Dispersion Interaction Energies between Chiral Molecules. *Phys. Rev. A* **1994**, *50*, 4767–4777. [CrossRef] [PubMed]
35. Craig, D.P.; Thirunamachandran, T. New Approaches to Chiral Discrimination in Coupling between Molecules. *Theor. Chem. Acc.* **1999**, *102*, 112–120. [CrossRef]
36. Salam, A. Intermolecular Energy Shifts between Two Chiral Molecules in Excited Electronic States. *Mol. Phys.* **1996**, *87*, 919–929. [CrossRef]
37. Barcellona, P.; Safari, H.; Salam, A.; Buhmann, S.Y. Enhanced Chiral Discriminatory van der Waals Interactions Mediated by Chiral Surfaces. *Phys. Rev. Lett.* **2017**, *118*, 193401. [CrossRef]
38. Safari, H.; Barcellona, P.; Buhmann, S.Y.; Salam, A. Medium-Assisted van der Waals Dispersion Interactions Involving Chiral Molecules. *New J. Phys.* **2020**, *22*, 053049. [CrossRef]
39. Salam, A. On the Effect of a Radiation Field in Modifying the Intermolecular Interaction between Two Chiral Molecules. *J. Chem. Phys.* **2006**, *124*, 014302. [CrossRef]
40. Andrews, D.L.; Thirunamachandran, T. On Three-Dimensional Rotational Averages. *J. Chem. Phys.* **1977**, *67*, 5026–5033. [CrossRef]
41. Thirunamachandran, T. Vacuum Fluctuations and Intermolecular Interactions. *Phys. Scr.* **1988**, *1988*, 123. [CrossRef]
42. Buhmann, S.Y.; Safari, H.; Scheel, S.; Salam, A. Body-Assisted Dispersion Potentials of Diamagnetic Atoms. *Phys. Rev. A* **2013**, *87*, 012507. [CrossRef]
43. Andrews, D.L.; Jones, G.A.; Salam, A.; Woolley, R.G. Perspective: Quantum Hamiltonians for Optical Interactions. *J. Chem. Phys.* **2018**, *148*, 040901. [CrossRef]
44. Salam, A. Quantum Electrodynamics Effects in Atoms and Molecules. *WIREs Comput. Mol. Sci.* **2015**, *5*, 178–201. [CrossRef]
45. Salam, A. On the Contribution of the Diamagnetic Coupling Term to the Two-Body Retarded Dispersion Interaction. *J. Phys. B At. Mol. Opt. Phys.* **2000**, *33*, 2181–2193. [CrossRef]
46. Salam, A. Molecular Quantum Electrodynamics in the Heisenberg Picture: A Field Theoretic Viewpoint. *Int. Rev. Phys. Chem.* **2008**, *27*, 405–448. [CrossRef]

Disclaimer/Publisher's Note: The statements, opinions and data contained in all publications are solely those of the individual author(s) and contributor(s) and not of MDPI and/or the editor(s). MDPI and/or the editor(s) disclaim responsibility for any injury to people or property resulting from any ideas, methods, instructions or products referred to in the content.

Communication

Finite-Size Effects of Casimir–van der Waals Forces in the Self-Assembly of Nanoparticles

Raul Esquivel-Sirvent

Instituto de Física, Universidad Nacional Autónomia de México, P.O. Box 20-364, Mexico City 01000, Mexico; raul@fisica.unam.mx

Abstract: Casimir–van der Waals forces are important in the self-assembly processes of nanoparticles. In this paper, using a hybrid approach based on Lifshitz theory of Casimir–van der Waals interactions and corrections due to the shape of the nanoparticles, it is shown that for non-spherical nanoparticles, the usual Hamaker approach overestimates the magnitude of the interaction. In particular, the study considers nanoplates of different thicknesses, nanocubes assembled with their faces parallel to each other, and tilted nanocubes, where the main interaction is between edges.

Keywords: self-assembly; Casimir force; van der Waals force; Hamaker constant; nanoparticles

1. Introduction

The self-assembly of nanoparticles has become an attractive area of research since it can be used to construct materials with novel properties by arranging nanoparticles of different geometries in arrays that mimic crystalline structures with a given periodicity [1]. Unlike usual crystals with an atom in each site, in self-assembled supracrystals, a nanoparticle is placed in the crystalline sites [2]. The assembly and bonding of the nanoparticles happen due to a combination of forces and, in many cases, ligands such as strands of nucleic acids [3]. Several physical properties can be modified in self-assembled systems, such as the optical response using plasmonic nanoparticles, as well as electrical and thermal conductivity properties, making self-assembly a practical way for building nanocomposites [4]. The potential use of nanocomposites as biosensors and nano-biomaterials has been studied [5], as well their therapeutic delivery of drugs at the nanoscale [6].

Self-assembly typically occurs in a solvent, such as water, and the interactions are described by the DLVO theory, named after Derjaguin, Landau, Verwey, and Overbeek. The DLVO theory includes the screened electrostatic interaction via the Poisson-Boltzman equation and the van der Waals interaction, assuming they are additive [7–10]. Depletion forces, that are attractive, due to the presence of micelles can also be present [11]The van der Waals force is usually calculated using the Hamaker approach, which assumes a pair-wise summation [12]. Based on the fluctuation–dissipation theorem and Rytov's theory of fluctuating electrodynamics [13], Lifshitz [14,15] developed the theory of generalized van der Waals forces between macroscopic bodies. The Lifshitz equation for the Casimir–van der Waals interaction energy can be written as in the Hamaker's approach; however, now the Hamaker constant can be explicitly calculated if the dielectric functions of the particles and the surrounding media are known.

Of interest to the calculations presented in this paper is the equilibrium formation of colloidal Au nanoprisms. Young et al. synthesized and self-assembled triangular nanoprisms in a one-dimensional periodic array [16]. The periodicity of these lamellar superlattices depends on the solution's temperature and ionic strength. The equilibrium condition comes from the balance of the attractive van der Waals force, the repulsive electrostatic potential from the solution of the Poisson–Boltzmann equation, and the attractive depletion force that comes from the formation of micelles, since surfactants are added to avoid aggregation. A similar work by Munkhbat et al. [17] designed tunable self-assembled

Citation: Esquivel-Sirvent, R. Finite-Size Effects of Casimir–van der Waals Forces in the Self-Assembly of Nanoparticles. *Physics* **2023**, *5*, 322–330. https://doi.org/10.3390/physics5010024

Received: 2 February 2023
Revised: 4 March 2023
Accepted: 6 March 2023
Published: 21 March 2023

Copyright: © 2023 by the author. Licensee MDPI, Basel, Switzerland. This article is an open access article distributed under the terms and conditions of the Creative Commons Attribution (CC BY) license (https://creativecommons.org/licenses/by/4.0/).

Casimir microcavities made of parallel Au nano palettes and showed that only the electrostatic and Casimir–van der Waals interactions play a dominant role. Both above-described systems yield equilibrium periodic structures. The possibility of having optical cavities with a high-quality factor by combining the repulsive Casimir and buoyancy forces has also been considered [18]. A suitable choice of the dielectric functions of the plates and the media between them yields a repulsive force.

Finite-size effects due to the shape of the interacting bodies have been a challenge to the precise calculation of the Casimir–van der Waals force. Hamaker introduced [12] a pair-wise summation to calculate the interaction energy between two spheres, resulting in an expression that includes the energy of interaction multiplied by a factor that depends on the geometry of the objects. Other geometries have also been considered, such as the interaction between spheres and shells, as well as shells and walls [19]. Dantchev and Valchev presented [20] a surface integration approach generalizing the Derjaguin or proximity theorem approximation for the interaction between a three-dimensional object and a half-space. In particular, Dantchev and Valchev considered the case of spheres, cylinders, and the interaction of liposomes and lipid bilayers. The problem of extending the original theory of Hamaker is well described in an extensive review by Rusanov and Brodskaya [21], where they present the interaction of many systems of interest in colloidal science, such as spherical particles, wedges, and cylinders of different lengths.

Furthermore, we are interested in the self-assembly of polymer-grafted metal nanocubes into arrays of one-dimensional strings with well-defined interparticle orientations and tunable electromagnetic properties [22]. The nanocubes are assembled in two configurations: one considering the edge–to-edge interactions of the nanocubes, and the second one considering the face-to-face interactions. Unlike spherical nanoparticles characterized by one dipolar plasmonic resonance, cubes have several dipolar modes [23].

Since Lifshitz theory provides a more accurate description than the simple Hamaker approximation, a hybrid approach is preferred. In this paper, the interaction energy between two parallel surfaces is calculated using Lifshitz theory, adjusting for geometrical effects [24]. Within this approach and using the results of de Rocco and Hoover [25], we evaluate the Casimir–van der Waals interaction in several systems of interest in self-assembly.

2. Lifshitz Theory and the Hamaker Constant

Lifshitz theory considers two parallel slabs separated by a distance L and a temperature T. The plates have lateral dimensions that are much larger than L. The dielectric function of the plates is $\varepsilon(\omega)$ in a medium with a dielectric function $\varepsilon_m(\omega)$. After making the rotation to imaginary frequencies ($\omega \to i\omega$) and introducing the Matsubara frequencies $\zeta_n = 2\pi n K_B T/\hbar$, where K_B is Boltzmann constant, $\hbar$ is the reduced Planck's constant, and n is a natural number, the Casimir–van der Waals energy per unit area is written as [14,15,26]

$$\mathcal{E}(T,L) = \frac{K_B T}{2\pi} {\sum_{n=0}^{\infty}}' \int_0^{\infty} dQ Q \ln[D_p(Q,\zeta_n,L) D_s(Q,\zeta_n,L)]. \quad (1)$$

The prime in the sum indicates that the $n = 0$ term has to be multiplied by $1/2$. The wave vector in the gap is $K_0 = (Q, k_0)$, with the z-component, $k_0 = \sqrt{\varepsilon_m \zeta_n^2/c_0^2 + Q^2}$, where c_0 denotes the speed of light in vacuum. Within the material, the corresponding z component of the wave vector is $k = \sqrt{\varepsilon \zeta_n^2/c_0^2 + Q^2}$. These definitions of the normal components of the wave vectors are evaluated in the Matsubara frequencies. The function $D_\nu(Q,\zeta,L)$ is

$$D_\nu(Q,\zeta_n,L) = 1 - r_\nu^2 e^{-2k_0 L}, \quad (2)$$

where r_ν are the reflection coefficients of the plates for either $\nu = p$ or $\nu = s$ polarization. For clarity, it is important to notice that the reflection coefficients depend on the frequency and parallel component of the wave vector Q through the definitions of k_0 and k, defined above. For a slab of thickness d, one has

$$r_\nu = \rho_\nu \frac{1 - e^{-2\delta}}{1 - \rho_\nu^2 e^{-2\delta}}, \qquad (3)$$

where the phase δ is defined as $\delta = (d/c_0)\sqrt{\zeta_n^2(\varepsilon(i\zeta_n) - 1) + c_0^2 k_0}$, and the Fresnel coefficients ρ_ν are

$$\rho_s = \frac{k_0 - k}{k_0 + k}, \qquad (4)$$

and

$$\rho_p = \frac{k - \varepsilon(i\zeta_n)k_0}{k + \varepsilon(i\zeta_n)k_0}. \qquad (5)$$

In general, Equation (1) is correct for half-spaces, finite-width plates, layered systems [27], or nonlocal dielectric functions [28] provided that the appropriate reflection coefficients are calculated [29]. The only restriction is that Lifshitz theory assumes that the plate extension is infinite.

The interaction energy given by Equation (1) can be rewritten in the form of the Hamaker formula. Defining the variable $x = 2k_0 L$, the energy per unit area reads:

$$\mathcal{E}(T, L) = -\frac{A_H(T)}{12\pi L^2}, \qquad (6)$$

where the Hamaker constant, $A_H(T)$, is

$$A_H(T) = -\frac{3kT}{2} \sum_{n=0}^{\infty}{}' \int_{x_0}^{\infty} dx\, x \ln[D_p(x, \zeta_n) D_s(x, \zeta_n)], \qquad (7)$$

and the lower limit of integration is $x_0 = \zeta_n \sqrt{\varepsilon_m} L/c_0$.

For two equal plates facing each other of surface area S, the total energy is $E_H(L) = \mathcal{E}(L)S$.

3. Finite-Size Effects

When dealing with finite-size effects, the energy of interaction between the bodies can be, in general, written as $E(L, T) = A_H K(a, b, d)$, where $K(a, b, d)$ is a geometric correction that depends on the dimensions of the body indicated by a, b, d. However, A_H is calculated from Equation (7), which, as explained before, is for parallel plates strictly of infinite length or with dimensions much larger than the separation L. In what follows the following notation is used: E is the total energy between the bodies, and $\mathcal{E}$ is the energy density (see Equation (6)).

For the case of finite-size plates, the geometric factor was derived by De Rocco and Hoover [25]. For two parallel plates of size $a \times b$ and thickness c, the geometric factor is

$$\begin{aligned}
K_{\text{pl}}(x, a, b) = {} & \frac{1}{4} \ln\left(\frac{x^4 + x^2 a^2 + x^2 b^2 + a^2 b^2}{x^4 + x^2 a^2 + x^2 b^2}\right) + \frac{x^2 - a^2}{4ax} \tan^{-1}\left(\frac{a}{x}\right) \\
& + \frac{x^2 - b^2}{4bx} \tan^{-1}\left(\frac{b}{x}\right) + \frac{x(a^2 + b^2)^{3/2}}{6a^2 b^2} \tan^{-1}\left(\frac{x}{\sqrt{a^2 + b^2}}\right) \\
& + \left(\frac{1}{6x^2} + \frac{1}{6a^2}\right) b\sqrt{x^2 + a^2} \tan^{-1}\left(\frac{b}{\sqrt{a^2 + x^2}}\right) \\
& + \left(\frac{1}{6x^2} + \frac{1}{6b^2}\right) a\sqrt{x^2 + a^2} \tan^{-1}\left(\frac{a}{\sqrt{b^2 + x^2}}\right).
\end{aligned} \qquad (8)$$

The variable x is a dummy variable that represents the position of the body where the geometric factor is evaluated. The corresponding interaction energy between the plates is

$$E(L,a,b,c) = -\frac{A_H}{\pi^2}[K_{pl}(L+2c) - 2K_{pl}(L+c) + K_{pl}(L)]. \qquad (9)$$

The other case of interest is the self-assembly of cubes (see Ref. [22]). Equations (8) and (9) are correct for two cubes with parallel faces, setting $a = b = c$. When the cubes are tilted and the interaction is edge-to-edge, the geometric factor of the interaction (denoted as K_{cb}) is

$$\begin{aligned}K(x,d,c)_{cb} =& \frac{1}{8}\ln\left(\frac{d^2+x^2}{c^2+d^2+x^2}\right) + \frac{1}{8}\left(\frac{x}{d}-\frac{d}{x}\right)\tan^{-1}\left(\frac{d}{x}\right) \\ &+ \frac{(c^2+d^2)^{3/2}x}{12c^2d^2}\tan^{-1}\left(\frac{x}{\sqrt{d^2+c^2}}\right) \\ &+ \frac{c\sqrt{d^2+x^2}}{12}\left(\frac{1}{d^2}+\frac{1}{x^2}\right)\tan^{-1}\left(\frac{c}{\sqrt{d^2+x^2}}\right) \\ &+ \frac{d(c^2+x^2)^{1/2}}{12}\left(\frac{1}{c^2}+\frac{1}{x^2}\right)\tan^{-1}\left(\frac{d}{\sqrt{c^2+x^2}}\right),\end{aligned} \qquad (10)$$

In this case, the separation L is between the edges of the cubes and $d = L/\sqrt{2}$, and the energy is

$$E(d,a,b,c) = -\frac{A_H}{\pi^2}(K_{cb}(d+2a,d,b,c) - 2K_{cb}(d+a,d,b,c) + K_{cb}(d,d,b,c)). \qquad (11)$$

4. Results

To evaluate the finite-size effects, consider that the nanoparticles are made of Au with a dielectric function given by a Drude model: $\varepsilon(i\zeta_n) = 1 + w_p^2/(\zeta_n^2 + \zeta_n\gamma)$, where $w_p = 9$ eV and $\gamma = 0.02$ eV. The plates (and cubes) are surrounded by water. The dielectric function of the water used here is the data reported in Ref. [30] calculated along the rotated frequency space ($\omega \to i\zeta_n$).

To understand the effect of finite-size effects, let us compare the energy between the plates E_H and the energy predicted using Equation (9) and Equation (11). As was stated above, the energy for the plates E_H is given by $E_H = S\mathcal{E} = SA_H/12\pi L^2$., where S is the surface of the plates. The value of the Hamaker constant in Refs. [16,17] was calculated assuming semi-infinite plates, which is $r_{s,p}$ in Equation (2)—the knownFresnel coefficients. Let us define the ratio $E_r = E(L;a,b,c)/E_H(L)$. If $E_r \sim 1$, then finite-size effects are not significant. Figure 1 shows E_r as a function of the separation for the case of parallel plates. The blue curve represents plates of size $a = b = 2000$ nm and thickness $c = 30$ nm, roughly the size reported in Ref. [17], and the red line corresponds to plates of size $a = b = 145$ nm and thickness $c = 7.5$ nm as in Ref. [16]. In both cases, one can observe that the Casimir–van der Waals interaction is underestimated when using Equation (6). As L decreases, the value of E_r increases, since $L \ll a,b$.

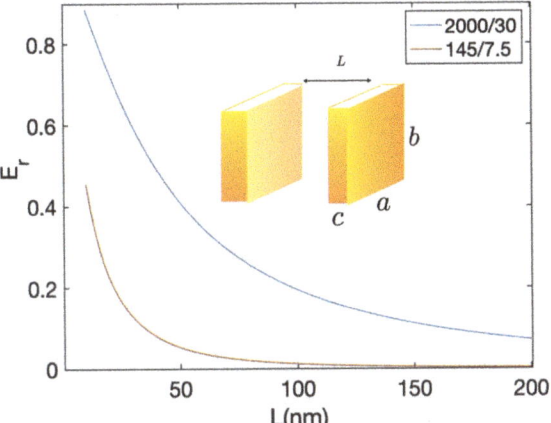

Figure 1. The energy ratio, E_r, for two sizes of square plates of size $a = b$ and thickness c. The lines correspond to the value a/c as indicated. The sizes correspond to those used in Refs. [16,17]. See text for details.

For completeness, Figure 2 shows how E_r changes with thickness c while the other dimensions are kept fixed ($a = b = 2000$ nm).

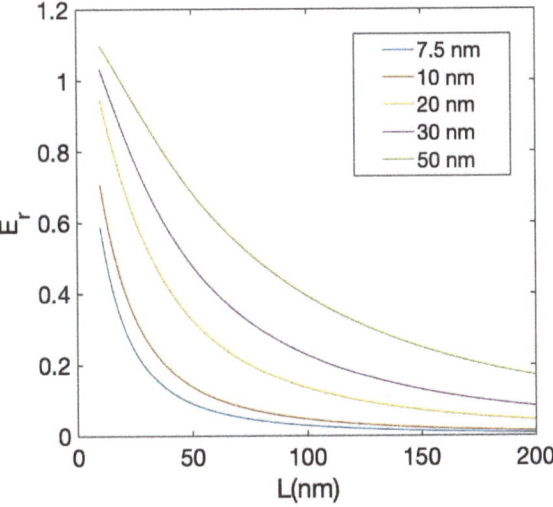

Figure 2. The variation of E_r as a function of the separation, L, of the two plates for different values of the thickness c of each as indicated, and $a = b = 2000$ nm. Even for large enough values of c, the energy ratio is less than unity.

At a fixed separation L, the dependence of E_r with the thickness, c, increases linearly for small values of c and levels off asymptotically to the value expected for half-spaces. Keeping the area of the plates constant and at two arbitrary separations of $L = 50$ nm and $L = 100$ nm, Figure 3 shows the variation with the thickness. Thus, one can quantify the correction needed for Equation (6).

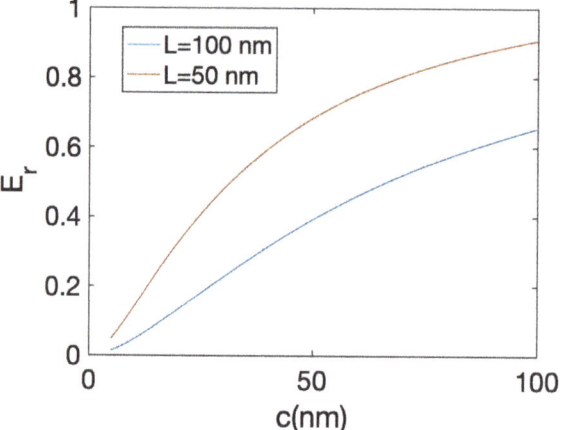

Figure 3. For two fixed separations between the plates, $L = 50$ nm and $L = 100$ nm, the ratio E_r increases with increasing the value of the thickness c, leveling-off asymptotically to the value expected for half-spaces. The dimensions of the plate are $a = b = 2000$ nm.

The interaction energy E_r for cubes is presented in Figure 4. The face–face and edge–edge interactions are considered with edge lengths of $a = b = c = 80$ nm as reported in Ref. [22]. The face–face interaction is just a particular case of parallel plates with $a = b = c$, and the behavior is the same. As L decreases, the ratio a/L increases, and E_r increases. For all separations L, since $E_r < 1$, one can see that using Equation (6) overestimates the interaction. The ratio E_r is calculated using Equation (11) for the edge–edge interaction. The behavior of E_r is different from the other cases. For the tilted cubes, the behavior is different, and the interaction increases with increasing separation until it reaches a maxima. To further understand this behavior, E_r is plotted Figure 5 for the edge–edge interaction for cubes of different sizes. The behavior of the curves is the same for different sizes except that, depending on the size of the cube, there are different values of the maxima, but it occurs when the separation between the edges is the same as the size of the cube $L = c$. It should also be noted that the maximum value attained by E_r is independent of the size of the cube. Whether this is an artifact of the Hamaker approach or due to the singularity of having the interaction between two edges is an issue that needs further exploration.

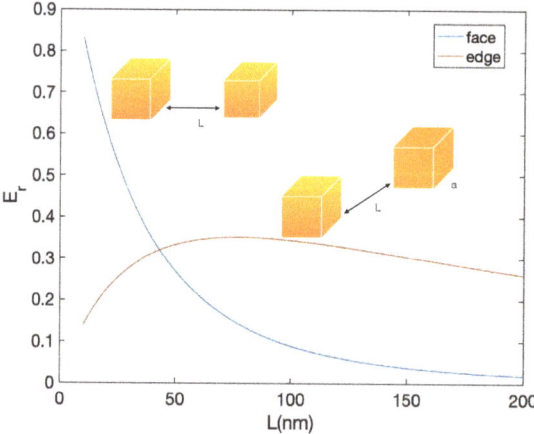

Figure 4. Energy ratio, E_r between two cubes facing each other and for two tilted cubes, as a function of the separation L. The nanocubes have dimensions $a = b = c = 80$ nm.

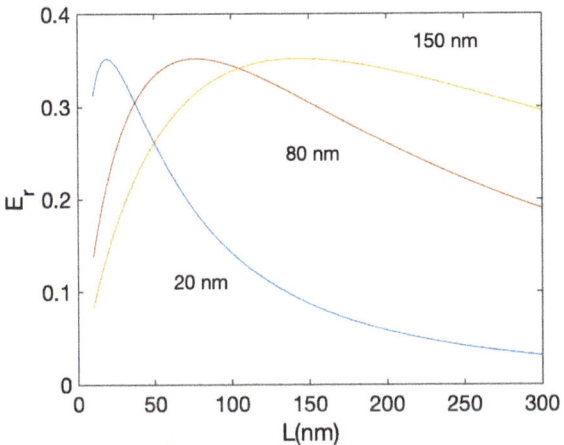

Figure 5. Energy ratio, E_r, between two tilted nanocubes of different sizes. The sizes considered are 20 nm (blue line), 80 nm (red line), and 150 nm (orange line). The maximum in each curve happens when the separation equals the size of the cube.

5. Discussion

The results presented in this study assumed that the surrounding medium was water. No effect of electrolytes that screen the van der Waals interaction [31,32] was considered. The screening will change the magnitude of the interaction energy. The dielectric function of the nanoparticle is assumed to be that of bulk Au. In the case of small metallic nanoparticles, the damping has to be corrected to consider the change in the electronmean-free path. For silver, this correction implies a change in the Hamaker constant of 138% nanoparticles [33].

In the case of plates and nanoparticles, there is another effect that has not been considered: spatial dispersion or nonlocal effects. For plates whose thickness is less than the electron mean-free path, or with nanoparticles with an average size smaller than the mean-free path, the dependence of the dielectric function with the wave vector has to be taken into account. As shown in Ref. [34], introducing spatial dispersion affects the Hamaker constant's value at short separations. The difference in the Hamaker constant between the local and nonlocal cases can be as large as two orders of magnitude. Nonlocal effects become relevant when the size of the bodies is of the order of magnitude of the skin depth. For Au (depending on the frequency), the skin depth is $\sim$40 nm. Thus, the nanoplates used for self-assembly in the literature [16,17] fall in the range of thickness where spatial dispersion has to be taken into account.

6. Conclusions

The ability to synthesize nanoparticles of different shapes and use them in self-assembly requires understanding all the interactions, particularly the Casimir–van der Waals interaction. The simplified approach not taking into account the shape overestimates the force. The use of Equations (6) and (7) is not the only procedure available in the literature for estimating the Casimir–van der Waals interaction. Numerical simulations for arbitrary 3D objects have been reported earlier [35] but require more computer-intensive calculations. The procedure presented in this paper can be considered a first approach to estimating the influence of the geometry for the case of nanoplates and nanocubes. The interaction energy obtained, considering finite-size effects, is smaller than that predicted by the conventional Hamaker approach. Geometric effects and other considerations, such as spatial dispersion, should provide a better prediction of the Casimir–van der Waals interaction for an accurate design of self-assembled structures.

Funding: This research received no external funding.

Data Availability Statement: Not applicable.

Acknowledgments: The author thanks Shunashi G. Castillo-López for helpful discussions and comments to the manuscript.

Conflicts of Interest: The author declares no conflict of interest.

References

1. Jones, M.R.; Macfarlane, R.J.; Lee, B.; Zhang, J.; Young, K.L.; Senesi, A.J.; Mirkin, C.A. DNA-nanoparticle superlattices formed from anisotropic building blocks. *Nat. Mat.* **2010**, *9*, 913–917. [CrossRef] [PubMed]
2. Macfarlane, R.J.; Lee, B.; Jones, M.R.; Harris, N.; Schatz, G.C.; Mirkin, C.A. Nanoparticle superlattice engineering with DNA. In *Spherical Nucleic Acids. Volume 2*; Mirkin, C.A., Ed.; Jenny Stanford Publishing/Taylor & Francis Group: New York, NY, USA, 2020; Chapter 23. [CrossRef]
3. Wu, C.; Han, D.; Chen, T.; Peng, L.; Zhu, G.; You, M.; Qiu, L.; Sefah, K.; Zhang, X.; Tan, W. Building a multifunctional aptamer-based DNA nanoassembly for targeted cancer therapy. *J. Am. Chem. Soc.* **2013**, *135*, 18644–18650. [CrossRef] [PubMed]
4. Genix, A.C.; Oberdisse, J. Nanoparticle self-assembly: From interactions in suspension to polymer nanocomposites. *Soft Matt.* **2018**, *14*, 5161–5179. [CrossRef] [PubMed]
5. Mendes, A.C.; Baran, E.T.; Reis, R.L.; Azevedo, H.S. Self-assembly in nature: Using the principles of nature to create complex nanobiomaterials. *WIREs (Wiley Interdiscip. Rev.) Nanomed. Nanobiotechnol.* **2013**, *5*, 582–612. [CrossRef] [PubMed]
6. Yadav, S.; Sharma, A.K.; Kumar, P. Nanoscale self-assembly for therapeutic delivery. *Front. Bioeng. Biotechnol.* **2020**, *8*, 127. [CrossRef] [PubMed]
7. Verwey, E.J.W. Theory of the stability of lyophobic colloids. *J. Phys. Chem.* **1947**, *51*, 631–636. [CrossRef]
8. Birdi, K.S. *Handbook of Surface and Colloid Chemistry*; CRC Press/Taylor & Francis Group: Boca Raton, FL, USA, 2008. [CrossRef]
9. Jose, N.A.; Zeng, H.C.; Lapkin, A.A. Hydrodynamic assembly of two-dimensional layered double hydroxide nanostructures. *Nat. Comm.* **2018**, *9*, 1–12. [CrossRef]
10. French, R.H.; Parsegian, V.A.; Podgornik, R.; Rajter, R.F.; Jagota, A.; Luo, J.; Asthagiri, D.; Chaudhury, M.K.; Chiang, Y.M.; Granick, S.; et al. Long range interactions in nanoscale science. *Rev. Mod. Phys.* **2010**, *82*, 1887–1944. [CrossRef]
11. Trokhymchuk, A.; Henderson, D. Depletion forces in bulk and in confined domains: From Asakura–Oosawa to recent statistical physics advances. *Curr. Opin. Colloid Interface Sci.* **2015**, *20*, 32–38. [CrossRef]
12. Hamaker, H.C. The London—van der Waals attraction between spherical particles. *Physica* **1937**, *4*, 1058–1072. [CrossRef]
13. Vinogradov, E.A.; Dorofeev, I.A. Thermally stimulated electromagnetic fields of solids. *Phys.-Usp.* **2009**, *52*, 425–459. [CrossRef]
14. Lifshitz, E.M. The theory of molecular attractive forces between solids. *Sov. Phys. JETP* **1956**, *2*, 73–83. Available onine: http://jetp.ras.ru/cgi-bin/e/index/e/2/1/p73?a=list (accessed on 1 March 2023).
15. Dzyaloshinskii, I.E.; Lifshitz, E.M.; Pitaevskii, L.P. General theory of Van der Waals' forces. *Sov. Phys. Usp.* **1961**, *4*, 153–176. [CrossRef]
16. Young, K.L.; Jones, M.R.; Zhang, J.; Macfarlane, R.J.; Esquivel-Sirvent, R.; Nap, R.J.; Wu, J.; Schatz, G.C.; Lee, B.; Mirkin, C.A. Assembly of reconfigurable one-dimensional colloidal superlattices due to a synergy of fundamental nanoscale forces. *Proc. Nat. Acad. Sci. USA* **2012**, *109*, 2240–2245. [CrossRef]
17. Munkhbat, B.; Canales, A.; Küçüköz, B.; Baranov, D.G.; Shegai, T.O. Tunable self-assembled Casimir microcavities and polaritons. *Nature* **2021**, *597*, 214–219. [CrossRef] [PubMed]
18. Esteso, V.; Carretero-Palacios, S.; Míguez, H. Casimir–Lifshitz force based optical resonators. *J. Phys. Chem. Lett.* **2019**, *10*, 5856–5860. [CrossRef]
19. Tadmor, R. The London-van der Waals interaction energy between objects of various geometries. *J. Phys. Cond. Matt.* **2001**, *13*, L195–L202. [CrossRef]
20. Dantchev, D.; Valchev, G. Surface integration approach: A new technique for evaluating geometry dependent forces between objects of various geometry and a plate. *J. Colloid Interface Sci.* **2012**, *372*, 148–163. [CrossRef]
21. Rusanov, A.I.; Brodskaya, E.N. Dispersion forces in nanoscience. *Russ. Chem. Rev.* **2019**, *88*, 837–874. [CrossRef]
22. Gao, B.; Arya, G.; Tao, A.R. Self-orienting nanocubes for the assembly of plasmonic nanojunctions. *Nat. Nanotech.* **2012**, *7*, 433–437. [CrossRef]
23. Langbein, D. Normal modes at small cubes and rectangular particles. *J. Phys. A Math. Gen.* **1976**, *9*, 627–644. [CrossRef]
24. Parsegian, V.A. *Van der Waals Forces. A Handbook for Biologists, Chemists, Engineers, and Physicists*; Cambridge University Press: New York, NY, USA, 2005. [CrossRef]
25. De Rocco, A.G.; Hoover, W.G. On the interaction of colloidal particles. *Proc. Nat. Acad. Sci. USA* **1960**, *46*, 1057–1065. [CrossRef] [PubMed]
26. French, R.H. Origins and applications of London dispersion forces and Hamaker constants in ceramics. *J. Am. Ceram. Soc.* **2000**, *83*, 2117–2146. [CrossRef]
27. Pinto, F. Computational considerations in the calculation of the Casimir force between multilayered systems. *Int. J. Mod. Phys. A* **2004**, *19*, 4069–4084. [CrossRef]

28. Esquivel-Sirvent, R.; Svetovoy, V. Nonlocal thin films in calculations of the Casimir force. *Phys. Rev.* **2005**, *72*, 045443. [CrossRef]
29. Mochán, W.L.; Villarreal, C.; Esquivel-Sirvent, R. On Casimir forces for media with arbitrary dielectric properties. *Rev. Mex. Fis.* **2002**, *48*, 339–342. Available online: https://www.scielo.org.mx/scielo.php?script=sci_arttext&pid=S0035-001X2002000400010 (accessed on 1 March 2023).
30. Roth, C.M.; Lenhoff, A.M. Improved parametric representation of water dielectric data for Lifshitz theory calculations. *J. Colloid Interface Sci.* **1996**, *179*, 637–639. [CrossRef]
31. Fiedler, J.; Walter, M.; Buhmann, S.Y. Effective screening of medium-assisted van der Waals interactions between embedded particles. *J. Chem. Phys.* **2021**, *154*, 104102. [CrossRef]
32. Nunes, R.O.; Spreng, B.; de Melo e Souza, R.; Ingold, G.L.; Maia Neto, P.A.; Rosa, F.S.S. The Casimir interaction between spheres immersed in electrolytes. *Universe* **2021**, *7*, 156. [CrossRef]
33. Pinchuk, A.O. Size-dependent Hamaker constant for silver nanoparticles. *J. Phys. Chem. C* **2012**, *116*, 20099–20102. [CrossRef]
34. Esquivel-Sirvent, R.; Schatz, G.C. Spatial nonlocality in the calculation of Hamaker coefficients. *J. Phys. Chem. C* **2012**, *116*, 420–424. [CrossRef]
35. Reid, M.H.; Rodriguez, A.W.; White, J.; Johnson, S.G. Efficient computation of Casimir interactions between arbitrary 3D objects. *Phys. Rev. Lett.* **2009**, *103*, 040401. [CrossRef] [PubMed]

Disclaimer/Publisher's Note: The statements, opinions and data contained in all publications are solely those of the individual author(s) and contributor(s) and not of MDPI and/or the editor(s). MDPI and/or the editor(s) disclaim responsibility for any injury to people or property resulting from any ideas, methods, instructions or products referred to in the content.

Article

The Asymmetric Dynamical Casimir Effect

Matthew J. Gorban [1,2,*], William D. Julius [1,2], Patrick M. Brown [1,2], Jacob A. Matulevich [1,2] and Gerald B. Cleaver [1,2]

1. Early Universe, Cosmology and Strings (EUCOS) Group, Center for Astrophysics, Space Physics and Engineering Research (CASPER), Baylor University, Waco, TX 76798, USA
2. Department of Physics, Baylor University, Waco, TX 76798, USA
* Correspondence: matthew_gorban1@baylor.edu

Abstract: A mirror with time-dependent boundary conditions will interact with the quantum vacuum to produce real particles via a phenomenon called the dynamical Casimir effect (DCE). When asymmetric boundary conditions are imposed on the fluctuating mirror, the DCE produces an asymmetric spectrum of particles. We call this the asymmetric dynamical Casimir effect (ADCE). Here, we investigate the necessary conditions and general structure of the ADCE through both a waves-based and a particles-based perspective. We review the current state of the ADCE literature and expand upon previous studies to generate new asymmetric solutions. The physical consequences of the ADCE are examined, as the imbalance of particles produced must be balanced with the subsequent motion of the mirror. The transfer of momentum from the vacuum to macroscopic objects is discussed.

Keywords: quantum vacuum; vacuum fluctuation; dynamical Casimir effect; asymmetry; asymmetric excitations; asymmetric dynamical Casimir effect

Citation: Gorban, M.J.; Julius, W.D.; Brown, P.M.; Matulevich, J.A.; Cleaver, G.B. The Asymmetric Dynamical Casimir Effect. *Physics* 2023, 5, 398–423. https://doi.org/10.3390/physics5020029

Received: 24 January 2023
Revised: 14 February 2023
Accepted: 13 March 2023
Published: 11 April 2023
Corrected: 15 March 2024

Copyright: © 2023 by the authors. Licensee MDPI, Basel, Switzerland. This article is an open access article distributed under the terms and conditions of the Creative Commons Attribution (CC BY) license (https://creativecommons.org/licenses/by/4.0/).

1. Introduction

In 1948, Hendrik Casimir introduced the notion that the macroscopic boundaries of enclosed cavities impose strict limitations on the quantum vacuum and restrict fundamental vacuum modes of the background free scalar field [1]. The physical interaction between the quantum mechanical vacuum and surfaces with various geometries and boundary conditions (or physically, the properties of the materials constituting that surface) is known as the Casimir effect. This is commonly referred to as a physical manifestation of the quantum vacuum [2–8]. Perhaps one of the most remarkable consequences of modern quantum theory is the extension of this phenomenon into the case of an open cavity with time-varying boundary conditions. When this occurs, the coupling between vacuum quantum fields and time-dependent boundaries results in particle production from the quantum vacuum. This was first introduced in Gerald T. Moore's 1969 doctoral thesis [9], in which he demonstrated that a moving cavity in one dimension produces nonzero energy photonic modes from the initial vacuum state. Over the following decade, this phenomenon would be more thoroughly examined by many others, including additional studies by DeWitt [10] and Fulling and Davies [11,12], although it was not until 1989 that the now commonplace name dynamical Casimir effect (DCE) was first introduced [13].

There is now an abundance of literature on the DCE; see [14–16] for several detailed reviews of this topic. In these, the following definition of the DCE is given: "a macroscopic phenomena caused by changes of vacuum quantum states of fields due to fast time variations of positions (or properties; e.g., plasma frequency or conductivity) of boundaries confining the fields (or other parameters)" [16]. Most notably, the DCE will result in the generation of quanta (photons) of the electromagnetic field directly due to the time-dependent interaction of a macroscopic process with the quantum vacuum.

While the DCE has also been investigated in various three-dimensional configurations, such as cylindrical waveguides [17], parallel plates [18], and spherical [19], cylindrical [20], and rectangular cavities [21,22], we focus our attention on a (spatially) one-dimensional model. Specifically, it is a (1+1)D (dimensional) spacetime permeated by a massless scalar quantum field in the presence of a point mirror with certain optical properties. The (1+1)D model provides an excellent proving ground by which the underlying fundamental physics can be explored. This lets us directly examine the effects of altering the properties and configurations of the mirrors and allows for the analysis of the general nature of the type of time fluctuations needed to induce particle production from the vacuum [22–30]. We avoid using a perfectly reflective mirror [9] as it produces an undesirable result: the renormalized energy can be negative when the mirror starts moving [11,31]. With this in mind, we are interested in the specific case of a partially reflective mirror, which has positive definite (renormalized) radiative energy [31,32]. For a review of the physics of partially reflective mirrors, see [33–42].

Our particular method of modeling a partially reflective mirror uses the established $\delta - \delta'$ potential [35,39,41,43]. When constructing a $\delta - \delta'$ mirror (here δ' is the spatial derivative of the Dirac δ) [44,45], spatial asymmetry is built in, causing the quantum vacuum to act unequally on either side of the mirror. Moving $\delta - \delta'$ mirrors [46] and $\delta - \delta'$ mirrors with time-dependent boundary conditions [47,48] all lead to the creation of an asymmetric distribution of particles due to the unequal vacuum interactions with either side of the mirror. Specifically, this is due to the combination of broken spatial symmetry and fast time fluctuations of the positions or properties of the mirror. We call this phenomenon the asymmetric dynamical Casimir effect (ADCE).

This paper sets out to review the relevant literature on this topic and to put forward a complete analysis of the necessary general conditions to generate this asymmetry, expanding on previous analyses of the $\delta - \delta'$ mirror. We compare several different models and examine the similarities between them to formulate a general approach to producing the ADCE. Specifically, we show that both the scattering-based approach for an asymmetric mirror in (1+1)D and the quantum-particles-based approach, in which we build in asymmetry into a known DCE solution via an asymmetric Bogoliubov transformation, both lead to remarkably similar asymmetric particle distributions. Lastly, we discuss some physical consequences of the ADCE. Specifically, that an asymmetric production of particles results in net motional forces on previously stationary objects.

Natural units are used throughout this paper, with $c = \hbar = 1$, where c denotes the speed of light and $\hbar$ is the reduced Planck constant. Here, we occasionally make use of the Einstein summation notation, where Greek indices run over time and 1D space coordinate pair, $\{t, x\}$. We normalize the Fourier transform following the wave propagation convention, keeping a $1/2\pi$ factor on the forward transform. We note that some of the literature cited here utilize other conventions and so caution is warranted when utilizing these transforms.

2. Scattering Approach for Mirror in 1+1 Vacuum

Here, we review the scattering framework used to analyze the effect of mirrors on quantum scalar fields [46]. We start with a massless scalar field, which we take to initially be interacting with a (partially reflecting, possibly time-varying) mirror. Since this mirror's position is allowed to vary in time, one must exercise caution when introducing coordinates. If the mirror is not moving relative to the laboratory frame, the laboratory and co-moving coordinates are identical, so one is safe to not distinguish them. In the case of moving mirrors, we introduce all of our formalism and fields in a frame co-moving with the mirror, then transform back to a laboratory frame when calculating physical quantities of interest. In this case, we denote the co-moving time coordinates with primes and the laboratory frame coordinates without primes. In the limited cases where we must work with moving objects in the frequency domain, we prime the functions themselves, so as to not confuse them with Green's function parameters.

The massless scalar field, $\phi(t,x)$, is a solution to the Klein–Gordon equation,

$$\left[\partial_t^2 - \partial_x^2 + 2U(t,x)\right]\phi(t,x) = 0, \tag{1}$$

where $U(t,x)$ is some general potential modeling a mirror with various properties and ∂_α denotes the partial derivative with respect to α. This has the corresponding Lagrangian,

$$\mathscr{L} = \mathscr{L}_0 - U(t,x)\phi^2(t,x), \tag{2}$$

where $\mathscr{L}_0$ is the (1+1)D scalar Lagrangian,

$$\mathscr{L}_0 = \frac{1}{2}\left[(\partial_t\phi(t,x))^2 - (\partial_x\phi(t,x))^2\right]. \tag{3}$$

The corresponding Euler–Lagrange equation is

$$\frac{\partial\mathscr{L}}{\partial\phi} - \partial_\nu\left(\frac{\partial\mathscr{L}}{\partial(\partial_\nu\phi)}\right) = 0. \tag{4}$$

The fields resulting from these equations may be decomposed as

$$\phi(t,x) = \Theta(x)\phi_+(t,x) + \Theta(-x)\phi_-(t,x), \tag{5}$$

where $\Theta(x)$ is the Heaviside step-function and $\phi_\pm$ is the solution on either side of the mirror. Since both of $\phi_\pm$ obey the Klein–Gordon equation individually, they can be represented by the sum of two freely counterpropagating fields in the frequency domain,

$$\phi_+(t,x) = \int \frac{dw}{\sqrt{2\pi}}\left[\varphi_{\text{out}}(\omega)e^{iwx} + \psi_{\text{in}}(\omega)e^{-iwx}\right]e^{-iwt} \tag{6}$$

and

$$\phi_-(t,x) = \int \frac{dw}{\sqrt{2\pi}}\left[\varphi_{\text{in}}(\omega)e^{iwx} + \psi_{\text{out}}(\omega)e^{-iwx}\right]e^{-iwt}, \tag{7}$$

where the amplitudes of the incoming and outgoing fields are labeled accordingly, and ω denotes the frequency.

The incoming fields, φ_{in} and ψ_{in}, are unaffected by the mirror and take the form

$$\varphi_{\text{in}}(\omega) = (2|w|)^{-1/2}\left[\Theta(\omega)a_L(\omega) + \Theta(-w)a_R^\dagger(-w)\right] \tag{8}$$

and

$$\psi_{\text{in}}(\omega) = (2|w|)^{-1/2}\left[\Theta(\omega)a_R(\omega) + \Theta(-w)a_L^\dagger(-w)\right], \tag{9}$$

where $a_j(\omega)$ and $a_j^\dagger(\omega)$ ($j = L, R$) are the annihilation and creation operators for the left (L) and right (R) sides of the mirror, which obey the commutation relation

$$[a_i(\omega), a_j^\dagger(\omega')] = \delta(\omega - \omega')\delta_{ij}, \tag{10}$$

where δ_{ij} is the Kronecker delta.

The ingoing and outgoing counterpropagating fields may be related using a scattering matrix with possibly frequency dependent reflection ($r_\pm(\omega)$) and transmission ($s_\pm(\omega)$) coefficients. In this case, the scattering matrix is

$$S(\omega) = \begin{pmatrix} s_+(\omega) & r_+(\omega) \\ r_-(\omega) & s_-(\omega) \end{pmatrix}, \tag{11}$$

with

$$\Phi_{\text{out}}(\omega) = S(\omega)\Phi_{\text{in}}. \tag{12}$$

Here, we are making use of the vectorized shorthand

$$\Phi_{\text{in}}(\omega) = \begin{pmatrix} \varphi_{\text{in}}(\omega) \\ \psi_{\text{in}}(\omega) \end{pmatrix} \text{ and } \Phi_{\text{out}}(\omega) = \begin{pmatrix} \varphi_{\text{out}}(\omega) \\ \psi_{\text{out}}(\omega) \end{pmatrix} \tag{13}$$

to represent ingoing and outgoing counterpropagating fields. In any situation where $\Phi(\omega)$ is used without a subscript, it can be assumed that the given relation holds for both ingoing and outgoing fields. The S-matrix is required to be unitary and causally consistent. For a complete analysis of the properties of the S-matrix, see [46,49,50]. Calculating the reflection and transmission coefficients determines the scattering system and completely defines the relationship between incoming and outgoing fields interacting with the mirror.

To solve for the components of the S-matrix, matching conditions between incoming and outgoing fields must be calculated. This gives a system of equations, which can be solved to obtain the reflection and transmission coefficients [44,51–53]. These matching conditions are found by minimizing a variation on the action, which is to say, the resulting system of equations is equivalent to solving the above Euler–Lagrange equation.

2.1. The Static Asymmetric $\delta - \delta'$ Mirror

The first step in adding in the necessary asymmetry needed to produce the ADCE is to introduce an asymmetric $\delta - \delta'$ potential,

$$U(x) = \mu\delta(x) + \lambda\delta'(x), \tag{14}$$

into the Lagrangian, where μ is related to the plasma frequency of the mirror and λ is a dimensionless factor. This potential models a partially reflective mirror [44,46,47]. The Lagrangian in this case becomes

$$\mathscr{L} = \mathscr{L}_0 - [\mu\delta(x) + \lambda\delta'(x)]\phi^2(t, x). \tag{15}$$

This potential results in the Klein–Gordon Equation (1), taking the form

$$\left[\partial_t^2 - \partial_x^2 + 2\mu\delta(x) + 2\lambda\delta'(x)\right]\phi(t, x) = 0. \tag{16}$$

In the frequency domain, this becomes

$$\left[-\partial_x^2 + 2\mu\delta(x) + 2\lambda\delta'(x)\right]\Phi(\omega, x) = \omega^2\Phi(\omega, x), \tag{17}$$

which can be used to find the matching conditions [46],

$$\Phi(w, 0^+) = \frac{1+\lambda}{1-\lambda}\Phi(w, 0^-) \tag{18}$$

and

$$\partial_x\Phi(w, 0^+) = \frac{1-\lambda}{1+\lambda}\partial_x\Phi(w, 0^-) + \frac{2\mu}{1-\lambda^2}\Phi_-(w, 0^-). \tag{19}$$

These matching conditions govern the relationship between $\Phi_\pm$, which can be written in terms of the reflection and transmission coefficients,

$$\Phi_+(\omega, x) = s_-(\omega)e^{-i\omega x}\Theta(-x) + (e^{-i\omega x} + r_-(\omega)e^{i\omega x})\Theta(x) \tag{20}$$

and

$$\Phi_-(\omega, x) = (e^{i\omega x} + r_+(\omega)e^{-i\omega x})\Theta(-x) + s_+(\omega)e^{i\omega x}\Theta(x). \tag{21}$$

Applying the matching conditions, the explicit forms for the components of the scattering matrix are

$$r_\pm(\omega) = \frac{-i\mu_0 \pm 2w\lambda_0}{i\mu_0 + w(1+\lambda_0^2)} \tag{22}$$

and
$$s_\pm(\omega) = \frac{w(1-\lambda_0^2)}{i\mu_0 + w(1+\lambda_0^2)},\tag{23}$$

where we now include the notations μ_0 and λ_0 to explicitly denote these as the zeroth-order terms. This distinction becomes important as we start to include perturbative effects below. The inequality between $r_+(\omega) \neq r_-(\omega)$ is due to the underlying asymmetry of the potential itself, i.e., it is a direct consequence of the δ' term.

Note that, when $\lambda_0 = 1$, the mirror is perfectly reflective and the left and right sides now possess Dirichlet and Robin boundary conditions, respectively. Additionally, the change $\lambda_0 \to -\lambda_0$ will swap these properties from one side of the mirror to the other.

2.2. The Time-Varying Asymmetric Mirror

2.2.1. Particle Creation from Fluctuations in Boundary Conditions

Here, we make a digression to address the mechanism for particle creation resulting from fluctuating boundary conditions. Thus far, we have not worried about such effects as it can easily be shown that it is necessary to introduce *time fluctuations* to generate particle production. Recall from Equation (12) that $\Phi_{\text{out}}(\omega) = S(\omega)\Phi_{\text{in}}$. Then, knowing $\Phi_{\text{out}}(\omega)$ allows for the computation of the spectrum of created particles as the spectral distribution of created particles is given by [37]

$$N(\omega) = 2\omega\, \text{Tr}\left[\langle 0_{\text{in}}|\Phi_{\text{out}}(-\omega)\Phi_{\text{out}}^{\text{T}}(\omega)|0_{\text{in}}\rangle\right],\tag{24}$$

where $\text{Tr}[M]$ denoted the trace of a some matrix M, and the number of created particles is

$$\mathcal{N} = \int_0^\infty d\omega\, N(\omega).\tag{25}$$

From Equation (24), one can see that, regardless of the asymmetry in $S(\omega)$, there are no zeroth-order contributions to particle creation. Thus, it is necessary to introduce some perturbation in time as the mechanism to cause particle production. One also sees that spatial asymmetry leads to asymmetry in the spectrum of created particles.

We quantify this asymmetry by splitting both the spectral distribution and total number of particles into their right (+) and left (−) components as

$$N(\omega) = N_+(\omega) + N_-(\omega)\tag{26}$$

and

$$\mathcal{N} = \mathcal{N}_+ + \mathcal{N}_-,\tag{27}$$

respectively. One can then make use of the quantities $N_\mp/N_\pm$, $\mathcal{N}_\mp/\mathcal{N}_\pm$, and $\Delta N = N_- - N_+$ as a means of comparing and quantifying the asymmetry between the two sides of the mirror. We refer to the quantities $N_\mp/N_\pm$, $\mathcal{N}_\mp/\mathcal{N}_\pm$, and ΔN as the spectral ratio, particle creation ratio, and spectral difference, respectively. Specifically, these quantities are useful in evaluating and understanding the functional form of the asymmetry present in the system. In particular, ΔN (and subsequently $\Delta \mathcal{N}$) can be used to calculate potential energy fluxes and force differentials that will play a part in the dynamics of the system. More on this point is discussed in Section 5. When the mirror no longer exhibits asymmetric interactions with the vacuum the ratios become unity and the difference vanishes.

Demonstrating and observing these physical quantities is an active area of research for experimentalists in search of better tools to understand and quantify the real-world limitations of the theory. While there have been experimental proposals of mechanically induced DCE [54–59], there are many difficulties to overcome in the creation of a physically realizable high-frequency mechanically oscillating mirror [16,60,61]. This issue has led to the proposal of alternate methods for observing the DCE [13,54,60,62–69] and experimental evidence supports the real production of particles from time-varying materials

[70–72]. Most notably, the first experimental DCE detection used a superconducting circuit whose electrical length is changed by modulating the inductance of a superconducting quantum interference device (SQUID) at high frequencies [61]. These experiments can be effectively modeled with a time-dependent $\mu(t)$ in a single δ mirror, with the entire mirror's properties varying in time. This was a motivating factor for the investigation into time-dependent material properties in the $\delta - \delta'$ mirror, specifically $\mu(t)$ [47,48], which we review in Section 2.2.3. In addition to this solution, it is also convenient to model a $\delta - \delta'$ mirror with perturbative fluctuations on λ, the scale factor attached to the δ' term that determines the magnitude of asymmetry. This is akin to altering the surface structure of the material, as opposed to effectively changing the bulk material properties with the time-varying μ. This solution provides a potentially better model for real-world applications and experimental setups. For example, Mott insulators that undergo metal–insulator transitions can have their surface properties change on picosecond timescales with a multiple-order magnitude change in surface conductivity [73,74]. Experimentally, this can be performed through the use of ultrahigh-frequency pulsed lasers to alter the surface structure on incredibly short timescales [75–78].

2.2.2. Fluctuations in Position: $q(t)$

One of the standard methods for inducing time fluctuations to generate the DCE is to have the position of a mirror change in time. From [46], there is a moving asymmetric $\delta - \delta'$ mirror, whose position is given by $x = q(t)$ in the laboratory frame. The movement is taken to be nonrelativistic ($|\dot{q}(t)| \ll 1$) and limited by a small value ϵ, such that $q(t) = \epsilon g(t)$ with $|g(t)| \leq 1$. Scattering is assumed to be

$$\Phi'_{\text{out}}(\omega) = S(\omega)\Phi'_{\text{in}}(\omega) \tag{28}$$

in the co-moving frame where the mirror is instantaneously at rest (tangential frames). To solve this in the laboratory frame, we use the relation

$$\Phi'(t',0) = \Phi(t,\epsilon g(t)) = [1 - \epsilon g(t)\eta \partial_t]\Phi(t,0) + \mathcal{O}(\epsilon^2), \tag{29}$$

where

$$\Phi(t,x) = \begin{pmatrix} \tilde{\varphi}(t-x) \\ \tilde{\psi}(t+x) \end{pmatrix}. \tag{30}$$

Here, $\tilde{\varphi}$ and $\tilde{\psi}$ are components of the field in the temporal domain and $\eta = \text{diag}(1,-1)$. Taking advantage of the fact that $dt = dt'$ to the second order, Equation (29) can be rewritten as

$$\Phi'(t,0) = [1 - \epsilon g(t)\eta \partial_t]\Phi(t,0). \tag{31}$$

One finds that applying this transform to Equation (28) in the frequency domain yields

$$\Phi_{\text{out}}(\omega) = S_0(\omega)\Phi_{\text{in}}(\omega) + \epsilon \int \frac{d\omega'}{2\pi} \delta S_q(\omega,\omega')\Phi_{\text{in}}(\omega'), \tag{32}$$

where we suppress the evaluation of $x = 0$ in $\Phi(\omega,0)$ going forward for compactness. One also has:

$$\delta S_q(\omega,\omega') = i\omega' \mathcal{G}(\omega - \omega')[S_0(\omega)\eta - \eta S_0(\omega')], \tag{33}$$

with $\mathcal{G}(\omega)$ being the Fourier transform of $g(t)$. We refer to δS_q as the delta-S matrix, a perturbative term that arises from the first-order perturbation in Equation (29) due to the time-varying fluctuations of the mirror's position. This term is of particular physical importance, as it carries the asymmetry that will result in the asymmetric production of particles on each side of the mirror.

Due to the introduction of the small deviation in mirror position $g(t)$, a first-order term emerges that will give rise to particle production. As it is shown below, the introduction of the $\delta - \delta'$ potential leads to an asymmetric production of particles about the two sides of the mirror.

We now prescribe a specific form to the motion,

$$g(t) = \cos(\omega_0 t)\exp(-|t|/\tau), \tag{34}$$

where τ is the effective oscillation lifetime and ω_0 is the characteristic frequency of oscillation. Only the monochromatic limit is considered, with $\omega_0 \tau \gg 1$. In this limit, the system undergoes (effectively) spatially symmetric motion about its starting position. The Fourier transform of Equation (34) is approximately

$$\frac{|\mathcal{G}(\omega)|^2}{\tau} \approx \frac{\pi}{2}[\delta(\omega + \omega_0) + \delta(\omega - \omega_0)]. \tag{35}$$

One can also obtain the right and left spectral distributions as

$$\frac{N_\pm(\omega)}{\tau} = \frac{\epsilon^2}{\pi}\omega(\omega_0 - \omega)\Lambda_\pm(\omega, \omega_0 - \omega)\Theta(\omega_0 - \omega), \tag{36}$$

where the asymmetry in the distribution of particles of the two sides can be seen in

$$\Lambda_\pm(\omega, \omega - \omega_0) = \frac{1}{4}\operatorname{Re}\left[\frac{8\lambda_0^2 \omega(\omega_0 - \omega) - 2\mu_0^2 + i\mu_0 \omega_0 (1 \mp \lambda_0)^2}{(i\mu_0 + \omega(1 + \lambda_0^2))[i\mu_0 + (\omega_0 - \omega)(1 + \lambda_0^2)]}\right]. \tag{37}$$

A change from $\lambda_0 \to -\lambda_0$ flips $\Lambda_\pm \to \Lambda_\mp$ and therefore also flips $N_\pm \to N_\mp$ [35,36].

For a detailed analysis of the spectrum of particles created and the interplay between different combinations of μ_0 and λ_0, see [46]. Highlighting a few key points, one can see that setting $\lambda_0 = 1$ produces the largest difference in magnitude between the spectra emitted by the two sides with a spectral ratio of

$$\frac{N_-}{N_+} = \frac{[\mu_0^2 + 4\omega(\omega_0 - \omega)]^2}{(\mu_0^2 + 4\omega^2)[\mu_0^2 + 4(\omega_0 - \omega)^2]}. \tag{38}$$

Additionally, when $\lambda_0 = 1$, $\Lambda_- \approx 1/2$, which corresponds to a Dirichlet spectrum. The maximum spectral difference occurs when $\mu_0/\omega_0 \approx 1$, where the mirror imposes perfectly reflecting Dirichlet and maximally suppressed Robin conditions on the field about the left and right sides of the mirror, respectively. The Robin side exhibits strong suppression at this point, corresponding to a value of $\gamma_0 \omega_0 \approx 2.2$, where γ_0 is the Robin parameter, $\gamma_0 = 2/\mu_0$ [79–81]. The vast majority of the particles are produced on the left side of the mirror. When $\lambda_0 = 0$, the asymmetry vanishes and the results simplify to those of a δ mirror [35,36]. This spectrum increases monotonically with μ_0. As $\mu_0 \to \infty$, the spectrum asymptotically approaches a Dirichlet spectrum.

The spectral difference for the moving $\delta - \delta'$ mirror obeying the oscillation function (34), shown in Figure 1, becomes

$$\frac{\Delta N}{\tau} = \frac{\epsilon^2}{\pi}\lambda_0 \omega_0^2 (1 + \lambda_0^2) Y(\omega) Y(\omega_0 - \omega)\Theta(\omega_0 - \omega), \tag{39}$$

with

$$Y(\omega) = \frac{\mu_0 \omega}{\mu_0^2 + \omega^2(1 + \lambda_0^2)^2}, \tag{40}$$

which again indicates that more particles are produced on the left side of the mirror ($\lambda_0 > 0$).

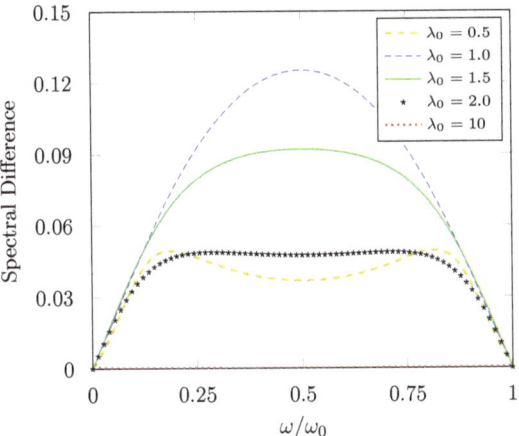

Figure 1. The plot of $(\epsilon^2 \tau/\pi)^{-1} \times \Delta N$, the difference between the spectral distributions of particles created on the two sides of a $\delta - \delta'$ mirror as a function of ω/ω_0 for different values of λ_0, with $\mu_0 = 1$. See text for details.

2.2.3. Fluctuations in Properties: $\mu(t)$

As discussed in Section 2.2.1 above, while it is theoretically possible to oscillate a thin mirror at high frequencies, current technological limitations prevent this from being experimentally viable. Thankfully, the oscillation of a boundary's position is not the only option for introducing time dependence in surface interactions. In the $\delta - \delta'$ model, it is possible to modify the fundamental properties of the mirror [47,48]. Now, we are interested in modifying the plasma frequency (through the modification of μ), such that $\mu \to \mu(t) = \mu_0[1 + \epsilon f(t)]$, where $\mu_0 \geq 1$ is a constant and $f(t)$ is an arbitrary function with $|f(t)| \leq 1$ and $\epsilon \ll 1$. As done in Section 2.1 when deriving the matching conditions in Equations (18) and (19), it is convenient to work in the frequency domain, where the derivative matching condition term (19) now becomes

$$\partial_x \Phi(\omega, 0^+) = \frac{1 - \lambda_0}{1 + \lambda_0} \partial_x \Phi(\omega, 0^-) + \frac{2}{1 - \lambda_0^2} \int \frac{d\omega'}{2\pi} \mathcal{M}(\omega - \omega') \Phi_-(\omega', 0^-), \quad (41)$$

where

$$\mathcal{M}(\omega) = \mu_0 (\delta(\omega) + \epsilon \mathcal{F}(\omega)) \quad (42)$$

is the Fourier transform of $\mu(t)$ and $\mathcal{F}(\omega)$ is the Fourier transform of $f(t)$. The matching conditions now contain perturbative terms that modify the S-matrix.

To the first order, the final form of $\Phi_{\text{out}}(\omega) = S(\omega)\Phi_{\text{in}}$ becomes

$$\Phi_{\text{out}}(\omega) = S_0(\omega)\Phi_{\text{in}}(\omega) + \int \frac{d\omega'}{2\pi} \delta S_\mu(\omega, \omega') \Phi_{\text{in}}(\omega'), \quad (43)$$

where S_0 is the zeroth-order scattering matrix found from Equations (22) and (23). The asymmetric correction that originates from the introduction of $f(t)$ takes the form $\delta S_\mu(\omega, \omega') = \epsilon \alpha(\omega, \omega') \mathbb{S}_\mu(\omega')$, where

$$\alpha_\mu(\omega, \omega') = -\frac{i\mu_0 \mathcal{F}(\omega - \omega')}{i\mu_0 + \omega(1 + \lambda_0^2)} \quad (44)$$

with

$$\mathbb{S}_\mu(\omega') = \begin{pmatrix} s_+(\omega') & 1 + r_+(\omega') \\ 1 + r_-(\omega') & s_-(\omega') \end{pmatrix}. \quad (45)$$

Using Equation (43), the spectrum of particles (24) can be calculated. The left and right components of this spectrum are

$$N_\pm(\omega) = \frac{\epsilon^2}{2\pi^2}(1 \pm \lambda_0)^2(1+\lambda_0^2)\int_0^\infty d\omega'\, n(\omega,\omega') + \mathcal{O}(\epsilon^2), \tag{46}$$

where

$$n(\omega,\omega') = Y(\omega)Y(\omega')|\mathcal{F}(\omega+\omega')|^2. \tag{47}$$

The spectral distribution ratio and particle creation ratio are

$$\frac{N_-}{N_+} = \frac{\mathcal{N}_-}{\mathcal{N}_+} = \left(\frac{1-\lambda_0}{1+\lambda_0}\right)^2. \tag{48}$$

Thus, one sees a constant, frequently independent difference between the spectrum of particles created when the asymmetric mirror with time-dependent properties interacts with the vacuum.

For positive (negative) values of λ_0, the right (left) side has a greater production of particles. When $\lambda_0 = \pm 1$, only one side of the mirror experiences the creation of particles. The asymmetry vanishes when $\lambda_0 = 0$ as expected, once again highlighting the necessary combination of spatial and temporal perturbations needed to produce the ADCE.

When $f(t)$ takes the form (34), the spectral distribution becomes

$$\frac{N_\pm}{\tau} = \frac{\epsilon^2}{4\pi}(1\pm\lambda_0)^2(1+\lambda_0^2)Y(\omega)Y(\omega_0-\omega)\Theta(\omega_0-\omega), \tag{49}$$

and the spectal difference between these two sides is now

$$\frac{\Delta N}{\tau} = -\frac{\epsilon^2}{\pi}\lambda_0(1+\lambda_0^2)Y(\omega)Y(\omega_0-\omega)\Theta(\omega_0-\omega). \tag{50}$$

This is, remarkably, identical to the the spectral difference of the moving $\delta - \delta'$ mirror (39) up to an overall minus sign and factor of ω_0^2. This is due to the fact that ΔN removes the symmetric background of the two fields and isolates the purely asymmetric component of the spectrum, which amounts to calculating the difference between $\mathrm{Re}[r_+]$ and $\mathrm{Re}[r_-]$.

More on this is the general form of the scattering is addressed.

2.3. General Form of Asymmetric Scattering

There are apparent similarities between the two given examples of time-dependent $\delta - \delta'$ mirrors; thus, one may propose a general form of asymmetric time-dependent perturbations on objects in (1+1)D that are capable of generating ADCE photons. The mechanism that drives the time-dependent perturbations is arbitrary, but we specify that it is bounded by $|f(t)| \leq 1$ where $f(t)$ is some (usually, but not necessarily, periodic) driving function of the fluctuation. There must also be some spatial delineation that manifests in the boundary conditions to produce the asymmetry on opposite sides of the object. This asymmetry will show itself in the transmission and reflection coefficients of the S-matrix, where either $r_+(\omega) \neq r_-(\omega)$ or $s_+(\omega) \neq s_-(\omega)$. Starting as before, we seek the first-order perturbative effects on the scattering matrix, governing the relationship between the incoming and outgoing fields interacting with an object $\Phi_{\mathrm{out}}(\omega) = S(\omega)\Phi_{\mathrm{in}}(\omega)$. Fluctuations in time yield

$$\Phi_{\mathrm{out}}(\omega) = S_0(\omega)\Phi_{\mathrm{in}}(\omega) + \epsilon\int \frac{d\omega'}{2\pi}\delta S(\omega,\omega')\Phi_{\mathrm{in}}(\omega') + \mathcal{O}(\epsilon^2), \tag{51}$$

where the S_0 is the zeroth-order, time-independent scattering matrix for the system. Here, the matrix δS takes the form

$$\delta S(\omega,\omega') = \alpha(\omega,\omega')\mathcal{F}(\omega-\omega')\mathbb{S}(\omega,\omega'), \tag{52}$$

where $\mathbb{S}(\omega,\omega')$ and $\alpha(\omega,\omega')$ are the first-order scattering matrix and amplitude, respectively, found by imposing the correct boundary conditions. While both terms can be functions of both ω and ω', this is not necessary, as is quite evident from the analysis on the fluctuations in properties from Section 2.2.3. Additionally, $\mathcal{F}(\omega)$ is the Fourier transform of $f(t)$.

The two examples just above follow this form. The same is true for a system that modifies the mirror's reflectivity by introducing a kinetic term in the $\delta - \delta'$ potential [48]. Adding the term $2\chi_0 \delta(x)(\partial_t \phi(t,x))^2$ to Equation (15), where χ_0 is a constant parameter, and varying the parameter $\mu(t)$ changes the transmission and reflection coefficients (22) and (23) such that $\mu_0 \to \mu_0 - \chi_0 \omega^2$ in the denominator of these terms. Solving for Equation (52) leads to

$$\alpha_\chi(\omega,\omega') = -\frac{i\mu_0}{i\mu_0 - i\chi_0 \omega^2 + \omega(1+\lambda_0^2)} \tag{53}$$

and

$$\mathbb{S}_\chi(\omega') = \begin{pmatrix} s_+(\omega') & 1+r_+(\omega') \\ 1+r_-(\omega') & s_-(\omega') \end{pmatrix}, \tag{54}$$

which are nearly identical to Equations (44) and (45). Additionally, while the spectrum of particles is slightly modified by the addition of the χ_0 term, the spectral ratio between the two sides of the object are the same as Equation (48). In Section 2.4, one again observes perturbations of the form (51) when we investigate what happens when the λ_0 term of the $\delta - \delta'$ mirror fluctuates in time.

To investigate the asymmetry of the particle production, we make use of the following formula:

$$\langle 0_{\rm in} | \Phi_{\rm in}(\omega) \Phi_{\rm in}^{\rm T}(\omega') | 0_{\rm in} \rangle = \frac{\pi}{\omega} \delta(\omega+\omega') \Theta(\omega). \tag{55}$$

The spectral distribution becomes

$$N(\omega) = \frac{1}{2\pi} \int_0^\infty \frac{d\omega'}{2\pi} \frac{\omega}{\omega'} \mathrm{Tr}\left[\delta S(\omega,-\omega') \delta S^\dagger(\omega,-\omega')\right]$$
$$= \frac{\epsilon^2}{2\pi} \int_0^\infty \frac{d\omega'}{2\pi} \frac{\omega}{\omega'} |\alpha(\omega,-\omega')|^2 |\mathcal{F}(\omega+\omega')|^2 \mathrm{Tr}\left[\mathbb{S}(\omega,-\omega')\mathbb{S}^\dagger(\omega,-\omega')\right], \tag{56}$$

which can be integrated over ω to find the total number of particles created, $\mathcal{N}$.

The decomposition of Equation (56) into its left and right pieces is

$$N_\pm(\omega) = \frac{\epsilon^2}{2\pi} \int_0^\infty \frac{d\omega'}{2\pi} \frac{\omega}{\omega'} |\alpha(\omega,-\omega')|^2 |\mathcal{F}(\omega+\omega')|^2 \Lambda_\pm(\omega,-\omega'), \tag{57}$$

where $\Lambda_\pm = \mathrm{Tr}[\mathbb{S}\mathbb{S}^\dagger]_\pm$.

Prescribing Equation (34) to Equation (57), we arrive at the general form of the spectral decomposition when the time fluctuations are in the approximately symmetric monochromatic limit,

$$N_\pm(\omega) = \frac{\epsilon^2}{8\pi} \left(\frac{\omega}{\omega_0 - \omega}\right) |\alpha(\omega,\omega_0-\omega)|^2 \Lambda_\pm(\omega,\omega_0-\omega) \Theta(\omega_0-\omega), \tag{58}$$

where one can see that the ratio N_-/N_+ is equal to Λ_-/Λ_+. The general spectral difference ΔN is now proportional to $\Delta\Lambda$, the difference between Λ_- and Λ_+. The quantity ΔN is useful not only because of its ability to isolate the difference in the asymmetric outputs of the mirror, but also because it corresponds to the physically meaningful quantities and this can manipulate the dynamics of the system. The asymmetry of the mirror is the foundational element that produces asymmetric quantum effects, whereby vacuum excitations give rise to a non-null mean final velocity and cause a stationary object to begin to move [47]. Keeping this in mind, within the framework of the scattering approach we can make some general comments on the form of ΔN when the fluctuation takes the form

$f(t) = \cos(\omega_0 t)\exp(-|t|/\tau)$. Since ΔN is proportional to $\Delta\Lambda$, one can see that its solution originates from the difference between the real part of $\text{Tr}[\mathbb{SS}^\dagger]$, which amounts to calculating the difference between the asymmetric components of the first-order scattering matrix. Specifically, since this matrix can be expressed in terms of the zeroth-order transmission and reflection coefficients, it is really the fundamental asymmetry of the unperturbed S-matrix that carries over into the asymmetry of the first-order fluctuations and thus into ΔN.

The specific form of the S-matrix can be constructed in such a way that its components possess some sort of asymmetry, such as what we have seen thus far with the $\delta - \delta'$ potentials in the Lagrangian. Actually, it is possible to analyze asymmetric systems without a pre-described Lagrangian. As long as the the scattering matrix obeys its necessary conditions [46,49,50], numerous asymmetric objects can be constructed. With the $\delta - \delta'$ mirrors in Sections 2.2.2 and 2.2.3, the asymmetry is present due to the inequality, $r_+ \neq r_-$. Thus, the quantity ΔN for the mirror will be some function of $\text{Re}[r_- - r_+]$. As remarked before in Section 2.2.3, this is the origin of the near equality between ΔN of the two $\delta - \delta'$ mirrors with fluctuations in the position $q(t)$ and the material property $\mu(t)$.

In general, there are three asymmetric forms of the S-matrix:

- when $r_+ \neq r_-$ and $s_+ = s_-$,
- when $s_+ \neq s_-$ and $r_+ = r_-$,
- when both $r_+ \neq r_-$ and $s_+ \neq s_-$.

Thus, ΔN can ultimately be expressed as a function of the following, for the previous forms:

- $\alpha\,\text{Re}[r_- \pm r_+]$,
- $\beta\,\text{Re}[s_- \pm s_+]$,
- $\alpha\,\text{Re}[r_- \pm r_+] + \beta\,\text{Re}[s_- \pm s_+]$,

where α and β are calculable scale factors with functional dependence on variables that define the S-matrix (μ_0 and λ_0 for the $\delta - \delta'$ mirror).

There is an important caveat we must address with regard to general scattering. These similarities only hold when the mechanism driving scattering affects the position or some material property related to the plasma frequency. This is because such mechanisms act by causing the strength of the δ function in the potential to become time-dependent. Such considerations do not extend straightforwardly to allowing the strength of the δ' term, which is addressed in Section 2.4 just below.

2.4. Fluctuations in Properties: $\lambda(t)$

Having already explored the consequences of making μ_0 time-dependent in the $\delta - \delta'$ mirror, we now calculate the effects of taking $\lambda_0 \to \lambda(t) = \lambda_0[1 + \epsilon f(t)]$. Starting with the field equation of the system,

$$\left[\partial_t^2 - \partial_x^2 + 2\mu\delta(x) + 2\lambda(t)\delta'(x)\right]\phi(t,x) = 0, \tag{59}$$

we take the Fourier transform as conducted in Section 2.1. Then, one has;

$$\left[-\partial_x^2 + 2\mu_0\delta(x)\right]\Phi(\omega,x) + 2\int\frac{d\omega'}{2\pi}\mathcal{L}(\omega-\omega')\delta'(x)\Phi(\omega',x) = \omega^2\Phi(\omega',x), \tag{60}$$

where $\mathcal{L}(\omega)$ is the Fourier transform of $\lambda(t)$. Using the same machinery as before, one arrives at the continuity equations needed to solve for the matching conditions,

$$-\partial_x\Phi(\omega,0^+) + \partial_x\Phi(\omega,0^-) + \mu[\Phi(\omega,0^+) + \Phi(\omega,0^-)]$$
$$-\int\frac{d\omega'}{2\pi}\mathcal{L}(\omega-\omega')(\partial_x\Phi(\omega',0^+) + \partial_x\Phi(\omega',0^-)) = 0, \tag{61}$$

and

$$-\Phi(\omega,0^+) + \Phi(\omega,0^-) + \int \frac{d\omega'}{2\pi}\mathcal{L}(\omega-\omega')(\Phi(\omega',0^+) + \Phi(\omega',0^-)) = 0. \quad (62)$$

From these continuity equations, it becomes understandable that unlike the matching conditions in Equations (18) and (19), general matching conditions for $\lambda(t)$ cannot be found using this approach. This is due to the presence of the convolution integral between $\mathcal{L}(\omega-\omega')$ and $\partial_x\Phi(\omega')$ in Equation (61). This convolution ultimately leads to nonlinear mixing of different frequency terms.

To illustrate this difficulty straightforwardly, the form of $f(t)$ used in prior Sections (see Equation (34)) was employed in the continuity equations to investigate the resulting scattering coefficients, assuming the preservation of linearity a priori. The result is that the scattering coefficients in the frequency domain become dependent on $\omega \pm \omega_0$ modes ($s_\pm(\omega \pm \omega_0)$, $r_\pm(\omega \pm \omega_0)$). A detailed derivation of these scattering terms can be seen in Appendix A. To that end, work is currently underway to apply the Bogoliubov approach to this problem; however, those results are reserved for a future paper.

3. Bogoliubov Approach for Mirror in 1+1 Vacuum

In contrast to the waves-based scattering approach of Section 2 whereby the perturbative effects of time fluctuations are present in higher-order terms of the S-matrix, in the particle-based framework the perturbative effects can be calculated by investigating the higher-order terms present in the Bogoliubov transform between the input and output creation and annihilation operators of the field. The scattering approach is convenient when looking at the consequences of adding a potential (i.e., mirror) to a background vacuum field in a Lagrangian (3). However, it is often of interest to understand how the vacuum interacts with mirrors that directly impose specific boundary conditions on the field. The Robin boundary condition (henceforth Robin b.c.) is a suitable example of this, as shown below. This approach allows for specific boundary conditions to be imposed on the underlying field itself without directly knowing or specifying a generating potential.

The particles-based perturbative procedure introduced by Ford [82] has been used extensively to describe the effects of small changes in simple mirror geometries that produce radiative effects. Here, we draw from two separate instances of perturbative corrections on a mirror with Robin boundary conditions: the first incorporates time-dependent changes in properties of the boundary [83] and the second uses a moving boundary with an oscillating position [79]. To illustrate how different manifestations of time-dependent fluctuations produce the same effect, we first review [79,83] side-by-side, deriving the Bogoliubov transformation for the different cases. These Bogoliubov transformations encode the difference between the input/output creation and annihilation operators and provide a parallel way of demonstrating the transformation of the scattering matrix seen in Section 2. Following this, we demonstrate the ability to build in asymmetry to generate ADCE photons from the originally symmetric moving Robin boundary in a similar manner to before.

3.1. Fluctuating Robin Boundary Condition

The Robin b.c. for a mirror in (1+1)D is

$$\gamma_0 \left[\frac{\partial \phi(t,x)}{\partial x} \right]_{x=0} = \phi(t,0), \quad (63)$$

where γ_0 is the parameter that allows for continuous interpolation between Dirichlet ($\gamma_0 \to 0$) and Neumann ($\gamma_0 \to \infty$) boundary conditions. The Robin b.c. is a useful tool for representing phenomenological models that describe penetrable surfaces [84] as the Robin parameter is related to the penetration depth into the metallic boundary by the field. The parameter γ_0^{-1} corresponds to the plasma frequency of the material and γ_0 acts as the plasma wavelength.

FLUCTUATIONS IN POSITION	**FLUCTUATIONS IN PROPERTIES**
For a moving mirror, the Robin b.c. only holds in the co-moving frame, where $\delta q(t)$ is the time-dependent position of the mirror. In the laboratory frame, this equation is $$\gamma_0 \left[\frac{\partial}{\partial x} + \delta\dot{q}(t)\frac{\partial}{\partial t}\right]\phi(t,\delta q(t)) = \phi(t,\delta q(t)), \quad (64)$$ where γ_0 is the zeroth-order time-independent Robin parameter.	A mirror with time-dependent boundary conditions modifies the Robin b.c. with first-order corrections to the Robin parameter, giving $$\left[\gamma_0 + \delta\gamma(t)\right]\frac{\partial\phi}{\partial x}(t,0) = \phi(t,0), \quad (65)$$ where $\delta\gamma(t)$ is a smooth time-dependent function satisfying the condition $|\delta\gamma(t)| \ll \gamma_0$.

Adopting a perturbative approach and following Ford [82], we take $\phi(t,x) = \phi_0(t,x) + \delta\phi(t,x)$, where ϕ_0 is the unperturbed field of a static, time-independent mirror at $x = 0$ and $\delta\phi$ is the small perturbation from the fluctuations on the static boundary.

This is equivalent to expansions in δq and its derivatives to the first order: $$\gamma_0\left[\frac{\partial\delta\phi(t,x)}{\partial x}\right]_{x=0} - \delta\phi(t,0) =$$ $$\delta q(t)\left[\frac{\partial\phi_0}{\partial x}(t,0) - \gamma_0\frac{\partial^2\phi_0}{\partial x^2}(t,0)\right] \quad (66)$$ $$-\delta\dot{q}(t)\gamma_0\frac{\partial\phi_0}{\partial t}(t,0).$$	Using the fact that both ϕ_0 and $\delta\phi$ satisfy the Klein–Gordon equation, we have $$\gamma_0\left[\frac{\partial\delta\phi(t,x)}{\partial x}\right]_{x=0} - \delta\phi(t,0) =$$ $$-\delta\gamma(t)\frac{\partial\phi_0}{\partial x}(t,0). \quad (67)$$

It is now useful to work in the frequency domain; thus, we employ the following Fourier transforms:

$$\Psi(\omega,x) = \int dt\,\psi(t,x)e^{i\omega t}, \quad \delta Q(\omega) = \int dt\,\delta q(t)e^{i\omega t},$$
$$\delta\Phi(\omega,x) = \int dt\,\delta\phi(t,x)e^{i\omega t}, \quad \delta\Gamma(\omega) = \int dt\,\delta\gamma(t)e^{i\omega t}. \quad (68)$$

The normal mode expansion of the unperturbed field for $x > 0$ is

$$\Phi_0 = \sqrt{\frac{4\pi}{|\omega|(1+\gamma_0^2\omega^2)}}\left[\sin(\omega x) + \gamma_0\omega\cos(\omega x)\right]\left[\Theta(\omega)a(\omega) - \Theta(-\omega)a^\dagger(-\omega)\right], \quad (69)$$

where $a(\omega)$ and $a^\dagger(\omega)$ are the bosonic annihilation and creation operators, respectively, which satisfy the commutation relation $[a(\omega), a^\dagger(\omega')] = 2\pi\delta(\omega - \omega')$. To solve for Φ, one must first calculate $\delta\Phi$, which can be found by introducing the following Green's function:

$$\left(\partial_x^2 - \omega^2\right)G(\omega,x,x') = \delta(x-x'). \quad (70)$$

By employing Green's theorem, one obtains the following as the solution for the outgoing field:

$$\Phi_{\text{out}}(\omega,x) = \Phi_{\text{in}}(\omega,x) + \left[G_R^{\text{ret}}(\omega,0,x) - G_R^{\text{adv}}(\omega,0,x)\right] \times \left[\frac{\partial\delta\Phi}{\partial x}(\omega,0) - \frac{\delta\Phi(\omega,0)}{\gamma_0}\right] \quad (71)$$

where G_R^{ret} (G_R^{adv}) is the retarded (advanced) Robin Green function, given by

$$G_R^{\text{ret}}(\omega,0,x) = \frac{\gamma_0}{1-i\gamma_0\omega}e^{i\omega t} \tag{72}$$

and

$$G_R^{\text{adv}}(\omega,0,x) = \frac{\gamma_0}{1+i\gamma_0\omega}e^{-i\omega t}. \tag{73}$$

Using the following equality

$$\gamma_0 \frac{\partial \delta\Phi}{\partial x}(\omega,0) - \delta\Phi(\omega,0) = \int \frac{d\omega'}{2\pi}\left[\frac{\partial}{\partial x}+\omega\omega'\right]\Phi_0(\omega',0) \times \delta Q(\omega-\omega') \tag{74}$$

in the equation for Φ_{out}, the resulting Bogoliubov transformation then becomes

$$a_{\text{out}} = a_{\text{in}} +$$
$$2i\sqrt{\frac{\omega}{1+\gamma_0^2\omega^2}}\int \frac{d\omega'}{2\pi}\sqrt{\frac{\omega'}{1+\gamma_0^2\omega'^2}} \tag{75}$$
$$\times\left[\Theta(\omega')a_{\text{in}}(\omega')-\Theta(-\omega')a_{\text{in}}^\dagger(-\omega')\right]$$
$$\times (1+\gamma_0^2\omega\omega')\delta Q(\omega-\omega').$$

Using the following equality

$$\gamma_0 \frac{\partial \delta\Phi}{\partial x}(\omega,0) - \delta\Phi(\omega,0) = -\int \frac{d\omega'}{2\pi}\frac{\partial \Phi_0}{\partial x}(\omega',0) \times \delta\Gamma(\omega-\omega'). \tag{76}$$

in the equation for Φ_{out}, the resulting Bogoliubov transformation then becomes

$$a_{\text{out}} = a_{\text{in}} -$$
$$2i\sqrt{\frac{\omega}{1+\gamma_0^2\omega^2}}\int \frac{d\omega'}{2\pi}\sqrt{\frac{\omega'}{1+\gamma_0^2\omega'^2}} \tag{77}$$
$$\times\left[\Theta(\omega')a_{\text{in}}(\omega')-\Theta(-\omega')a_{\text{in}}^\dagger(-\omega')\right]$$
$$\times \delta\Gamma(\omega-\omega').$$

Thus one finds a relationship between the input/output Bogoliubov transforms of the moving and time-dependent Robin b.c., whereby they differ by an overall minus sign and an additional factor of $(1+\gamma_0^2\omega\omega')$. Note that the two representations coincide when the boundary reduces to the purely Dirichlet boundary condition ($\gamma_0 \to 0$), with the difference between a_{out} and a_{in} reducing to

$$a_{\text{out}} - a_{\text{in}} = \pm 2i \int \frac{d\omega'}{2\pi}\sqrt{\omega\omega'}\left[\Theta(\omega')a_{\text{in}}(\omega')-\Theta(-\omega')a_{\text{in}}^\dagger(-\omega')\right]\delta\mathcal{F}(\omega-\omega') \tag{78}$$

where $\delta\mathcal{F}$ is the Fourier transform of the parameter that drives the small perturbation. Here, the difference between a_{out} and a_{in} isolates the terms that encode particle production and highlights the similarities between different methods of creating particles via unique ways of generating time-varying perturbations.

3.2. Moving Asymmetric Robin Boundary

Just as it was examined in the Section 2 in order to induce the ADCE, the system must be set up in such a way that the boundary divides the space *and* imposes an asymmetry. This was accomplished by introducing the asymmetric $\delta - \delta'$ potential into the Lagrangian for the free scalar field to simulate a mirror whose two sides possess different properties. One must be mindful when building asymmetry into these field solutions, as it is possible for mathematical inconsistencies to arise if the asymmetry is not carefully introduced [85].

Here, we introduce asymmetry into the moving Robin boundary [79] analyzed in Section 3.1. An asymmetric perturbation on the moving Robin b.c. begins the same way as the standard moving Robin mirror, with

$$\left[\frac{\partial}{\partial x}+\delta\dot{q}(t)\frac{\partial}{\partial t}\right]\phi(t,\delta q(t)) = \frac{1}{\gamma_0}\phi(t,\delta q(t)) \tag{79}$$

being the Robin boundary condition in the laboratory frame for a small deviation $\delta q(t)$ about $x = 0$.

Following the same procedure as Ref. [79], one finds the first-order field ($\phi = \phi_0 + \delta\phi$) satisfies the following equation at $x = 0$:

$$\left[\frac{\partial \delta\phi(t,x)}{\partial x}\right]_{x=0} - \frac{1}{\gamma_0}\delta\phi(t,0)$$
$$= \delta q(t)\frac{1}{\gamma_0}\left[\frac{\partial \phi_0(t,x)}{\partial x} - \gamma_0\frac{\partial^2 \phi_0(t,x)}{\partial x^2}\right]_{x=0} - \delta\dot{q}(t)\left[\frac{\partial \phi_0(t,x)}{\partial t}\right]_{x=0}. \quad (80)$$

It is here that we impose the asymmetry of the mirror. Motivated by the use of the δ'-potential in the $\delta - \delta'$ examples from the scattering section, we take advantage of the properties of the δ'-potential and incorporate it into Equation (80). Recall the definition of $\delta'(x)$ from Ref. [46],

$$\delta'(x)f(x) = \delta'(x)\frac{f(0^+) + f(0^-)}{2} - \delta(x)\frac{f'(0^+) + f'(0^-)}{2}. \quad (81)$$

Using the symmetry of the time-independent Robin solution, one finds that

$$\delta'(x)\frac{\partial \phi_0(t,x)}{\partial x} = \delta'(x)\left[\frac{\partial \phi_0(t,x)}{\partial x}\right]_{x=0} - \delta(x)\left[\frac{\partial^2 \phi_0(t,x)}{\partial x^2}\right]_{x=0}. \quad (82)$$

Thus, to build asymmetry into the moving Robin b.c., while at the same time remaining mathematically consistent with the definition of δ', we incorporate a $\delta - \delta'$ term into the spatial derivatives at zero in Equation (80) giving the new equality,

$$\left[\frac{\partial \delta\phi(t,x)}{\partial x}\right]_{x=0} - \frac{1}{\gamma_0}\delta\phi(t,0)$$
$$= \delta q(t)\frac{1}{\gamma_0}\left[\delta'(x)\left[\frac{\partial \phi_0(t,x)}{\partial x}\right]_{x=0} - \gamma_0\delta(x)\left[\frac{\partial^2 \phi_0(t,x)}{\partial x^2}\right]_{x=0}\right] - \delta\dot{q}(t)\left[\frac{\partial \phi_0(t,x)}{\partial t}\right]_{x=0}. \quad (83)$$

This manifests in there being two separate solutions about $x = 0$,

$$\left[\frac{\partial \delta\phi(t,x)}{\partial x}\right]_{x=0^{\pm}} - \frac{1}{\gamma_0}\delta\phi(t,0^{\pm})$$
$$= \delta q(t)\frac{1}{\gamma_0}\left[\pm\frac{\partial \phi_0(t,x)}{\partial x} - \gamma_0\frac{\partial^2 \phi_0(t,x)}{\partial x^2}\right]_{x=0^{\pm}} - \delta\dot{q}(t)\left[\frac{\partial \phi_0(t,x)}{\partial t}\right]_{x=0^{\pm}}. \quad (84)$$

Following the same derivation as in Section 3.1, one arrives at the Bogoliubov transform for the relationship between annihilation operators a_{out} and a_{in}, appropriately labeled with a positive (negative) superscript for the $x > 0$ ($x < 0$) region,

$$a_{\text{out}}^{(\pm)} = a_{\text{in}}^{(\pm)} + 2i\sqrt{\frac{\omega}{(1+\gamma_0^2\omega^2)}}\int\frac{d\omega'}{2\pi}\left(\pm 1 + \gamma_0^2\omega\omega'\right)\sqrt{\frac{\omega'}{(1+\gamma_0^2\omega'^2)}}$$
$$\times\left[\Theta(\omega')a_{\text{in}}^{(\pm)}(\omega') - \Theta(-\omega')a_{\text{in}}^{(\pm)}(-\omega')^{\dagger}\right]\delta Q(\omega - \omega'), \quad (85)$$

where we see that the positive solution is the same as in Section 3.1 Note that the vacua solution only accounts for the outgoing solution about either side of the mirror since $\delta\phi(t,x)$ must describe the contribution from the mirror and not the incoming waves moving towards the mirror [83].

3.2.1. Spectral Distribution

The infinitesimal spectral distribution of the particles created on either side of the mirror, between ω and $\omega + d\omega$ ($\omega \geq 0$), is given by

$$N_\pm d\omega = \langle 0_{in} | a_{in}^{(\pm)}(\omega)^\dagger a_{in}^{(\pm)}(\omega) | 0_{in} \rangle \frac{d\omega}{2\pi}. \tag{86}$$

The complete spectrum is found by using Equations (85) in (86), giving

$$N_\pm = \frac{2\omega}{\pi(1+\gamma_0^2\omega^2)} \int_0^\infty \frac{d\omega'}{2\pi} \frac{\omega'[1 \mp \gamma_0 \omega \omega']^2}{(1+\gamma_0^2\omega'^2)} |\delta\Gamma(\omega+\omega')|^2. \tag{87}$$

One may once again assign a specific form to the time-dependent function that drives the motion of the mirror. Following the same procedure from Refs. [46,79,83], implemented for the moving $\delta - \delta'$ system in Section 2, we use

$$\delta\gamma(t) = \epsilon \cos(\omega_0 t) \exp(-|t|/\tau), \tag{88}$$

where, as before in Equation (34), τ is the oscillation lifetime and ω_0 is the characteristic frequency of the oscillation with $\omega_0 \tau \gg 1$. We denote the Fourier transform of $\gamma(t)$ with $\delta\Gamma(\omega)$. The function $\delta\Gamma(\omega)$ contains two extremely narrow peaks around $\omega = \pm \omega_0$ and can therefore be approximated as

$$\frac{|\delta\Gamma(\omega)|^2}{\tau} \approx \epsilon^2 \frac{\pi}{2} [\delta(\omega - \omega_0) + \delta(\omega + \omega_0)]. \tag{89}$$

The new definition of $\delta\Gamma(\omega)$ in Equation (89) allows us to explicitly compute the spectrum on either side of the mirror, which becomes

$$\frac{N_\pm}{\tau} = \frac{\epsilon^2}{2\pi} \frac{\omega(\omega_0 - \omega)[1 \mp \gamma_0^2(\omega_0 - \omega)\omega]^2}{(1+\gamma_0^2\omega^2)[1+\gamma_0^2(\omega_0-\omega)^2]} \Theta(\omega_0 - \omega), \tag{90}$$

where one sees, as in Refs. [79,83], that no particles are created for frequencies higher than the characteristic frequency ω_0 of the time-dependent perturbation on the Robin b.c. As expected, the spectrum is invariant under the replacement $\omega \to \omega_0 - \omega$ and is symmetric about $\omega = \omega_0/2$. This indicates that particles are created in pairs such that the sum of their frequencies is ω_0.

Once again, one may calculate physically relevant quantities that give us more insight into the dynamics of the system. The spectral ratio is

$$\frac{N_-}{N_+} = \left(\frac{1 + \gamma_0^2 \omega(\omega_0 - \omega)}{1 - \gamma_0^2 \omega(\omega_0 - \omega)} \right)^2, \tag{91}$$

and the spectral difference is

$$\frac{\Delta N}{\tau} = \frac{\epsilon^2}{\pi} \frac{2[\gamma_0 \omega(\omega_0 - \omega)]^2}{(1+\gamma_0^2 \omega^2)[1+\gamma_0^2(\omega_0-\omega)^2]} \Theta(\omega_0 - \omega). \tag{92}$$

One can see that the spectral ratio and difference for the newly calculated moving asymmetric Robin boundary closely resembles those found in Section 2.2.2 for the time-dependent moving $\delta - \delta'$ mirror. One sees from Equation (91) that the left half of the mirror always produces a larger number of particles than the right half, excluding the points $\omega = 0$ and $\omega = \omega_0$ where the spectrum vanishes. This is also apparent in spectral difference, as it is positive for all values outside the end points. As expected, in the Dirichlet limit when $\gamma_0 = 0$ the asymmetry vanishes. For a closer look at the difference between spectral outputs by the two sides of the moving asymmetric Robin b.c., including the influence of difference values of γ_0; see Figure 2.

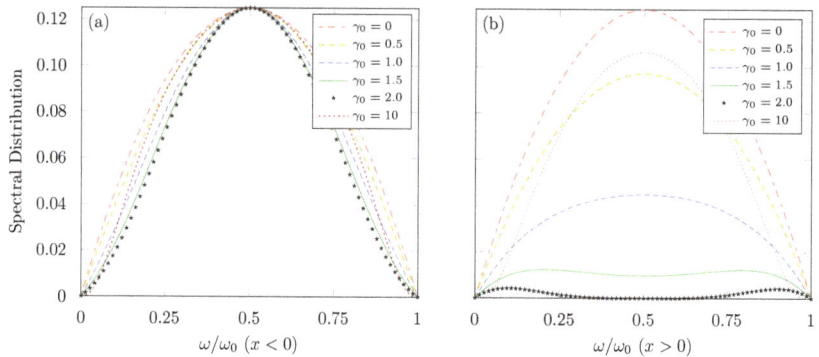

Figure 2. The spectral distribution of particles created on the two sides of the mirror as a function of ω/ω_0 for different values of γ_0: (**a**) the plot of $(\epsilon^2\tau/\pi)^{-1} \times N_-$; (**b**) the plot of $(\epsilon^2\tau/\pi)^{-1} \times N_+$. See text for details.

3.2.2. Particle Creation Rate

The total number of particles created, effectively the (average) particle creation rate, is

$$R_\pm = \frac{N_\pm}{\tau} = \left(\frac{\epsilon^2}{2\pi}\right) \int_0^{\omega_0} \frac{\omega(\omega_0-\omega)\left[1 \mp \gamma_0^2(\omega_0-\omega)\omega\right]^2}{(1+\gamma_0^2\omega^2)\left[1+\gamma_0^2(\omega_0-\omega)^2\right]} d\omega$$
$$= \left(\frac{\epsilon^2\omega_0^3}{2\pi}\right) F_\pm(\xi) \tag{93}$$

where $\xi = \gamma_0 w_0$ with

$$F_+(\xi) = \frac{\xi(4\xi + \xi^3 + 12\arctan(\xi)) - 6(2+\xi^2)\ln(1+\xi^2)}{6\xi^4(\xi^2+4)} \tag{94}$$

and

$$F_-(\xi) = \frac{\xi(24\xi + \xi^3 - 36\arctan(\xi)) - 6(-2+\xi^2)\ln(1+\xi^2)}{6\xi^4}. \tag{95}$$

This particle creation rate is the physically meaningful quantity that can be experimentally measured. One can see that N is proportional to τ (a result of the open geometry of the cavity). The particle creation rate in the limits of $\gamma_0\omega_0 \ll 1$ (Dirichlet) and $\gamma_0\omega_0 \gg 1$ (Neumann) converge to the same value:

$$R_\pm \approx \left(\frac{\epsilon^2\omega_0^2}{12\pi}\right), \tag{96}$$

which matches what is found in the literature [23,26,37].

4. Comparison between the Different Approaches

The moving asymmetric Robin boundary solution that we constructed in Section 3.2 bears a striking resemblance to the moving $\delta - \delta'$ mirror that originates from the scattering approach. One can examine the two solutions alongside each other by looking at their respective spectral distributions in Figures 2 and 3. For the sake of comparison, let us consider the maximally asymmetric cases for the different solutions, which correspond to $\lambda_0 = 1$ and $\gamma_0 = 2$ (taking $\mu_0 = 1$ with $\gamma_0 = 2/\mu_0$). The spectrum N_+, on right side of the mirror, is the same as both the original unperturbed moving Robin b.c. spectrum [79] and the spectrum produced by the right side of the moving $\delta - \delta'$ mirror [46]. This Robin spectrum, in Figure 2b, is associated with the highest degree of asymmetry as it is maximally suppressed when $\gamma_0 = 2$ (or $\lambda_0 = 1$), with the spectrum completely vanishing at $\omega_0/2$. From Figures 2a and 3a, the spectrum produced by the left half, $\mathcal{N}_-$, is a purely reflective Dirichlet spectrum when $\lambda_0 = 1$ and $\gamma_0 = 0$ for the moving $\delta - \delta'$ and asymmetric Robin mirror, respectively. However, in the maximally asymmetric case of the moving asymmetric Robin mirror, when $\gamma_0 = 2$, there is an inhibition of modes away from $\omega_0 = 2$ that sharpens the purely reflective Dirichlet peak and leaves the maximum value at $\omega_0/2$ unchanged.

The slight inhibition of modes away from $\omega_0/2$ in the maximally asymmetric case of the moving asymmetric Robin b.c. solution, seen in Figure 4b, is what leads to the difference between the particle production ratio $\mathcal{N}_+/\mathcal{N}_-$ of the asymmetric Robin and $\delta - \delta'$ mirrors. This is well seen in the increased asymmetry in the $\delta - \delta'$ solution for different values of w_0 when compared to the asymmetric Robin solution. Particle production is maximally suppressed for $\gamma_0 \omega_0 \approx 2.2$, the frequency of maximal asymmetric particle production, which gives rise to approximately the same minimum in the particle creation ratios seen in Figure 5. Both minima occur at $\omega_0 \approx 1.1$, where $\mathcal{N}_+/\mathcal{N}_- \approx 0.016$ for the asymmetric Robin and $\mathcal{N}_+/\mathcal{N}_- \approx 0.013$ for the $\delta - \delta'$ mirrors. From another view, the left side produces about 60 and 75 times that of the right side, respectively.

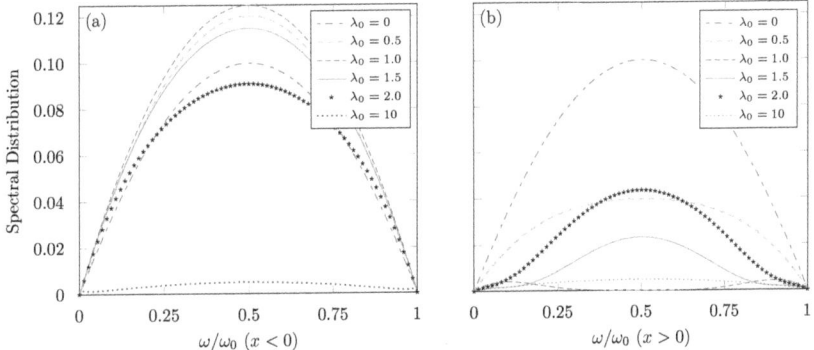

Figure 3. The spectral distribution of particles created on the two sides of a moving $\delta - \delta'$ mirror as a function of ω/ω_0 for different values of λ_0, with $\mu_0 = 1$: (**a**) the plot of $(\epsilon^2 \tau/\pi)^{-1} \times dN_-/d\omega$; (**b**) the plot of $(\epsilon^2 \tau/\pi)^{-1} \times dN_+/d\omega$. See text for details. Figure is generated from results within Ref. [46].

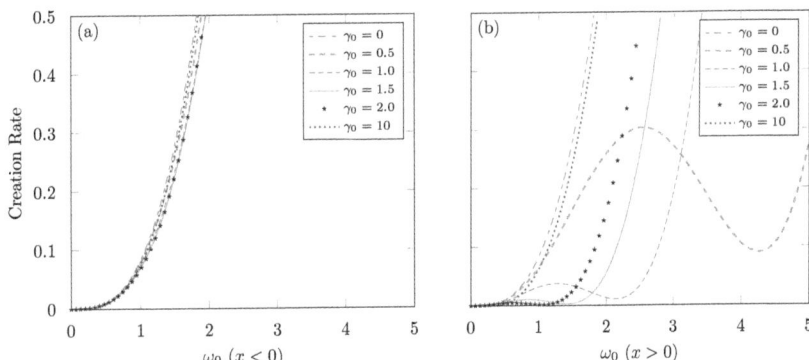

Figure 4. The creation rate of particles on the two sides of the mirror as a function of ω_0 for different values of γ_0: (**a**) the plot of $(\epsilon^2 T/\pi)^{-1} \times R_-$; (**b**) the plot of $(\epsilon^2 T/\pi)^{-1} \times R_+$. See text for details.

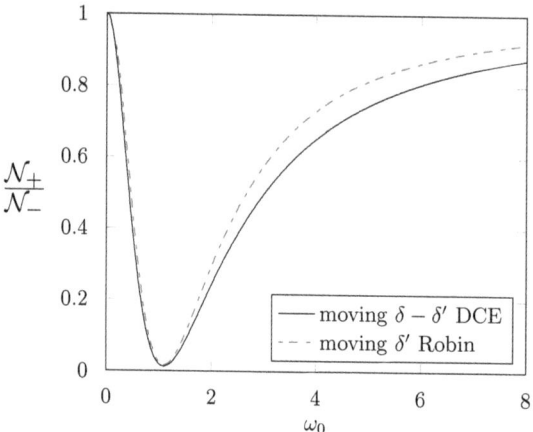

Figure 5. The particle creation ratio of $\mathcal{N}_+/\mathcal{N}_-$ for both the moving asymmetric Robin and $\delta - \delta'$ boundaries. For comparison, the asymmetric Robin mirror with $\gamma_0 = 2$ and the $\delta - \delta'$ mirror with $\lambda_0 = \mu_0 = 1$ are shown.

5. Discussion

The means by which macroscopic systems interacting with the quantum vacuum are able to produce the ADCE are apparent; it is necessary to generate solutions that include both fluctuations in time and explicitly broken spatially symmetry. Without fluctuations in time, be it on the object's position, material properties, etc., the production of particles vanishes. Unless asymmetric boundary conditions are imposed on either side of an object in (1+1)D, the production of particles will always be symmetric about the two sides of the object and the ADCE will not exist. The appearance of the ADCE is independent of the method used to generate the boundary that interacts with the vacuum. Whether an asymmetric system is solved in a waves based scattering interaction framework or with a particle-based calculation of the creation/annihilation operators, the same asymmetric effect is present in the solutions of these two approaches. This is especially evident in our newly constructed moving asymmetric Robin b.c. solution, where the introduction of spatial asymmetry to an otherwise symmetric mirror obeying the Robin b.c. induced a change in the particle output of one side of the mirror.

One of the more remarkable consequences of the ADCE is that the unbalanced production of particles will cause an otherwise stationary system to be perturbed via its interaction with the vacuum and induce motion as momentum is "extracted" from the vacuum [47]. The initial state of the object, for $t < -\tau$ ($\tau > 0$), is that of a stationary, time-independent object interacting with a field. It is completely described by the quantum vacuum state as there are no quantum interactions before the time fluctuations occur. The characteristic oscillations of the time-dependent boundary begin at $-\tau$, i.e., some generic variable of the system $\varepsilon_0 \to \varepsilon(t)$, after which the object is free to move. Note that, once the object is able to move, the quantum field will cause the object to experience Brownian motion [86–89]. Assuming the object is large enough, this motion can be neglected. At this point, if the object possesses no spatial asymmetry while undergoing time fluctuations, the object remains in its starting position as the symmetric production of particles applies an equal and opposite response to the object. For an asymmetric object, particle production is favored to one side, which results in a net force on the object, a transfer of momentum to the previously stationary system, and a dissipation of energy from the mirror. This is expected from the underlying symmetries of quantum field theory (translational invariance, locality, and unitarity). A nonzero vacuum momentum, and a nonvanishing total force, are to be found in any asymmetrically excited system [90].

The total energy of the created particles, $\mathcal{E} = \mathcal{E}_+ + \mathcal{E}_-$, is the sum of the two sides where $\mathcal{E}_\pm = \int_0^\infty d\omega N_\pm(\omega)\omega$. The momentum is now $\mathcal{P} = \mathcal{P}_+ + \mathcal{P}_-$, where $\mathcal{P}_\pm = \pm\mathcal{E}_\pm$. The quantity that determines the asymmetric dynamics is ΔN, as one now has $\Delta \mathcal{E}, \Delta \mathcal{P}$, and $\Delta F \neq 0$. If the system is closed, the energy of the particles emitted comes at the expense of the internal energy of the object, as energy is needed to drive the time fluctuations, and the mass of the object will now change in time. To ensure the total momentum of the system is conserved, the object experiences a net force and now has a nonzero final momentum since the total momentum of the particles no longer vanishes for asymmetric objects. For a detailed analysis of the forces and dynamical evolution of an asymmetric object with time-dependent material properties interacting with the vacuum, see [47]. Here, it is necessary to not only include the motional contribution from the vacuum's interaction with the time-dependent properties of object, but also the interaction due to its newly perturbed fluctuation in position. Thus, to perform a detailed analysis of the motional corrective terms introduced in [47], one must account for the interaction term between the time dependence on $\mu(t)$ and the position $q(t)$ in the $\delta - \delta'$ example that was explored in Section 2 (see [91] for this process conducted on a symmetric Robin boundary). Accounting for every form of time fluctuations is necessary to understand the full dynamics of the system, an analysis we intend to perform in the future.

Understanding the fundamental mechanisms of asymmetric vacuum interactions provides the basis to investigate an abundance of vacuum interactions that seek to probe the extreme limits of physical theory. Already, we have seen an otherwise stationary object gain momentum out of seemingly nothing, due to its interaction with the vacuum, a surprising result that actually arises from the conservation of momentum. This is not the only time that asymmetric systems have gained momentum from vacuum interactions. It has been shown that a net transfer of linear momentum can occur in a system composed of two excited, dissimilar atoms [90]. Just as it was seen throughout this paper, a quantum system with asymmetric excitations leads to an imbalanced production of emitted particles and gives rise to a net force and transfer of momentum from the vacuum. Linear asymmetry is not the only means by which to generate some motive force from the vacuum: chiral particles can also achieve a similar effect. These particles, which do not posses mirror invariance, can gain kinetic "Casimir" momentum when subjected to a magnetic field [92,93]. There are claims, albeit controversial [94–97], that the vacuum can impart momentum asymmetrically on magnetoelectric materials [94]. Asymmetric momentum transfer is said to arise from the magnetoelectric molecular structure, as it possesses optical anisotropy since the structure breaks the temporal and spatial symmetries of electromagnetic modes. Even though the details still need to be fully worked out, it is clear that asymmetric vacuum interactions play a role in understanding magnetoelectric and other anistropic materials.

6. Conclusions

We reviewed past studies on the $\delta - \delta'$ mirror and showed that, regardless of the mechanism and form of the time-dependent fluctuations, the ADCE is produced. Fluctuations on λ_0 were explored and we discussed obstructions to analyzing linear scattering in this case. Experimental motivations were discussed. We showed, in the scattering framework, that physically relevant quantities originate purely from the difference between right- and left-half asymmetric transmission and reflection coefficients. A newly formulated solution using the Bogoliubov transform introduces an asymmetric formulation of the moving Robin boundary. This solution bears a striking resemblance to the moving $\delta - \delta'$ mirror, demonstrating the ability to break symmetric boundary solutions and build up new forms of ADCE configurations. Byproducts of the ADCE were explored, namely the transfer of momentum to otherwise stationary systems, causing an object to move through the vacuum without any addition external forces beyond the vacuum interactions. Remarkably, momentum transfer here emerges from the enforcement of conservation laws, not a violation of them.

Within the framework of objects interacting in (1+1)D with the massless scalar quantum field, we have explored the effects of introducing asymmetry to time-dependent systems interacting with the quantum vacuum and demonstrated general consequences that asymmetric boundary conditions impart upon these systems. Whether the problem is approached from the perspective of quantum particles or quantum fields, the end result is the same: an asymmetric production of photons between the two sides of an object. An explicit breaking of mirror symmetry about the two sides of an object is necessary to generate the asymmetry needed to produce different spectra and quantities of particles about the two sides of the object. Additionally, without time-dependent fluctuations of object–vacuum interactions the particle production vanishes. It is necessary to have perturbations on both the spatial and temporal domains of the system to break the underlying symmetry of vacuum interactions.

Author Contributions: Conceptualization, M.J.G.; methodology, M.J.G.; software, M.J.G. and W.D.J.; validation, M.J.G., W.D.J., P.M.B. and J.A.M.; formal analysis, M.J.G., W.D.J., P.M.B. and J.A.M.; investigation, M.J.G. and W.D.J.; resources, G.B.C.; data curation, M.J.G. and W.D.J.; writing—original draft preparation, M.J.G. and W.D.J.; writing—review and editing, P.M.B., J.A.M. and G.B.C.; supervision, G.B.C.; project administration, G.B.C. All authors have read and agreed to the published version of the manuscript.

Funding: This research received no external funding.

Data Availability Statement: Not applicable.

Acknowledgments: The authors would like to thank Ramesh Radhakrishnan, Cooper Watson, and Eric Davis for beneficial discussions and reviews. The authors would also like to thank the reviewers.

Conflicts of Interest: The authors declare no conflict of interest.

Abbreviations

The following abbreviations are used in this manuscript:

ADCE	asymmetric dynamical Casimir effect
b.c.	boundary condition
D	dimension
DCE	dynamical Casimir effect
SQUID	superconducting quantum interference device

Appendix A. $\lambda(t)$ Linear Scattering

Here, we provide a derivation of the scattering terms for $f(t)$ chosen such that the resulting expressions for matching conditions are as simple as possible. This allows us straightforward illustration of the way in which we are obstructed from deriving scattering matrix elements as we did in the rest of this paper.

Starting from Equations (61) and (62),

$$-\partial_x \Phi(\omega 0^+) + \partial_x \Phi(\omega, 0^-) + \mu[\Phi(\omega, 0^+) + \Phi(\omega, 0^-)]$$
$$- \int \frac{d\omega'}{2\pi} \mathcal{L}(\omega - \omega')(\partial_x \Phi(\omega', 0^+) + \partial_x \Phi(\omega', 0^-)) = 0$$

and

$$-\Phi(\omega, 0^+) + \Phi(\omega, 0^-) + \int \frac{d\omega'}{2\pi} \mathcal{L}(\omega - \omega')(\Phi(\omega', 0^+) + \Phi(\omega', 0^-)) = 0,$$

it becomes seen that a general form of the matching conditions cannot be derived due to convolution Fourier transforms. To demonstrate the difficulty these integrals provide for the matching conditions, we take specific form $\lambda(t) = \lambda_0[1 + \epsilon f(\omega_0 t, |t|/\tau)]$, where f is assumed for now to have the same type of functional dependence found in Equation (34). We note, though, that we do not specify an explicit functional definition for f. Instead, making the general assumption that in the limit where $\tau \to \infty$, one has a "monochromatic-like" limit where its Fourier transform satisfies

$$\lim_{\tau \to \infty} \mathcal{F}(\omega) = b[\delta(\omega + \omega_0) + \delta(\omega - \omega_0)], \tag{A1}$$

where b is some normalization constant for the Dirac delta distributions. Using this, one then has the Fourier transform of $\lambda(t)$ as

$$\mathcal{L}(\omega) = \lambda_0[\delta(\omega) + \epsilon \mathcal{F}(\omega)], \tag{A2}$$

where in what follows we assume we already computed the limit on τ whenever evaluating integrals.

Substituting $\mathcal{L}(\omega - \omega')$ into Equations (61) and (62), one has:

$$-\partial_x \Phi(\omega 0^+) + \partial_x \Phi(\omega, 0^-) + \mu[\Phi(\omega, 0^+) + \Phi(\omega, 0^-)]$$
$$-\frac{\lambda_0}{2\pi}[\partial_x \Phi(\omega, 0^+) + \partial_x \Phi(\omega, 0^-)] - \lambda_0 \epsilon \int \frac{d\omega'}{2\pi} \mathcal{F}(\omega - \omega')(\partial_x \Phi(\omega', 0^+) + \partial_x \Phi(\omega', 0^-)) = 0 \tag{A3}$$

and

$$-\Phi(\omega, 0^+) + \Phi(\omega, 0^-) + \frac{\lambda_0}{2\pi}(\Phi(\omega, 0^+) + \Phi(\omega, 0^-))$$
$$+\lambda_0 \epsilon \int \frac{d\omega'}{2\pi} \mathcal{F}(\omega - \omega')(\Phi(\omega', 0^+) + \Phi(\omega', 0^-)) = 0. \tag{A4}$$

Now, explicitly evaluating these integrals under the above limits and assumptions, one obtains:

$$-\partial_x \Phi(\omega 0^+) + \partial_x \Phi(\omega, 0^-) + \mu[\Phi(\omega, 0^+) + \Phi(\omega, 0^-)] - \frac{\lambda_0}{2\pi}[\partial_x \Phi(\omega, 0^+) + \partial_x \Phi(\omega, 0^-)]$$
$$-\frac{\lambda_0 \epsilon b}{2\pi}[\partial_x \Phi(\omega - \omega_0, 0^+) + \partial_x \Phi(\omega - \omega_0, 0^-) + \partial_x \Phi(\omega + \omega_0, 0^+) + \partial_x \Phi(\omega + \omega_0, 0^-)] = 0 \tag{A5}$$

and

$$-\Phi(\omega, 0^+) + \Phi(\omega, 0^-) + \frac{\lambda_0}{2\pi}(\Phi(\omega, 0^+) + \Phi(\omega, 0^-))$$
$$+\frac{\lambda_0 \epsilon b}{2\pi}[\Phi(\omega - \omega_0, 0^+) + \Phi(\omega - \omega_0, 0^-) + \Phi(\omega + \omega_0, 0^+) + \Phi(\omega + \omega_0, 0^-)] = 0. \tag{A6}$$

Next, we further assume that the ingoing and outgoing fields are linearly related as before, giving

$$\Phi_+(\omega, x) = s_-(\omega)e^{-i\omega x}\Theta(-x) + (e^{-i\omega x} + r_-(\omega)e^{i\omega x})\Theta(x)$$

and

$$\Phi_-(\omega, x) = (e^{i\omega x} + r_+(\omega)e^{-i\omega x})\Theta(-x) + s_+(\omega)e^{i\omega x}\Theta(x).$$

Now, Equations (A5) and (A6) can be re-expressed explicitly in terms of transmission and reflection coefficients, offering
Φ_+:

$$-i\omega(1 + \frac{\lambda_0}{2\pi})(r_-(\omega) - 1) - i\omega(1 - \frac{\lambda_0}{2\pi})s_-(\omega)$$
$$+\mu[1 + r_-(\omega) + s_-(\omega)] = \frac{\lambda_0 \epsilon b}{2\pi}[i(\omega - \omega_0)(r_-(\omega - \omega_0) - 1) \tag{A7}$$
$$+i(\omega - \omega_0)s_-(\omega - \omega_0) + i(\omega + \omega_0)(r_-(\omega + \omega_0) - 1) + i(\omega + \omega_0)s_-(\omega + \omega_0)],$$

and

$$(\frac{\lambda_0}{2\pi} - 1)(1 + r_-(\omega)) + (\frac{\lambda_0}{2\pi} + 1)s_-(\omega)$$
$$= -\frac{\lambda_0 \epsilon b}{2\pi}[2 + r_-(\omega - \omega_0) + s_-(\omega - \omega_0) + r_-(\omega + \omega_0) + s_-(\omega + \omega_0)]. \quad (A8)$$

Φ_-:

$$-i\omega(1 + \frac{\lambda_0}{2\pi})s_+(\omega) + i\omega(1 - \frac{\lambda_0}{2\pi})(1 - r_+(\omega)) + \mu[1 + s_+(\omega) + r_+(\omega)]$$
$$= \frac{\lambda_0 \epsilon b}{2\pi}[i(\omega - \omega_0)s_+(\omega - \omega_0) + i(\omega - \omega_0)(1 - r_+(\omega - \omega_0))$$
$$+ i(\omega + \omega_0)s_+(\omega + \omega_0) + i(\omega + \omega_0)(1 - r_+(\omega + \omega_0))], \quad (A9)$$

and

$$(\frac{\lambda_0}{2\pi} - 1)s_+(\omega) + (\frac{\lambda_0}{2\pi} + 1)(1 + r_+(\omega))$$
$$= -\frac{\lambda_0 \epsilon b}{2\pi}[2 + s_+(\omega - \omega_0) + r_+(\omega - \omega_0) + s_+(\omega + \omega_0) + r_+(\omega + \omega_0)]. \quad (A10)$$

Equations (A7)–(A10) provide four coupled equations, with 12 unknown terms: four scattering terms for each frequency argument appearing ($\omega, \omega \pm \omega_0$). Therefore, there are not enough constraints on the fields to produce a definitive solution to the $\lambda(t)$ perturbation for the (1 + 1)D mirror in this scattering approach. The authors are not aware of any technique within this linear scattering framework that would allow for one to solve problems of this type. Additionally, this result seems to suggest that there may be some general obstruction that prevents this type of linear scattering framework from solution when the potential contains a δ' potential with time-dependent strength. This is because potentials in this form typically couple different frequencies together in a way that prevents the matching conditions from being solvable. The authors are still optimistic than an approach based upon Bogoliubov transformations may be more successful, but such an approach requires substantial development which is reserved for future work.

References

1. Casimir, H.B.G. On the attraction between two perfectly conducting plates. *Proc. Kon. Ned. Akad. Wetensch.* **1948**, *51*, 793–795. Available online: https://dwc.knaw.nl/DL/publications/PU00018547.pdf (accessed on 11 March 2023).
2. Milton, K.A. *The Casimir Effect: Physical Manifestations of Zero-Point Energy*; World Scientific Publishing Co. Pte. Ltd.: Singapore, 2001. [CrossRef]
3. Milonni, P.W.; Shih, M.L. Casimir forces. *Contemp. Phys.* **1992**, *33*, 313–322. [CrossRef]
4. Milton, K.A. The Casimir effect: Recent controversies and progress. *J. Phys. Math. Gen.* **2004**, *37*, R209–R277. [CrossRef]
5. Lamoreaux, S.K. The Casimir force: Background, experiments, and applications. *Rep. Prog. Phys.* **2004**, *68*, 201–236. [CrossRef]
6. Bordag, M.; Klimchitskaya, G.L.; Mohideen, U.; Mostepanenko, V.M. *Advances in the Casimir Effect*; Oxford University Press: Oxford, UK, 2009. [CrossRef]
7. Simpson, W.M.; Leonhardt, U.; Simpson, W.M. *Forces of the Quantum Vacuum*; World Scientific Publishing Co. Pte. Ltd.: Singapore, 2015. [CrossRef]
8. Palasantzas, G.; Dalvit, D.A.; Decca, R.; Svetovoy, V.B.; Lambrecht, A. Casimir Physics. *J. Phys. Condens. Matter* **2015**, *27*, 210301. [CrossRef]
9. Moore, G.T. Quantum theory of the electromagnetic field in a variable-length one-dimensional cavity. *J. Math. Phys.* **1970**, *11*, 2679–2691. [CrossRef]
10. DeWitt, B.S. Quantum field theory in curved spacetime. *Phys. Rep.* **1975**, *19*, 295–357. [CrossRef]
11. Fulling, S.A.; Davies, P.C. Radiation from a moving mirror in two dimensional space-time: conformal anomaly. *Proc. R. Soc. Lond. Math. Phys. Sci.* **1976**, *348*, 393–414. [CrossRef]
12. Davies, P.C.; Fulling, S.A. Radiation from moving mirrors and from black holes. *Proc. R. Soc. Lond. Math. Phys. Sci.* **1977**, *356*, 237–257. [CrossRef]
13. Yablonovitch, E. Accelerating reference frame for electromagnetic waves in a rapidly growing plasma: Unruh-Davies-Fulling-DeWitt radiation and the nonadiabatic Casimir effect. *Phys. Rev. Lett.* **1989**, *62*, 1742. [CrossRef]

14. Dodonov, V. Dynamical Casimir effect: Some theoretical aspects. *J. Phys. Conf. Ser.* **2009**, *161*, 012027. [CrossRef]
15. Dodonov, V. Current status of the dynamical Casimir effect. *Phys. Scr.* **2010**, *82*, 038105. [CrossRef]
16. Dodonov, V. Fifty years of the dynamical Casimir effect. *Physics* **2020**, *2*, 67–104. [CrossRef]
17. Maia Neto, P.A. The dynamical Casimir effect with cylindrical waveguides. *J. Opt. B Quantum Semiclass. Opt.* **2005**, *7*, S86–S88. [CrossRef]
18. Mundarain, D.F.; Maia Neto, P.A. Quantum radiation in a plane cavity with moving mirrors. *Phys. Rev. A* **1998**, *57*, 1379–1390. [CrossRef]
19. Eberlein, C. Sonoluminescence as quantum vacuum radiation. *Phys. Rev. Lett.* **1996**, *76*, 3842–3845. . [CrossRef]
20. Crocce, M.; Dalvit, D.A.; Lombardo, F.C.; Mazzitelli, F.D. Hertz potentials approach to the dynamical Casimir effect in cylindrical cavities of arbitrary section. *J. Opt. B Quantum Semiclass. Opt.* **2005**, *7*, S32–S39. [CrossRef]
21. Crocce, M.; Dalvit, D.A.; Mazzitelli, F.D. Resonant photon creation in a three-dimensional oscillating cavity. *Phys. Rev. A* **2001**, *64*, 013808. [CrossRef]
22. Dodonov, V.V.; Klimov, A.B. Generation and detection of photons in a cavity with a resonantly oscillating boundary. *Phys. Rev. A* **1996**, *53*, 2664–2682. [CrossRef]
23. Alves, D.T.; Farina, C.; Maia Neto, P.A. Dynamical Casimir effect with Dirichlet and Neumann boundary conditions. *J. Phys. A Math. Gen.* **2003**, *36*, 11333–11342. [CrossRef]
24. Alves, D.T.; Farina, C.; Granhen, E.R. Dynamical Casimir effect in a resonant cavity with mixed boundary conditions. *Phys. Rev. A* **2006**, *73*, 063818. [CrossRef]
25. Alves, D.T.; Granhen, E.R. Energy density and particle creation inside an oscillating cavity with mixed boundary conditions. *Phys. Rev. A* **2008**, *77*, 015808. [CrossRef]
26. Alves, D.T.; Granhen, E.R.; Lima, M.G. Quantum radiation force on a moving mirror with Dirichlet and Neumann boundary conditions for a vacuum, finite temperature, and a coherent state. *Phys. Rev. D* **2008**, *77*, 125001. [CrossRef]
27. Alves, D.T.; Granhen, E.R.; Silva, H.O.; Lima, M.G. Exact behavior of the energy density inside a one-dimensional oscillating cavity with a thermal state. *Phys. Lett. A* **2010**, *374*, 3899–3907. [CrossRef]
28. Good, M.R.R.; Anderson, P.R.; Evans, C.R. Mirror reflections of a black hole. *Phys. Rev. D* **2016**, *94*, 065010. [CrossRef]
29. Good, M.R.R.; Zhakenuly, A.; Linder, E.V. Mirror at the edge of the universe: Reflections on an accelerated boundary correspondence with de Sitter cosmology. *Phys. Rev. D* **2020**, *102*, 045020. [CrossRef]
30. Good, M.R.R.; Lapponi, E.; Luongo, O.; Mancini, S. Quantum communication through a partially reflecting accelerating mirror. *Phys. Rev. D* **2021**, *104*, 105020. [CrossRef]
31. Haro, J.; Elizalde, E. Hamiltonian approach to the dynamical Casimir effect. *Phys. Rev. Lett.* **2006**, *97*, 130401. [CrossRef]
32. Haro, J.; Elizalde, E. Physically sound Hamiltonian formulation of the dynamical Casimir effect. *Phys. Rev. D* **2007**, *76*, 065001. [CrossRef]
33. Jackel, M.-T.; Reynaud, S. Fluctuations and dissipation for a mirror in vacuum. *Quantum Opt. J. Eur. Opt. Soc. Part B* **1992**, *4*, 39–53. [CrossRef]
34. Barton, G.; Eberlein, C. On quantum radiation from a moving body with finite refractive index. *Ann. Phys.* **1993**, *227*, 222–274. [CrossRef]
35. Barton, G.; Calogeracos, A. On the quantum electrodynamics of a dispersive mirror. I. Mass shifts, radiation, and radiative reaction. *Ann. Phys.* **1995**, *238*, 227–267. [CrossRef]
36. Barton, G.; Calogeracos, A. On the quantum electrodynamics of a dispersive mirror. II. The Boundary condition and the applied force via Dirac's theory of constraints. *Ann. Phys.* **1995**, *238*, 268–285. [CrossRef]
37. Lambrecht, A.; Jaekel, M.-T.; Reynaud, S. Motion induced radiation from a vibrating cavity. *Phys. Rev. Lett.* **1996**, *77*, 615–618. [CrossRef] [PubMed]
38. Obadia, N.; Parentani, R. Notes on moving mirrors. *Phys. Rev. D* **2001**, *64*, 044019. [CrossRef]
39. Nicolaevici, N. Quantum radiation from a partially reflecting moving mirror. *Class. Quant. Grav.* **2001**, *18*, 619–628. [CrossRef]
40. Haro, J.; Elizalde, E. Black hole collapse simulated by vacuum fluctuations with a moving semitransparent mirror. *Phys. Rev. D* **2008**, *77*, 045011. [CrossRef]
41. Nicolaevici, N. Semitransparency effects in the moving mirror model for Hawking radiation. *Phys. Rev. D* **2009**, *80*, 125003. [CrossRef]
42. Fosco, C.D.; Giraldo, A.; Mazzitelli, F.D. Dynamical Casimir effect for semitransparent mirrors. *Phys. Rev. D* **2017**, *96*, 045004. [CrossRef]
43. Dalvit, D.A.R.; Maia Neto, P.A. Decoherence via the dynamical Casimir effect. *Phys. Rev. Lett.* **2000**, *84*, 798–801. [CrossRef]
44. Muñoz-Castañeda, J.M.; Guilarte, J.M. δ-δ' generalized Robin boundary conditions and quantum vacuum fluctuations. *Phys. Rev. D* **2015**, *91*, 025028. [CrossRef]
45. Braga, A.N.; Silva, J.D.L.; Alves, D.T. Casimir force between δ-δ' mirrors transparent at high frequencies. *Phys. Rev. D* **2016**, *94*, 125007. [CrossRef]
46. Silva, J.D.L.; Braga, A.N.; Alves, D.T. Dynamical Casimir effect with δ-δ' mirrors. *Phys. Rev. D* **2016**, *94*, 105009. [CrossRef]
47. Silva, J.D.L.; Braga, A.N.; Rego, A.L.; Alves, D.T. Motion induced by asymmetric excitation of the quantum vacuum. *Phys. Rev. D* **2020**, *102*, 125019. [CrossRef]

48. Rego, A.L.; Braga, A.N.; Silva, J.D.L.; Alves, D.T. Dynamical Casimir effect enhanced by decreasing the mirror reflectivity. *Phys. Rev. D* **2022**, *105*, 025013. [CrossRef]
49. Jaekel, M.T.; Reynaud, S. Casimir force between partially transmitting mirrors. *J. Phys. I* **1991**, *1*, 1395–1409. [CrossRef]
50. Maghrebi, M.F.; Golestanian, R.; Kardar, M. Scattering approach to the dynamical Casimir effect. *Phys. Rev. D* **2013**, *87*, 025016. [CrossRef]
51. Kurasov, P.B.; Scrinzi, A.; Elander, N. δ' potential arising in exterior complex scaling. *Phys. Rev. A* **1994**, *49*, 5095–5097. [CrossRef]
52. Kurasov, P. Distribution theory for discontinuous test functions and differential operators with generalized coefficients. *J. Math. Anal. Appl.* **1996**, *201*, 297–323. [CrossRef]
53. Gadella, M.; Negro, J.; Nieto, L. Bound states and scattering coefficients of the- $a\delta(x) + b\delta'(x)$ potential. *Phys. Lett. A* **2009**, *373*, 1310–1313. [CrossRef]
54. Kim, W.J.; Brownell, J.H.; Onofrio, R. Detectability of dissipative motion in quantum vacuum via superradiance. *Phys. Rev. Lett.* **2006**, *96*, 200402. [CrossRef] [PubMed]
55. Brownell, J.H.; Kim, W.J.; Onofrio, R. Modelling superradiant amplification of Casimir photons in very low dissipation cavities. *J. Phys. A Math. Theor.* **2008**, *41*, 164026. [CrossRef]
56. Motazedifard, A.; Dalafi, A.; Naderi, M.; Roknizadeh, R. Controllable generation of photons and phonons in a coupled Bose–Einstein condensate-optomechanical cavity via the parametric dynamical Casimir effect. *Ann. Phys.* **2018**, *396*, 202–219. [CrossRef]
57. Sanz, M.; Wieczorek, W.; Gröblacher, S.; Solano, E. Electro-mechanical Casimir effect. *Quantum* **2018**, *2*, 91. [CrossRef]
58. Qin, W.; Macrì, V.; Miranowicz, A.; Savasta, S.; Nori, F. Emission of photon pairs by mechanical stimulation of the squeezed vacuum. *Phys. Rev. A* **2019**, *100*, 062501. [CrossRef]
59. Butera, S.; Carusotto, I. Mechanical backreaction effect of the dynamical Casimir emission. *Phys. Rev. A* **2019**, *99*, 053815. [CrossRef]
60. Nation, P.B.; Johansson, J.R.; Blencowe, M.P.; Nori, F. Colloquium: Stimulating uncertainty: Amplifying the quantum vacuum with superconducting circuits. *Rev. Mod. Phys.* **2012**, *84*, 1–24. [CrossRef]
61. Wilson, C.M.; Johansson, G.; Pourkabirian, A.; Simoen, M.; Johansson, J.R.; Duty, T.; Nori, F.; Delsing, P. Observation of the dynamical Casimir effect in a superconducting circuit. *Nature* **2011**, *479*, 376–379. [CrossRef] [PubMed]
62. Schützhold, R.; Plunien, G.; Soff, G. Quantum radiation in external background fields. *Phys. Rev. A* **1998**, *58*, 1783–1793. [CrossRef]
63. Dodonov, V.V.; Klimov, A.B.; Nikonov, D.E. Quantum phenomena in nonstationary media. *Phys. Rev. A* **1993**, *47*, 4422–4429. [CrossRef]
64. Braggio, C.; Bressi, G.; Carugno, G.; Del Noce, C.; Galeazzi, G.; Lombardi, A.; Palmieri, A.; Ruoso, G.; Zanello, D. A novel experimental approach for the detection of the dynamical Casimir effect. *Europhys. Lett. (EPL)* **2005**, *70*, 754–760. [CrossRef]
65. De Liberato, S.; Ciuti, C.; Carusotto, I. Quantum vacuum radiation spectra from a semiconductor microcavity with a time-modulated vacuum Rabi frequency. *Phys. Rev. Lett.* **2007**, *98*, 103602. [CrossRef] [PubMed]
66. Günter, G.; Anappara, A.A.; Hees, J.; Sell, A.; Biasiol, G.; Sorba, L.; De Liberato, S.; Ciuti, C.; Tredicucci, A.; Leitenstorfer, A.; et al. Sub-cycle switch-on of ultrastrong light–matter interaction. *Nature* **2009**, *458*, 178–181. [CrossRef]
67. Johansson, J.R.; Johansson, G.; Wilson, C.M.; Nori, F. Dynamical Casimir effect in a superconducting coplanar waveguide. *Phys. Rev. Lett.* **2009**, *103*, 147003. [CrossRef]
68. Wilson, C.M.; Duty, T.; Sandberg, M.; Persson, F.; Shumeiko, V.; Delsing, P. Photon generation in an electromagnetic cavity with a time-dependent boundary. *Phys. Rev. Lett.* **2010**, *105*, 233907. [CrossRef] [PubMed]
69. Dezael, F.X.; Lambrecht, A. Analogue Casimir radiation using an optical parametric oscillator. *EPL (Europhys. Lett.)* **2010**, *89*, 14001. [CrossRef]
70. Lähteenmäki, P.; Paraoanu, G.S.; Hassel, J.; Hakonen, P.J. Dynamical Casimir effect in a Josephson metamaterial. *Proc. Natl. Acad. Sci. USA* **2013**, *110*, 4234–4238. [CrossRef]
71. Schneider, B.H.; Bengtsson, A.; Svensson, I.M.; Aref, T.; Johansson, G.; Bylander, J.; Delsing, P. Observation of broadband entanglement in microwave radiation from a single time-varying boundary condition. *Phys. Rev. Lett.* **2020**, *124*, 140503. [CrossRef]
72. Vezzoli, S.; Mussot, A.; Westerberg, N.; Kudlinski, A.; Dinparasti Saleh, H.; Prain, A.; Biancalana, F.; Lantz, E.; Faccio, D. Optical analogue of the dynamical Casimir effect in a dispersion-oscillating fibre. *Commun. Phys.* **2019**, *2*, 84. . [CrossRef]
73. Torricelli, G.; van Zwol, P.J.; Shpak, O.; Binns, C.; Palasantzas, G.; Kooi, B.J.; Svetovoy, V.B.; Wuttig, M. Switching Casimir forces with phase-change materials. *Phys. Rev. A* **2010**, *82*, 010101. [CrossRef]
74. Banishev, A.A.; Chang, C.-C.; Castillo-Garza, R.; Klimchitskaya, G.L.; Mostepanenko, V.M.; Mohideen, U. Modifying the Casimir force between indium tin oxide film and Au sphere. *Phys. Rev. B* **2012**, *85*, 045436. [CrossRef]
75. Wegkamp, D.; Stähler, J. Ultrafast dynamics during the photoinduced phase transition in VO_2. *Prog. Surf. Sci.* **2015**, *90*, 464–502. [CrossRef]
76. Mogunov, I.A.; Fernández, F.; Lysenko, S.; Kent, A.J.; Scherbakov, A.V.; Kalashnikova, A.M.; Akimov, A.V. Ultrafast insulator-metal transition in VO_2 nanostructures assisted by picosecond strain pulses. *Phys. Rev. Appl.* **2019**, *11*, 014054. [CrossRef]
77. Sood, A.; Shen, X.; Shi, Y.; Kumar, S.; Park, S.J.; Zajac, M.; Sun, Y.; Chen, L.-Q.; Ramanathan, S.; Wang, X.; et al. Universal phase dynamics in VO_2 switches revealed by ultrafast operando diffraction. *Science* **2021**, *373*, 352–355. [CrossRef] [PubMed]

78. Shabanpour, J.; Beyraghi, S.; Cheldavi, A. Ultrafast reprogrammable multifunctional vanadium-dioxide-assisted metasurface for dynamic THz wavefront engineering. *Sci. Rep.* **2020**, *10*, 8950. . [CrossRef] [PubMed]
79. Mintz, B.; Farina, C.; Maia Neto, P.A.; Rodrigues, R.B. Particle creation by a moving boundary with a Robin boundary condition. *J. Phys. A Math. Gen.* **2006**, *39*, 11325–11333. [CrossRef]
80. Mintz, B.; Farina, C.; Maia Neto, P.A.; Rodrigues, R.B. Casimir forces for moving boundaries with Robin conditions. *J. Phys. A Math. Gen.* **2006**, *39*, 6559–6565. [CrossRef]
81. Rego, A.L.; Mintz, B.; Farina, C.; Alves, D.T. Inhibition of the dynamical Casimir effect with Robin boundary conditions. *Phys. Rev. D* **2013**, *87*, 045024. [CrossRef]
82. Ford, L.H.; Vilenkin, A. Quantum radiation by moving mirrors. *Phys. Rev. D* **1982**, *25*, 2569–2575. [CrossRef]
83. Silva, H.O.; Farina, C. Simple model for the dynamical Casimir effect for a static mirror with time-dependent properties. *Phys. Rev. D* **2011**, *84*, 045003. [CrossRef]
84. Mostepanenko, V.M.; Trunov, N.N. Quantum field theory of the Casimir effect for real media. *Sov. J. Nucl. Phys.* **1985**, *42*, 818–822.
85. Farina, C.; Silva, H.O.; Rego, A.L.; Alves, D.T. Time-dependent Robin boundary conditions in the dynamical Casimir effect. *Int. J. Mod. Phys. Conf. Ser* **2012**, *14*, 306–315. [CrossRef]
86. Sinha, S.; Sorkin, R.D. Brownian motion at absolute zero. *Phys. Rev. B* **1992**, *45*, 8123–8126. . [CrossRef] [PubMed]
87. Jaekel, M.T.; Reynaud, S. Quantum fluctuations of position of a mirror in vacuum. *J. Phys. I France* **1993**, *3*, 1–20. [CrossRef]
88. Stargen, D.J.; Kothawala, D.; Sriramkumar, L. Moving mirrors and the fluctuation-dissipation theorem. *Phys. Rev. D* **2016**, *94*, 025040. [CrossRef]
89. Wang, Q.; Zhu, Z.; Unruh, W.G. How the huge energy of quantum vacuum gravitates to drive the slow accelerating expansion of the Universe. *Phys. Rev. D* **2017**, *95*, 103504. [CrossRef]
90. Donaire, M. Net force on an asymmetrically excited two-atom system from vacuum fluctuations. *Phys. Rev. A* **2016**, *94*, 062701. [CrossRef]
91. Silva, J.D.L.; Braga, A.N.; Rego, A.L.; Alves, D.T. Interference phenomena in the dynamical Casimir effect for a single mirror with Robin conditions. *Phys. Rev. D* **2015**, *92*, 025040. [CrossRef]
92. Donaire, M.; van Tiggelen, B.; Rikken, G.L.J.A. Casimir momentum of a chiral molecule in a magnetic field. *Phys. Rev. Lett.* **2013**, *111*, 143602. [CrossRef] [PubMed]
93. Donaire, M.; Van Tiggelen, B.A.; Rikken, G.L.J.A. Transfer of linear momentum from the quantum vacuum to a magnetochiral molecule. *J. Phys. Cond. Matter* **2015**, *27*, 214002. [CrossRef]
94. Feigel, A. Quantum vacuum contribution to the momentum of dielectric media. *Phys. Rev. Lett.* **2004**, *92*, 020404. [CrossRef] [PubMed]
95. Croze, O.A. Does the Feigel effect break the first law? *arXiv* **2013**, arXiv:1304.3338. https://doi.org/10.48550/arXiv.1304.3338.
96. Croze, O.A. Alternative derivation of the Feigel effect and call for its experimental verification. *Proc. R. Soc. A Math. Phys. Eng. Sci.* **2012**, *468*, 429–447. [CrossRef]
97. Birkeland, O.J.; Brevik, I. Feigel effect: Extraction of momentum from vacuum? *Phys. Rev. E* **2007**, *76*, 066605. [CrossRef] [PubMed]

Disclaimer/Publisher's Note: The statements, opinions and data contained in all publications are solely those of the individual author(s) and contributor(s) and not of MDPI and/or the editor(s). MDPI and/or the editor(s) disclaim responsibility for any injury to people or property resulting from any ideas, methods, instructions or products referred to in the content.

Article

Fluctuations-Induced Quantum Radiation and Reaction from an Atom in a Squeezed Quantum Field

Matthew Bravo [1], Jen-Tsung Hsiang [2],*and Bei-Lok Hu [3]

[1] Department of Physics, University of Maryland, College Park, MD 20742, USA; mbravo@terpmail.umd.edu
[2] Center for High Energy and High Field Physics, National Central University, Taoyuan 320317, Taiwan
[3] Maryland Center for Fundamental Physics and Joint Quantum Institute, University of Maryland, College Park, MD 20742, USA; blhu@umd.edu
* Correspondence: cosmology@gmail.com

Abstract: In this third of a series on quantum radiation, we further explore the feasibility of using the memories (non-Markovianity) kept in a quantum field to decipher certain information about the early universe. As a model study, we let a massless quantum field be subjected to a parametric process for a finite time interval such that the mode frequency of the field transits from one constant value to another. This configuration thus mimics a statically-bounded universe, where there is an 'in' and an 'out' state with the scale factor approaching constants, not a continuously evolving one. The field subjected to squeezing by this process should contain some information of the process itself. If an atom is coupled to the field after the parametric process, its response will depend on the squeezing, and any quantum radiation emitted by the atom will carry this information away so that an observer at a much later time may still identify it. Our analyses show that (1) a remote observer cannot measure the generated squeezing via the radiation energy flux from the atom because the net radiation energy flux is canceled due to the correlation between the radiation field from the atom and the free field at the observer's location. However, (2) there is a chance to identify squeezing by measuring the constant radiation energy density at late times. The only restriction is that this energy density is of the near-field nature and only an observer close to the atom can use it to unravel the information of squeezing. The second part of this paper focuses on (3) the dependence of squeezing on the functional form of the parametric process. By explicitly working out several examples, we demonstrate that the behavior of squeezing does reflect essential properties of the parametric process. Actually, striking features may show up in more complicated processes involving various scales. These analyses allow us to establish the connection between properties of a squeezed quantum field and details of the parametric process which performs the squeezing. Therefore, (4) one can construct templates to reconstitute the unknown parametric processes from the data of measurable quantities subjected to squeezing. In a sequel paper these results will be applied to a study of quantum radiations in cosmology.

Keywords: parametric creation of particles; squeezed state; fluctuations-induced quantum radiation; radiation reaction; non-Markovianity

1. Introduction

This paper is the third of a series on quantum radiation, in the form of emitted radiation with an energy flux (distinct from thermal radiance felt by an accelerated atom, as in the Unruh effect [1]) from the internal degrees of freedom (idf) of an atom, tracing its origin to the vacuum fluctuations of a quantum field, including its backreaction on the idf dynamics of the atom, in the form of quantum dissipation. In Ref. [2], two of us considered how vacuum fluctuations in the field act on the idf of an atom (we may call this the 'emitter'), modeled by a harmonic oscillator. We first showed how a stochastic component of the internal dynamics of the atom arises from the vacuum fluctuations of the field, resulting in

Citation: Bravo, M.; Hsiang, J.-T.; Hu, B.-L. Fluctuations-Induced Quantum Radiation and Reaction from an Atom in a Squeezed Quantum Field. *Physics* **2023**, *5*, 554–589. https://doi.org/10.3390/physics5020040

Received: 13 March 2023
Accepted: 10 May 2023
Published: 24 May 2023

Copyright: © 2023 by the authors. Licensee MDPI, Basel, Switzerland. This article is an open access article distributed under the terms and conditions of the Creative Commons Attribution (CC BY) license (https://creativecommons.org/licenses/by/4.0/).

the emittance of quantum radiation. We then showed how the backreaction of this quantum radiation induces quantum dissipation in the atom's idf dynamics. We explicitly identified the different terms representing these processes in the Langevin equations of motion. Then, using the example of a stationary atom, we showed how, in this case, the absence of radiation at a far-away observation point where a probe (or detector—note the so-called Unruh-DeWitt 'detector' [1,3] is an emitter in the present context) is located is actually a result of complex cancellations of the interference between emitted radiation from the atom's idf and the local fluctuations in the free field. By this, we pointed out that the entity which enters into the duality relation with vacuum fluctuations is not radiation reaction (in the quantum optics literature, e.g., [4–7], the relation between quantum fluctuations and radiation reaction is often mentioned without emphasizing the difference between classical radiation reaction [8,9] and quantum dissipation, which exist at two separate theoretical levels; only quantum dissipation enters in the fluctuations–dissipation relation with quantum fluctuations, not classical radiation reaction), which can exist as a classical entity, but quantum dissipation [10,11]. In the second paper [12], we considered the idf of the atom interacting with a quantum scalar field initially in a coherent state. We showed how the deterministic mean field drives the internal classical mean component to emit classical radiation and receive classical radiation reaction. Both components are statistically distinct and fully decoupled. It is clearly seen that the effects of the vacuum fluctuations of the field are matched with those of quantum radiation reaction, not with classical radiation reaction, as the folklore states, even promulgated in some textbooks. Furthermore, we identified the reason why quantum radiation from a stationary emitter is not observed, and a probe located far away only sees classical radiation.

1.1. Quantum Radiation from an Atom in a Squeezed Quantum Field

In this paper, we treat quantum radiation from an atom's idf interacting with a quantum field in a squeezed state. The squeezed state is probably the most important quantum state, next to the vacuum state, with both rich theoretical meanings and broad practical applications, as well known in quantum optics (see, e.g., [13]). We are particularly interested in its role in cosmology of the early universe.

Quantum Field Squeezed by an Expanding Universe

Squeezed states of a quantum field are naturally produced in an expanding universe in fundamental processes which involve the parametric amplification of quantum fluctuations, such as particle creation [14,15], either spontaneous production from the vacuum or stimulated production from n-particle states, and structure formation [16] from quantum fluctuations of the inflaton field. From relics such as primordial radiation and matter contents observed today, with the help of theoretical models governing their evolution, one attempts to deduce the state of the early universe at different stages of development. In addition to particle creation and structure formation, we add here another fundamental quantum process, namely, quantum radiation. One can ask questions such as: If such radiation of a quantum nature is detected, how can one use it to reveal certain quantum aspects of the early universe? Uncovering secrets of the early universe by digging out information stored in the quantum field is in a similar spirit to the quest we initiated about the non-Markovianity of the universe through memories kept in the quantum field [17].

1.2. Three Components: Radiation, Squeeze, Drive

Towards such a goal we carry out this investigation which involves three components: (1) quantum radiation, (2) squeezed state, and (3) driven dynamics, either under some external force or as provided by an expanding universe, which describes how the squeeze parameter changes in time. Component (1) was initiated in Ref. [2] and continued in Ref. [12], where the required technical tools for the present investigation can be found. Component (2) is performed here: we consider a quantum field in a squeezed state with a fixed squeeze parameter, regarded as the end state of the quantum field after being squeezed under some drive protocol or cosmological evolution. For the description of squeezed states we follow the descriptions in our earlier paper [18]. Under the classification of the three types of squeezing described there, we adhere to the first type, namely, with fixed squeezing. The second type of dynamical parametric squeezing will be treated in our sequel cosmological papers. We set aside the third type of squeezing due to finite coupling between the system (here, the atom's idf) and the environment (here, the quantum field) completely.

To see more explicitly what component (3) entails, consider a situation where the atom at the initial time t_i is in a quantum field in a squeezed state with a fixed squeeze parameter ζ_i and the same atom at the final time, t_f, in a squeezed field with a different squeeze parameter, ζ_f. If an atom emits quantum radiation at either or both the initial and the final times, comparing the signals from both should tell us something about how the drive had affected the atom through the quantum radiation emitted from the atom, or in cosmology, how the universe had evolved as measured by the parametric squeezing of the field. The more challenging situation is if it turns out that there is no emitted quantum radiation from an atom in a squeezed field. This is what our next work intends to find out.

1.3. Our Objectives, in Two Stages

We wish to ask questions in the same spirit as in Ref. [17]: Can we extract information about the history of the quantum field which had undergone a parametric process from the responses of an atom coupled to it in the epoch after the parametric process? In particular, for observers (receivers) situated far away from the atom (emitter), whether they can detect radiations emitted from the atom of a quantum nature. Quantum radiation is of special interest as it originates from the vacuum fluctuations of a quantum field, and is expected to keep some memories of the parametric process it went through. In a cosmological context, it acts as a carrier of quantum information about the early universe.

We divide our program of quantum radiation in cosmology into two stages, phrased as two questions. (A): Is there emitted quantum radiation from a stationary atom in a quantum field with a fixed squeeze parameter? If there is no emitted quantum radiation, then question (B): Is there emitted quantum radiation from a stationary atom in a quantum field subjected to a changing squeeze parameter, such as in an expanding universe? If so, what kind of evolutionary dynamics would produce what types of emitted radiation and with what magnitudes? The first stage addresses components (1) and (2) listed above, and the second stage, components (2) and (3), which will be continued in a follow-up paper.

1.3.1. Radiation Pattern as Template for Squeezing

This paper operates in the first stage, with a setup meant for a statically-bounded universe, not the continuously evolving type such as in the Friedmann–Lemaître–Robertson–Walker (FLRW) or the inflationary universe. The first half answers the first question (A), and the second half of this paper examines the response of the atom coupled to a field that had undergone a parametric process earlier. This evokes the template idea, i.e., one can use the response of the atom, coupled to the field in the out-region, to identify the dependence of the squeeze parameter on the earlier parametric process before the out-region. In the cosmology context, this offers a way for a late-time observer of such quantum phenomena to uncover how the universe had evolved in much earlier times.

1.3.2. Stress–Energy Tensor of Squeezed Field

To answer the central question (A), we use a simplified model in which the quantum field undergoes a parametric process of finite duration, such as under an external drive for a definite period of time, or in an asymptotically stationary (statically bounded) universe. The field could be a quantized matter or graviton field, or an inflaton field whose quantum fluctuations engender cosmological structures. Then, we calculate the expectation value of the stress–energy tensor of the radiation field emitted by an atom coupled to a free field in a fixed two-mode squeeze state. In order to identify the unambiguous signals and conform to the typical settings, we focus on the late-time results. We learned from our earlier invetsigation [2] for a quantum field in the vacuum state that the procedure for checking this is quite involved, as it entails both the radiation flux emitted from the atom as detected at the spacetime point of the probe, and an incoming flux from infinity.

1.4. Key Steps and Major Findings

The answer we found after calculating all relevant contributions for an atom in a quantum field in a squeezed state shows that there is no net radiation flux, the same as in Ref. [2]. This is a consequence of relaxation dynamics of the atom's internal dynamics when it is coupled to the squeezed field [19]. However, if the probe can measure the radiation energy density, it should obtain a residual constant radiation energy density at late time. Its value will fall off similar to the inverse cubic power of the distance between the probe and the atom. Thus, it is more similar to a near-field effect. Nonetheless, this energy density has an interesting characteristic: it depends on the squeeze parameter, which is what we are after.

Thus, our investigation turns to whether and how the squeeze parameter of the field after the parametric process would depend on the details of the process. We first show that the squeeze parameter can formally be expressed by the fundamental solutions of the parametrically driven field; the latter contains useful information about the process. Then, by working out several examples numerically we can make the following observations:

(1) For a monotonically varying process, the squeeze parameter has a monotonic dependence on the duration of this process; it does not depend on when the process starts, if we fix the duration.
(2) The magnitude of squeezing is related to the rate of change in the process. That is, large squeezing can be induced from a nonadiabatic transition. This is consistent with our understanding of spontaneous particle creation from parametric amplification of vacuum fluctuations [20] and that copious particles can be produced at the Planck time under rapid expansion of the universe [21]. Thus, we expect that nonadiabatic processes may contribute to larger residual radiation energy density around the atom.
(3) For a nonmonotonic parametric process, various scales in the process induce richer structure to the behavior of the squeeze parameter. In particular, if the parametric process changes with time sinusoidally at some frequency range, it may induce parametric resonance and yield exceptionally large squeezing in the out-region.

Similar considerations can likewise be applied to the frequency spectrum of the squeeze parameter of the field in the out-region. One can then examine its dependence on the parametric process which the field has undergone. This illustrates the way to obtain templates in how the squeeze parameter is related to the parametric process, and how certain information of the unknown parametric process can be inferred from these templates.

1.5. Organization

The paper is organized as follows. In Section 2, as a prerequisite, we briefly summarize our earlier results on the relaxation process of the atom–field interaction, and pose the questions we would like to answer in this paper. In Section 3, we lay out the formalism and the essential tools for detailing the nonequilibrium evolution of the atom's internal dynamics and the squeezed field when they are coupled together. In Section 4, we study the general spatial–temporal behavior of the energy flux and the energy density of the radiation field and examine their late-time behaviors. In Section 5, we turn to the functional dependence of the squeeze parameter on the functional form of the parametric process. Several examples are provided to illustrate the formal analysis. In Section 6, we give some concluding remarks. Appendix A offers a succinct summary of the two-mode squeezed state. In Appendices B and C, we offer more details on the late-time, large-distance behaviors of the energy flux and the energy density of the radiation field. In Appendix D, we discuss the time-translational invariance of the squeeze parameter in the out-region.

2. Scenario: Quantum Radiation in Atom–Field Systems

Classical radiation is a familiar subject, but what is its quantum field origin? Can one trace back its link all the way to quantum fluctuations? We obtained partial answers in our last two papers to this question for a harmonic atom interacting with a quantum field in a vacuum and in a coherent state. This paper deals with a squeezed field, necessary for treating cosmological quantum processes. Since fluctuation-induced quantum radiation is not a household topic, yet it might be useful to first provide a physical picture of the global landscape of quantum radiation based on our understanding from previous studies, setting the stage for the current paper.

It has been shown that [19] when the internal degree of freedom, modeled as a harmonic oscillator, of an atom (called a harmonic atom) in any initial state is linearly coupled to a massless linear scalar field in the stationary state, its motion will settle down to an equilibrium state. The presence of this equilibrium state, from the perspective of the atom, implies a balance of energy exchange between the atom and the environmental field. Expectantly, the radiation generated by the nonequilibrium motion of the atom's internal degree of freedom propagates outward to spatial infinity. It is lesser known whether and how, from the perspective of the field, the radiation energy from the atom is balanced. In our previous study [2], we demonstrated that the correlation between the outward radiation field and the local free field at spatial infinity constitutes an inward energy flow to balance the outward radiation energy flow. Furthermore, this inward flux serves to supplement the needed energy around the atom for the field to drive the atom's internal motion. Thus, we clearly see how energy flows from the atom to spatial infinity and then backflows to the atom. This is a consequence of the nonequilibrium fluctuation–dissipation relations [2,22] for the atom's internal degree of freedom and the free field, which is a stronger condition than the conservation of energy. This point is better appreciated once we take the global view of the total system, composed of the atom and field. The entire system is closed and the total energy is conserved, but this does not guarantee that the energy exchange between two subsystems is balanced unless the reduced dynamics is fully relaxed to an equilibrium state.

If, instead, the internal degree of freedom of the atom is initially coupled to the quantum field in a nonstationary state, similar to the squeezed state, then will the nonstationary nature of the field prevent the internal degree of freedom from approaching an equilibrium state? We showed that [19] when the (mode-dependent) squeeze parameter is time-independent, the atom's internal degree of freedom still settles down to an equilibrium state in the long run. The stationary components of the covariance matrix elements or the energy flows decay exponentially fast to time-independent constants as in the previous case, but the additional nonstationary components fall off to zero. The mechanisms that account for the late-time behavior of the stationary and the nonstationary components seem to be quite distinct. For the stationary components, it is a consequence of dissipation dynamically adapting to the driving fluctuations from the environment, but for the nonstationary components, cancellation due to fast phase variations in the nonstationary terms plays the decisive role. In contrast, from the viewpoint of the squeezed field, what is the nature of radiation emitted from the atom, and can this outward energy flux at distances far away from the atom find a corresponding inward flux at late times such that there is no net energy output to spatial infinity, as proven for the case of a stationary field? If so, what makes it possible?

These questions are of particular interest in a cosmological setting for the consideration of fundamental issues described in Section 1. There, evolution of the universe parametrically drives the ambient quantum field into a squeezed state. The extent of squeezing may depend on the characteristics of the parametric process actuated by the evolution of the background universe. When an atom is coupled to such a field, emitted radiation from the atom should carry the information of squeezing of the field, which in turn may reveal the evolution history of the universe.

As a prerequisite to addressing these issues, in this paper we are going to examine a simpler configuration that may cover the essential features. Consider, in Minkowski spacetime, a massless scalar field that undergoes a parametric process such that each mode frequency changes smoothly from one constant value, for example, ω_i to another ω_f, as shown in Figure 1. The parametric process occurs during the time interval $t_a \leq t \leq t_b$. In the out-region ($t \geq t_b$) of the process, an atom is coupled to the squeezed field at time t_0. The nonequilibrium evolution of the internal degree of freedom of the atom then generates outward radiation, which after time r will reach a detector (such as a satellite around the Earth) at a distance r away from the atom. The detector may measure the time variation of the energy flux. The corresponding signal will be extracted and amplified to sieve out the information regarding the parametric process of the field occurred earlier.

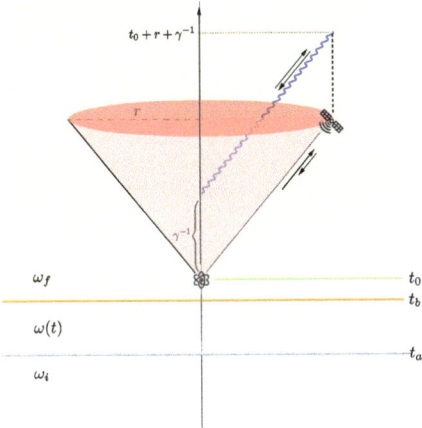

Figure 1. The configuration of the parametric process of the field. The process changes one of the mode frequencies from ω_i at time $t \leq t_a$ to ω_f at $t \geq t_b$. The transition is assumed to happen continuously. The interaction between an atom and the field is turned on at $t_0 > t_a$ and radiation is generated by the internal motion of the atom. For a receiver at distance r away from the atom, it will take another time r to meet the radiation. The relaxation time of the atom's internal motion is of the order of the inverse of the damping constant, γ^{-1}.

3. Massless Scalar Field Interacting with a Harmonic Atom

For the case of a massless scalar field ϕ coupled to a static atom, with its internal degree of freedom, χ, represented by a harmonic oscillator, one has the following Heisenberg equations of motion:

$$\ddot{\hat{\chi}}(t) + \omega_B^2 \hat{\chi}(t) = \frac{e}{m} \hat{\phi}(0, t), \tag{1}$$

$$(\partial_t^2 - \nabla^2) \hat{\phi}(x, t) = e \, \delta^{(3)}(x) \hat{\chi}(t), \tag{2}$$

where the atom is located at the origin of the spatial coordinates x in a $3+1$ Minkowski spacetime. The parameter e denotes the coupling strength between the internal degree of freedom of the atom and the field, and ω_B and m is the bare frequency and the mass of the oscillator, respectively. The overhead dot of a variable represents its time derivative with respect to t; if this variable is promoted to an operator, we add an overhead hat. $\delta^{(3)}(x)$ is the three-dimensional delta function, and we often use the shorthand notation, $\partial_a = \partial/\partial a$, for the partial derivative with respect to a.

Solving the equation of motion of the field (2) yields

$$\hat{\phi}(x, t) = \hat{\phi}_h(x, t) + e \int_0^t ds \, G_{R,0}^{(\phi)}(x, t; 0, s) \hat{\chi}(s), \tag{3}$$

where the homogeneous part $\hat{\phi}_h(x,t)$ gives the free field while the integral expression represents the radiation field emitted by the atom. The retarded Green's function $G_{R,0}^{(\phi)}(x,x')$ of the free field is defined by

$$G_{R,0}^{(\phi)}(x,x') = i\theta(t-t')\left[\hat{\phi}_h(x),\hat{\phi}_h(x')\right] = \frac{1}{2\pi}\theta(\tau)\delta(\tau^2 - R^2), \tag{4}$$

where $x = (x,t)$, $\tau = t-t'$, $R = |x-x'|$, and $\theta(\tau)$ is the Heaviside unit-step function. In the special case $R \to 0$, which is met for a single Brownian oscillator, we instead have

$$G_{R,0}^{(\phi)}(t,t') = -\frac{1}{2\pi}\theta(\tau)\delta'(\tau). \tag{5}$$

Here, we suppressed the trivial spatial coordinates $\mathbf{0}$, and the prime of a function denotes the derivative with respect to its argument. The superscript of the Green's function indicates the operator by which the Green's function is constructed, while the 0 in the subscript reminds us that the Green's function of interest is associated with a free operator; otherwise it is an interacting operator. For example, $G_{R,0}^{(\phi)}(\tau)$ refers to the retarded Green's function of the free field, while $G_R^{(\chi)}(\tau)$ is the retarded Green's function of the interacting χ.

Plugging Equation (3) into the equation of motion (2) for the atom's internal degree of freedom, one obtains:

$$\ddot{\hat{\chi}}(t) + \omega_B^2 \hat{\chi}(t) - \frac{e^2}{m}\int_0^t ds\, G_{R,0}^{(\phi)}(t,s)\hat{\chi}(s) = \frac{e}{m}\hat{\phi}_h(\mathbf{0},t). \tag{6}$$

This is the generalized quantum Langevin equation for $\hat{\chi}(t)$—it includes backreactions of the environmental field in terms of the noise force on the righthand side and the integral expression on the left hand side. As was identified in Equation (3), the latter is the reaction force due to emitted radiation, which retards the motion of the atom's internal degree of freedom.

Supposing the free field has a Markovian spectrum, we can reduce Equation (6) into a local form:

$$\ddot{\hat{\chi}}(t) + 2\gamma\dot{\hat{\chi}}(t) + \omega_R^2\,\hat{\chi}(t) = \frac{e}{m}\hat{\phi}_h(\mathbf{0},t), \tag{7}$$

and the oscillation frequency is renormalized to a physical value ω_R. Since we want to examine the nature of radiation and its late-time behavior in relation to the atom's internal dynamics in a self-consistent way, we need the field expression (3) written in terms of the solution to the equation of the reduced dynamics (7):

$$\hat{\chi}(t) = \hat{\chi}_h(t) + e\int_0^t ds\, G_R^{(\chi)}(t-s)\hat{\phi}_h(\mathbf{0},s), \tag{8}$$

where the retarded Green's function associated with $\hat{\chi}$ has the form

$$G_R^{(\chi)}(\tau) = \frac{1}{m\Omega}e^{-\gamma\tau}\sin\Omega\tau, \tag{9}$$

with $\Omega = \sqrt{\omega_R^2 - \gamma^2}$, γ is the damping constant, and $\hat{\chi}_h(t)$ is the homogeneous part. Equation (7) always implies that

$$\frac{d}{dt}\left[\frac{m}{2}\langle\dot{\hat{\chi}}(t)\rangle + \frac{m\omega_R}{2}\langle\hat{\chi}^2(t)\rangle\right] = P_{\hat{\xi}}(t) + P_\gamma(t), \tag{10}$$

where $P_{\hat{\xi}}(t)$ is the power delivered by the free field fluctuations:

$$P_{\hat{\xi}}(t) = \frac{e}{2}\langle\{\hat{\phi}_h(\mathbf{0},t),\dot{\hat{\chi}}(t)\}\rangle, \tag{11}$$

and $P_\gamma(t)$ represents the rate of energy lost due to dissipation:

$$P_\gamma(t) = -2m\gamma \langle \dot{\chi}^2(t) \rangle. \tag{12}$$

We showed [19] that even if the scalar field is initially in a nonstationary squeezed thermal state, the internal dynamics of the atom will still settle down to an asymptotic equilibrium state, where, in particular,

$$P_\xi(t) + P_\gamma(t) = 0. \tag{13}$$

This is a stronger condition than Equation (10), which reflects energy conservation of the reduced dynamics.

From Equations (3) and (8), the full interacting field is then given by

$$\hat{\phi}(x) = \hat{\phi}_h(x) + e \int_0^t dt'\, G_{R,0}^{(\phi)}(x,t;0,t') \left[\hat{\chi}_h(t') + e \int_0^{t'} ds\, G_R^{(\chi)}(t'-s) \hat{\phi}_h(0,s) \right]. \tag{14}$$

It turns out convenient to split the full interacting field into three physically distinct components, $\hat{\phi}(x) = \hat{\phi}_h(x) + \hat{\phi}_{TR}(x) + \hat{\phi}_{BR}(x)$, with

$$\hat{\phi}_{TR}(x) = e \int_0^t dt'\, G_{R,0}^{(\phi)}(x,t;0,t')\, \hat{\chi}_h(t'), \tag{15}$$

$$\hat{\phi}_{BR}(x) = e^2 \int_0^t dt'\, G_{R,0}^{(\phi)}(x,t;0,t') \int_0^{t'} ds\, G_R^{(\chi)}(t'-s) \hat{\phi}_h(0,s), \tag{16}$$

where $\hat{\phi}_{TR}(x)$ is the transient term associated with the atom's transient internal dynamics and $\hat{\phi}_{BR}(x)$ is the backreaction field correlated with $\hat{\phi}_h(x)$ everywhere.

Hadamard Function

At this point, we have not specified the state of the field. For the reason that is explained in Section 5, we assume that the initial state of the field the atom interacts with is a time-independent two-mode squeezed thermal state, for which the field's density matrix has the form

$$\hat{\rho}_{TMST} = \prod_k \hat{S}_2(\zeta_k) \hat{\rho}_\beta \hat{S}_2^\dagger(\zeta_k) \tag{17}$$

where $\hat{\rho}_\beta$ is the thermal density matrix of the free field at temperature $T_B = \beta^{-1}$, and $\hat{S}_2(\zeta_k)$ is the two-mode squeeze operator, with the squeeze parameter, $\zeta_k = \eta_k e^{i\theta_k}$, defined by

$$\hat{S}_2(\zeta_k) = \exp\left[\zeta_k^* \hat{a}_{+k} \hat{a}_{-k} - \zeta_k \hat{a}_{+k}^\dagger \hat{a}_{-k}^\dagger \right]. \tag{18}$$

Here, the † sign denotes the hermitian conjugate, $\eta_k \geq 0$, $2\pi > \theta_k \geq 0$, and k is the wave vector. For convenience, we collected the essential properties of the two-mode squeezed state in Appendix A. Here, we assume that the squeeze parameter is mode-independent to simplify the arguments.

If the free field has the plane-wave expansion,

$$\hat{\phi}_h(x) = \int \frac{d^3 k}{(2\pi)^{\frac{3}{2}}} \frac{1}{\sqrt{2\omega}} \left[\hat{a}_k e^{ik\cdot x} + \hat{a}_k^\dagger e^{-ik\cdot x} \right], \tag{19}$$

with $k \cdot x = -\omega t + \mathbf{k} \cdot \mathbf{x}$, $k = (\omega, \mathbf{k})$, and $\omega = |\mathbf{k}|$, then the free field's Hadamard function for this two-mode squeezed thermal state has the form,

$$G_{H,0}^{(\phi)}(x,x') = \int \frac{d^3 k}{(2\pi)^3} \frac{1}{4\omega} \coth \frac{\beta\omega}{2}\, e^{i\mathbf{k}\cdot(\mathbf{x}-\mathbf{x}')} \left[\cosh 2\eta\, e^{-i\omega(t-t')} - \sinh 2\eta\, e^{i\theta} e^{-i\omega(t+t')} \right] + \text{C.C.}. \tag{20}$$

Here, "C.C." represents the complex conjugate term. After carrying out the angular integration, one arrives at the following decomposition:

$$G_{H,0}^{(\phi)}(x,x') = G_{H,0}^{(\phi),ST}(x,x') + G_{H,0}^{(\phi),NS}(x,x'), \tag{21}$$

with

$$G_{H,0}^{(\phi),ST}(x,x') = \int_{-\infty}^{\infty} \frac{d\omega}{2\pi} \cosh 2\eta \coth \frac{\beta\omega}{2} \frac{\sin(\omega|x-x'|)}{4\pi|x-x'|} e^{-i\omega(t-t')}, \tag{22}$$

$$G_{H,0}^{(\phi),NS}(x,x') = -\int_{0}^{\infty} \frac{d\omega}{2\pi} \sinh 2\eta \coth \frac{\beta\omega}{2} \frac{\sin(\omega|x-x'|)}{4\pi|x-x'|} e^{-i\omega(t+t')+i\theta} + \text{C.C.} \tag{23}$$

Observe that the Hadamard function has an extra component that is not invariant under time translation, so we call it the nonstationary component of the Hadamard function. This is distinct from the case we consider in Ref. [2]. The emergence of this component is a consequence of the field being squeezed. We discuss its dynamical origin in Section 5. This nonstationary nature of the free field's Hadamard function will impact on the dynamical evolution of the atom's internal dynamics because the Hadamard function governs the statistics of the noise force sourcing (6), which naturally raises the concern as to whether the internal dynamics can ever equilibrate. As was mentioned earlier, it turns out the atom's internal motion can still relax to an equilibrium, meaning a time-translation-invariant state. Details of derivations and discussions on this point can be found in Ref. [19].

Equations (20) or (21) are the essential ingredient used to express the Hadamard function $G_H^{(\phi)}(x,x')$ of the interacting field, which is needed to evaluate the stress–energy tensor. Following the decomposition (15), the interacting field's Hadamard function is given by

$$G_H^{(\phi)}(x,x') = \frac{1}{2}\langle\{\hat{\phi}_h(x),\hat{\phi}_h(x')\}\rangle + \frac{1}{2}\Big[\langle\{\hat{\phi}_h(x),\hat{\phi}_{BR}(x')\}\rangle + \langle\{\hat{\phi}_{BR}(x),\hat{\phi}_h(x')\}\rangle\Big]$$
$$+ \frac{1}{2}\langle\{\hat{\phi}_{BR}(x),\hat{\phi}_{BR}(x')\}\rangle + \frac{1}{2}\langle\{\hat{\phi}_{TR}(x),\hat{\phi}_{TR}(x')\}\rangle, \tag{24}$$

where

$$\frac{1}{2}\langle\{\hat{\phi}_h(x),\hat{\phi}_h(x')\}\rangle = G_{H,0}^{(\phi)}(x,x'), \tag{25}$$

$$\frac{1}{2}\langle\{\hat{\phi}_h(x),\hat{\phi}_{BR}(x')\}\rangle = e^2 \int_0^{t'} ds_1'\, G_{R,0}^{(\phi)}(x',t';0,s_1') \int_0^{s_1'} ds_2'\, G_R^{(\chi)}(s_1'-s_2') G_{H,0}^{(\phi)}(x,t;0,s_2'), \tag{26}$$

$$\frac{1}{2}\langle\{\hat{\phi}_{BR}(x),\hat{\phi}_{BR}(x')\}\rangle = e^4 \int_0^t ds_1 \int_0^{t'} ds_1'\, G_{R,0}^{(\phi)}(x,t;0,s_1) G_{R,0}^{(\phi)}(x',t';0,s_1')$$
$$\times \int_0^{s_1} ds_2 \int_0^{s_1'} ds_2'\, G_R^{(\chi)}(s_1-s_2) G_R^{(\chi)}(s_1'-s_2') G_{H,0}^{(\phi)}(0,s_2;0,s_2'), \tag{27}$$

$$\frac{1}{2}\langle\{\hat{\phi}_{TR}(x),\hat{\phi}_{TR}(x')\}\rangle = \frac{e^2}{2} \int_0^t ds \int_0^{t'} ds'\, G_{R,0}^{(\phi)}(x,t;0,s) G_{R,0}^{(\phi)}(x',t';0,s') \langle\{\hat{\chi}_h(s),\hat{\chi}_h(s')\}\rangle. \tag{28}$$

The second group inside the square brackets in Equation (24) is of special interest because it describes the correlation between the radiation field and the free field at any location outside the atom. We do not see such counterpart in classical electrodynamics because in classical field theory, there is no vacuum state to establish any correlation with the radiation field. Here, the correlation must be present because (1) the internal dynamics of the atom, which emits the radiation, is driven by the free field at the atom's location and (2) the free field has nonvanishing correlation among any spacetime interval. Note that we are in the Heisenberg picture, so the expectation values are evaluated with respect to the initial state of both subsystems. The contribution in Equation (28) can be ignored at late times due to its transient nature. In addition, we also note that these four components are at most linearly proportional to $G_{H,0}^{(\phi)}(x,x')$. Thus, at least before $t - r \gg \gamma^{-1}$, where $r = |x|$, the interaction field is expected to behave similar to a squeezed field to some extent.

For later convenience, we introduce a shorthand notation for an expression which comes up frequently:

$$L_\omega(x,t) = \int_0^t ds_1 \, G_{R,0}^{(\phi)}(x, t - s_1) \int_0^{s_1} ds_2 \, G_R^{(\chi)}(s_1 - s_2) e^{-i\omega s_2} . \tag{29}$$

After we carry out the integrations, we find its explicit form is given by

$$L_\omega(x,t) = \theta(t-r) \widetilde{G}_R^{(\chi)}(\omega) \left\{ \widetilde{G}_{R,0}^{(\phi)}(x;\omega) e^{-i\omega t} - \frac{e^{-\gamma(t-r)}}{4\pi\Omega r} \left[(\omega + i\gamma) \cos\Omega(r - t) + i\Omega \sin\Omega(r - t) \right] \right\}, \tag{30}$$

where we used the identity for the Fourier transform of $G_{R,0}^{(\phi)}(x, \tau; 0, 0)$:

$$\widetilde{G}_{R,0}^{(\phi)}(x;\omega) = \frac{e^{i\omega r}}{4\pi r}. \tag{31}$$

The convention of the Fourier transformation we adopt is

$$f(t) = \int_{-\infty}^{\infty} \frac{d\omega}{2\pi} \, \tilde{f}(\omega) e^{-i\omega t}. \tag{32}$$

At times $t \gg r$, we find that the dominant term in Equation (30) is given by

$$L_\omega(x,t) = \theta(t-r) \widetilde{G}_R^{(\chi)}(\omega) \widetilde{G}_{R,0}^{(\phi)}(x;\omega) e^{-i\omega t}, \tag{33}$$

because in this case, the factor $e^{-\gamma(t-r)}$ causes the second term in the curly brackets in Equation (30) to be exponentially small. These are convenient snippets that greatly simplifies the analysis of the late-time behavior of the expectation values of the energy–momentum stress tensor of the interacting field. Henceforth, we proceed with the analysis separately for the contributions due to the stationary and the nonstationary components of $G_{H,0}^{(\phi)}(x,x')$.

At late times, the explicit expressions for the stationary and nonstationary components of the Hadamard functions of the free field and the interacting field ϕ become relatively simple. The interacting field's Hadamard function can likewise be decomposed into a stationary and a nonstationary component, $G_H^{(\phi)}(x,x') = G_H^{(\phi),ST}(x,x') + G_H^{(\phi),NS}(x,x')$, and

$$G_H^{(\phi),ST}(x,x') = G_{H,0}^{(\phi),ST}(x,x') + e^2 \int_{-\infty}^{\infty} \frac{d\omega}{2\pi} \left[\widetilde{G}_{H,0}^{(\phi),ST}(x,0;\omega) \widetilde{G}_R^{(\chi)*}(\omega) \widetilde{G}_{R,0}^{(\phi)*}(x';\omega) \right. \tag{34}$$

$$\left. + \widetilde{G}_{H,0}^{(\phi),ST}(x',0;\omega) \widetilde{G}_R^{(\chi)}(\omega) \widetilde{G}_{R,0}^{(\phi)}(x;\omega) \right] e^{-i\omega(t-t')} + \text{C.C.}$$

$$+ e^2 \int_{-\infty}^{\infty} \frac{d\omega}{2\pi} \cosh 2\eta \coth \frac{\beta\omega}{2} \widetilde{G}_{R,0}^{(\phi)}(x,\omega) \widetilde{G}_{R,0}^{(\phi)*}(x',\omega) \, \text{Im}\, \widetilde{G}_R^{(\chi)}(\omega) e^{-i\omega(t-t')} + \text{C.C.},$$

and

$$G_H^{(\phi),NS}(x,x') = G_{H,0}^{(\phi),NS}(x,x') - e^2 \int_0^{\infty} \frac{d\omega}{2\pi} \sinh 2\eta \coth \frac{\beta\omega}{2} \widetilde{G}_R^{(\chi)}(\omega) \tag{35}$$

$$\times \left[\frac{\sin(\omega|x|)}{4\pi|x|} \widetilde{G}_{R,0}^{(\phi)}(x';\omega) + \frac{\sin(\omega|x'|)}{4\pi|x'|} \widetilde{G}_{R,0}^{(\phi)}(x;\omega) \right] e^{-i\omega(t+t')+i\theta} + \text{C.C.}$$

$$- e^4 \int_0^{\infty} \frac{d\omega}{2\pi} \frac{\omega}{4\pi} \sinh 2\eta \coth \frac{\beta\omega}{2} \widetilde{G}_R^{(\chi)2}(\omega) \widetilde{G}_{R,0}^{(\phi)}(x;\omega) \widetilde{G}_{R,0}^{(\phi)}(x';\omega) e^{-i(\omega(t+t')+i\theta} + \text{C.C.},$$

where

$$\widetilde{G}_{H,0}^{(\phi),ST}(x,0;\omega) \equiv \widetilde{G}_{H,0}^{(\phi),ST}(x;\omega) = \cosh 2\eta \coth \frac{\beta\omega}{2} \frac{\sin(\omega|x|)}{4\pi|x|} = \cosh 2\eta \coth \frac{\beta\omega}{2} \operatorname{Im} \widetilde{G}_{R,0}^{(\phi)}(x;\omega). \qquad (36)$$

The energy–momentum stress tensor of the interacting field is then constructed via the Hadamard function, $G_H^{(\phi)}(x,x')$. Analyses of the late-time energy flow of the field at distances far away from the atom resulting from the atom–field interaction are contained in Appendices B and C. A general discussion of the spatial–temporal behavior of the stress–energy tensor due to the radiation field emitted from the atom follows in the next section.

4. Stress–Energy Tensor Due to the Radiation Field

The expectation value of the energy–momentum stress tensor due to the radiation field is defined by

$$\langle \Delta\hat{T}_{\mu\nu}(x) \rangle = \lim_{x' \to x} \left(\frac{\partial^2}{\partial x^\mu \partial x'^\nu} - \frac{1}{2} g_{\mu\nu} g^{\alpha\beta} \frac{\partial^2}{\partial x^\alpha \partial x'^\beta} \right) \left[G_H^{(\phi)}(x,x') - G_{H,0}^{(\phi)}(x,x') \right] \qquad (37)$$

with the signature of the metric $g_{\mu\nu}$ being $(-,+,+,+)$. Here, we subtracted off the contribution purely from the free field, which is irrelevant to the atom–field interaction. Due to symmetry, it has no dependence on the azimuthal angle, ϑ, and the polar angle, φ, of the spherical coordinate system, so the stress–energy tensor already reduces to

$$\langle \Delta\hat{T}_{\mu\nu} \rangle = \begin{pmatrix} \langle \Delta\hat{T}_{tt} \rangle & \langle \Delta\hat{T}_{tr} \rangle & 0 & 0 \\ \langle \Delta\hat{T}_{rt} \rangle & \langle \Delta\hat{T}_{rr} \rangle & 0 & 0 \\ 0 & 0 & 0 & 0 \\ 0 & 0 & 0 & 0 \end{pmatrix}. \qquad (38)$$

Then, we focus on the components $\langle \Delta\hat{T}_{tr} \rangle$ and $\langle \Delta\hat{T}_{tt} \rangle$ that are relevant to our following discussions.

Here, the analysis is not limited to the large-distance and the late-time limits, so that we can have a more complete, global view of the energy flow and energy density of the field caused by the atom–field interaction. We will place the analysis in those specific regimes in Appendices B and C.

Since in our configuration the coupling between the harmonic atom and the massless scalar field takes a different form from that between the point charge and the electromagnetic fields, it would be more illustrative if we first quickly cover the elementary derivations of the continuity equation and the total atom–field Hamiltonian in the classical regime to highlight their differences.

From the equation of motion of the scalar field (2), if we multiply both sides by $\partial_t \phi$, we obtain a local form of the continuity equation

$$\partial_t u_\phi + \nabla \cdot S_\phi = \rho \, \partial_t \phi, \qquad (39)$$

where $\rho(x,t) = e\,\delta^{(3)}(x)\chi(t)$ serves as the point dipole associated with the atom, and we identify the field energy density, u_ϕ, and the field momentum density, S_ϕ, respectively, by

$$u_\phi(x,t) = \frac{1}{2}\left[\partial_t \phi(x,t)\right]^2 + \frac{1}{2}\left[\nabla \phi(x,t)\right]^2, \qquad S_\phi(x,t) = -\partial_t \phi(x,t)\,\nabla \phi(x,t). \qquad (40)$$

The unusual issue with the right hand side of the continuity Equation (39) is best appreciated when we compare with the corresponding equation for the electromagnetic fields:

$$\partial_t u + \nabla \cdot S = -J \cdot E, \qquad \text{with} \qquad u = \frac{1}{2}E^2 + \frac{1}{2}B^2, \qquad S = E \times B, \qquad (41)$$

where J is the current density and the corresponding current is equal to $e\dot{x}$ for a point charge at the location x. The electric field, E, is defined by $E = -\nabla A^0 - \partial_t A$ and the magnetic field B by $B = \nabla \times A$. Here, A^0 is the scalar potential and A is the vector potential. Thus, the right hand side of Equation (39) should bear a similar interpretation to $-J \cdot E$, which is usually understood as the dissipation of the field energy due to the work performed by the field on the charge. That is, it is the opposite of the power delivered by the Lorentz force. However, there are a few subtle catches: (1) In classical field theory, we seldom consider the free field component of the field equation, and we never have vacuum field fluctuations, so the electromagnetic fields in Equation (41) are often assumed to be the field generated by the charge. In contrast, the quantum scalar field ϕ under our consideration is a full interacting field, comprising the free field and the radiation field emitted by the atom. Thus, the right hand side of Equation (39) will contain an extra contribution associated with the free field fluctuations. (2) The right hand side of Equation (39) does not correspond to the net power delivered by the field to the internal degree of freedom of the atom, which is $e\phi(\mathbf{0}, t) \partial_t \chi(t)$, apart from a renormalization contribution, according to the right hand side of Equation (1). Then the interpretation of the right hand of Equation (39) is not so straightforward, even though it has a form of the dipole interaction, similar to the right hand side of Equation (41).

We may trace such a feature to the Hamiltonian. The Hamiltonian $H_{\chi\phi}$ associated with the Lagrangian of the atom–field interacting system,

$$L_{\chi\phi} = \frac{m}{2}\dot{\chi}^2 - \frac{m\omega_B^2}{2}\chi^2 + \int_V dV\, e\,\delta(x)\chi(t)\phi(x,t) + \int_V d^3x \left\{\frac{1}{2}[\partial_t\phi(x,t)]^2 - \frac{1}{2}[\nabla\phi(x,t)]^2\right\}, \quad (42)$$

under our consideration is given by

$$H_{\chi\phi} = \frac{m}{2}\dot{\chi}^2 + \frac{m\omega_B^2}{2}\chi^2 - e\chi(t)\phi(\mathbf{0},t) + \int_V d^3x \left\{\frac{1}{2}[\partial_t\phi(x,t)]^2 + \frac{1}{2}[\nabla\phi(x,t)]^2\right\}, \quad (43)$$

where we implicitly assume that $\dot{\chi}(t)$ and $\dot{\phi}(x,t)$ are functions of the respective canonical momenta, $p(t)$ and $\pi(x,t)$. The quantity V denotes the whole spatial volume on a fixed time slice. Since the whole atom–field system forms a closed system, the total energy is conserved, so one has $dH/dt = 0$,

$$\frac{d}{dt}H_{\chi\phi}(t) = 0 = \dot{\chi}(t)\left\{m\ddot{\chi}(t) + m\omega_B^2\chi(t) - e\phi(\mathbf{0},t)\right\} \quad (44)$$
$$+ \int_V d^3x \left[-e\,\delta(x)\chi(t)\,\partial_t\phi(x,t) + \frac{d}{dt}\left\{\frac{1}{2}[\partial_t\phi(x,t)]^2 + \frac{1}{2}[\nabla\phi(x,t)]^2\right\}\right].$$

Note how the derivative of the interaction term is distributed, and further observe that the contribution inside of the first pair of curly brackets gives zero, following the equation of motion (1). It actually has an attractive interpretation. Following the derivations of Equations (6) and (7), it accounts for the reduced dynamics of the atom's internal degree of freedom, and, from the atom's perspective, describes the energy exchange between the atom and the surrounding quantum field.

The integral on the right hand side of Equation (44), on the other hand, describes the energy exchange from the field's perspective, and thus gives zero too. In the integrand, we see the presence of the same dipole interaction term $e\,\delta(x)\chi(t)\,\partial_t\phi(x,t)$ that appears on the right hand side of Equation (39). At first sight, it may seem odd that the dipole interaction term is solely responsible for the change of the field energy; however, we can further show that

$$\frac{d}{dt}H_{\chi\phi}(t) = 0 = \int_V d^3x\, \nabla \cdot [\nabla\phi(x,t)\,\partial_t\phi(x,t)] = -\oint_{\partial V} d\mathcal{A} \cdot S(x,t), \quad (45)$$

where $d\mathcal{A}$ is the area element on the boundary ∂V. It says that there is no flux entering from or leaving to spatial infinity because the boundary ∂V of the space volume V lies at spatial infinity. It is a rephrase of energy conservation in a closed system, so there is no inconsistency. On the other hand, Equation (45) is too restrictive because V is the total volume. The differential form of the continuity Equation (39) is more suitable to observe the energy distribution in the field.

In contrast to Equation (43), the Hamiltonian for the interacting system of a point charge and the electromagnetic field is

$$H_{xA} = \frac{m_B}{2} \dot{x}^2 + \frac{m_B \omega^2}{2} x^2 + \frac{1}{2} \int_V d^3x \, (E^2 + B^2). \tag{46}$$

It does not have the interaction term manifestly if the Hamiltonian is not expressed in terms of the canonical momentum of the charge; however, it is consistent with the observation that $-J \cdot E$ is opposite to the power delivered by the Lorentz force.

Now, let us discuss the general result of $\langle \Delta \hat{T}_{\mu\nu} \rangle$ that will be valid for $t > r$ at any location distance r away from the atom, so we come back to the manipulation of quantum operators. We first compute $\langle \Delta \hat{T}_{tr} \rangle$.

4.1. General Behavior of Field Energy Flux $\langle \Delta \hat{T}_{tr} \rangle$

If we write $\hat{\phi}_R(x,t)$ as

$$\hat{\phi}_R(x,t) = e \int_0^t ds \, G_{R,0}^{(\phi)}(x,t;0,s) \, \hat{x}(s) = \frac{e}{4\pi r} \theta(t-r) \hat{x}(t-r), \tag{47}$$

which explicitly shows that the radiation field is a retarded Coulomb-like field emitted by a source at an earlier time $t - r$. Note that the quite simple expression of the radiation field such as Equation (47) is not obtainable if the field has a nonMarkovian spectrum.

With the help of Equation (47), one obtains:

$$\frac{1}{2} \langle \{ \partial_t \hat{\phi}_R(x,t), \partial_r \hat{\phi}_R(x,t) \} \rangle = -\left(\frac{e}{4\pi r}\right)^2 \langle \hat{x}'^2(t-r) \rangle - \left(\frac{e}{4\pi r}\right)^2 \frac{1}{2r} \langle \{\hat{x}'(t-r), \hat{x}(t-r)\} \rangle, \tag{48}$$

for $t > r$, where a prime denotes taking the derivative with respect to the function's argument. If we compute the energy flow associated (48), across any spherical surface of radius r centered at the atom, we obtain

$$\int_{\partial V} dA \, \frac{1}{2} \langle \{ \partial_t \hat{\phi}_R(x,t), \partial_r \hat{\phi}_R(x,t) \} \rangle = -2m\gamma \langle \hat{x}'^2(t-r) \rangle - \frac{m\gamma}{r} \langle \{\hat{x}'(t-r), \hat{x}(t-r)\} \rangle$$

$$= P_\gamma(t-r) - \frac{m\gamma}{r} \partial_t \langle \hat{x}^2(t-r) \rangle, \tag{49}$$

where we used the substitution $e^2 = 8\pi\gamma m$. The first term has a special significance. From Refs. [2,18,22], one knows that $P_\gamma(t)$ is the energy that the atom's internal degree of freedom loses to the surrounding field due to damping. Thus, the first term tells us that the energy lost by the atom at time $t - r$ takes time r to propagate to a location at a distance r away from the atom. It is the only contribution in Equation (49) that may survive at spatial infinity.

The contribution from the cross-terms need a little more algebraic manipulations, but it gives:

$$\frac{1}{2} \langle \{ \partial_t \hat{\phi}_h(x,t), \partial_r \hat{\phi}_R(x,t) \} \rangle + \frac{1}{2} \langle \{ \partial_t \hat{\phi}_R(x,t), \partial_r \hat{\phi}_h(x,t) \} \rangle$$

$$= \frac{1}{4\pi r^2} P_\xi(t-r) - \frac{e^2}{4\pi r^2} \frac{\partial}{\partial t} \int_0^{t-r} ds \, G_R^{(\chi)}(t-r-s) \, G_{H,0}^{(\phi)}(x,t;0,s), \tag{50}$$

in which the integral in the last expression is nothing but

$$e^2 \int_0^{t-r} ds \, G_R^{(\chi)}(t-r-s) \, G_{H,0}^{(\phi)}(x,t;0,s) = \frac{e}{2} \langle \{\hat{x}(t-r), \hat{\phi}_h(x,t)\} \rangle. \tag{51}$$

Here, $P_\xi(t)$ is the power that the free quantum field fluctuations deliver to the atom's internal dynamics at time t at the atom's location. Then we find the incoming energy flow given by

$$\int_{\partial V} dA \left[\frac{1}{2} \langle \{\partial_t \hat{\phi}_h(\mathbf{x},t), \partial_r \hat{\phi}_R(\mathbf{x},t)\} \rangle + \frac{1}{2} \langle \{\partial_t \hat{\phi}_R(\mathbf{x},t), \partial_r \hat{\phi}_h(\mathbf{x},t)\} \rangle \right]$$
$$= P_\xi(t-r) - \frac{\partial}{\partial t} \left[\frac{e}{2} \langle \{\hat{\chi}(t-r), \hat{\phi}_h(\mathbf{x},t)\} \rangle \right]. \quad (52)$$

Combining Equations (49) and (52), one thus has:

$$\langle \Delta \hat{T}_{tr} \rangle = \frac{1}{4\pi r^2} \left\{ P_\xi(t-r) + P_\gamma(t-r) - \frac{\partial}{\partial t} \left[\frac{m\gamma}{r} \langle \hat{\chi}^2(t-r) \rangle + \frac{e}{2} \langle \{\hat{\chi}(t-r), \hat{\phi}_h(\mathbf{x},t)\} \rangle \right] \right\}, \quad (53)$$

From the relaxation of the reduced dynamics of the atom's internal degree of freedom, outlined in the beginning of Section 3, we learn that the internal dynamics will reach equilibrium, where $P_\xi(t) + P_\gamma(t) = 0$ for $t \gg \gamma^{-1}$. Hence, from Equation (53), we easily conclude that, for a fixed r, when $t-r$ is much greater than the relaxation time scale, the sum of the dominant terms $P_\xi(t-r) + P_\gamma(t-r)$ vanishes. At large distances away from the atom, the remaining terms are quite small, decaying at least similar to $1/r^3$. It is consistent with our results in Appendix B, where we explicitly show that they actually give zero at late times. Alternatively, we may deduce the same conclusion here, since (1) $\langle \hat{\chi}^2(t) \rangle$ will approach a constant at late times as the atom's internal dynamics asymptotically reach an equilibrium, and (2) with the help of the explicit expression,

$$\frac{e}{2} \langle \{\hat{\chi}(t-r), \hat{\phi}_h(\mathbf{x},t)\} \rangle = -\frac{e^2}{4\pi r} \int_0^\infty \frac{d\omega}{2\pi} \coth \frac{\beta \omega}{2} \cosh 2\eta \int_0^{t-r} ds \, G_R^{(\chi)}(t-r-s) \frac{\sin \omega r}{4\pi r^2} \cos \omega(t-s),$$

it is not hard to see that the time derivative of the last term inside the square brackets in Equation (53) vanishes at late times.

Thus, the detector away from the atom will not measure any radiation energy flux from the stationary atom at late times even though the atom is coupled to a nonstationary squeezed state.

4.2. General Behavior of Field Energy Density $\langle \Delta \hat{T}_{tt} \rangle$

It is also of interest to examine the change of the field energy density at distances far away from the atom resulting from their mutual interactions. Conceptually, the atom's internal dynamics will send out spatial infinity spherical waves centered at the location of the atom. Due to the quantum nature of the internal dynamics, this radiation wave, in general, has a random phase and its magnitude is inversely proportional to the distance to the atom. Following our previous discussion, a detector at a large but fixed distance away from the atom will receive a net outward energy flux in the beginning, and then it will find that the magnitude of the flux rapidly falls off to zero with time. After all these activities settle down, will the earlier energy flux leave any footprint in the space surrounding the atom, say, by shifting the local field energy density, albeit almost imperceptibly? This is what we try to find out in this section.

In the same manner, we rewrite $\langle \Delta T_{tt} \rangle$, following Equation (47). Other than the overall factor $1/2$, we first show:

$$\frac{1}{2} \langle \{\partial_t \hat{\phi}_R(\mathbf{x},t), \partial_t \hat{\phi}_R(\mathbf{x},t)\} \rangle + \frac{1}{2} \langle \{\partial_r \hat{\phi}_R(\mathbf{x},t), \partial_r \hat{\phi}_R(\mathbf{x},t)\} \rangle = -2 \times \frac{1}{4\pi r^2} \left\{ P_\gamma(t-r) + \frac{\partial}{\partial r} \left[\frac{m\gamma}{r} \langle \hat{\chi}^2(t-r) \rangle \right] \right\},$$

where $\partial_t \chi(t-r) = -\partial_r \chi(t-r)$. Here, due to the minus sign up front, this contribution tends to grow to a positive value at late times. We further note that the same expression $P_\gamma(t-r)$ also appears in the component (49) of $\langle \Delta \hat{T}_{tr}(t) \rangle$ that is purely caused by the radiation field. Thus, we see the outward energy flux due to Equation (49) that imparts field energy into the space around the atom.

For the cross-terms, we obtain:

$$2 \times \frac{1}{2} \langle \{\partial_t \hat{\phi}_h(x,t), \partial_t \hat{\phi}_R(x,t)\} \rangle + 2 \times \frac{1}{2} \langle \{\partial_r \hat{\phi}_h(x,t), \partial_r \hat{\phi}_R(x,t)\} \rangle$$

$$= -2 \times \frac{1}{4\pi r^2} \left\{ P_\xi(t-r) + \frac{\partial}{\partial r} \left[\frac{e}{2} \langle \{\hat{x}(t-r), \hat{\phi}_h(x,t)\} \rangle \right] \right\}. \tag{54}$$

Following the same argument, the inward energy flux from Equation (50), on the other hand, is prone to take away the field energy in the surrounding space.

Altogether, we thus find that the net field energy density outside the atom is given by

$$\langle \Delta \hat{T}_{tt} \rangle = -\frac{1}{4\pi r^2} \left\{ P_\xi(t-r) + P_\gamma(t-r) + \frac{\partial}{\partial r} \left[\frac{m\gamma}{r} \langle \hat{x}^2(t-r) \rangle + \frac{e}{2} \langle \{\hat{x}(t-r), \hat{\phi}_h(x,t)\} \rangle \right] \right\}, \tag{55}$$

for $t > r > 0$. This looks similar to Equation (53), and can be the consequence of the continuity equation. The dominant term in Equation (55) will vanish at late times, as a consequence of the relaxation of the internal dynamics of the atom, and the behavior of Equation (53) due to the appearance of $t - r$. Following our earlier arguments, the remaining term, $\langle \{\hat{x}(t-r), \hat{\phi}_h(x,t)\} \rangle$, on the other hand, becomes a time-independent constant at late times, which falls off at least similar to $1/r^3$, as is explicitly shown in Appendix B.2.

Indeed, Equations (53) and (55) enable us to verify the continuity equation,

$$\frac{1}{r^2} \frac{\partial}{\partial r} \left(r^2 \langle \Delta \hat{T}_{tr} \rangle \right) = \frac{1}{4\pi r^2} \left\{ \partial_r P_\xi(t-r) + \partial_r P_\gamma(t-r) - \frac{\partial^2}{\partial t \partial r} \left[\frac{m\gamma}{r} \langle \hat{x}^2(t-r) \rangle + \frac{e}{2} \langle \{\hat{x}(t-r), \hat{\phi}_h(x,t)\} \rangle \right] \right\}$$

$$= \frac{1}{4\pi r^2} \left\{ -\partial_t P_\xi(t-r) - \partial_t P_\gamma(t-r) - \frac{\partial^2}{\partial t \partial r} \left[\frac{m\gamma}{r} \langle \hat{x}^2(t-r) \rangle + \frac{e}{2} \langle \{\hat{x}(t-r), \hat{\phi}_h(x,t)\} \rangle \right] \right\}$$

$$= \frac{\partial}{\partial t} \langle \Delta \hat{T}_{tt} \rangle, \tag{56}$$

for $r > 0$. To include $r = 0$, the location of the atom, we need the form (39), which also takes into account the energy flow into and out of the atom, from the scalar field perspective.

From the considerations presented so far we may now see better how the radiation flux, generated by the internal dynamics of the atom, propagates outward and at the same time intakes the field energy in space outside the atom. Meanwhile, due to the remarkable correlation between the quantum radiation field and the free quantum field, there exists an inward flux. On its way toward the atom, it pulls out field energy stored in the proximity of the atom. From this hindsight, it can be understood that the net power $e\,\dot{\phi}(0,t)\dot{\chi}(t)$ delivered by the quantum field to the atom is not equal to the rate of work performed on the field $e\,\dot{\phi}(0,t)\chi(t)$, as can be inferred from Equation (44). Otherwise, $\langle \Delta \hat{T}_{tr} \rangle$ in Equation (53) will just be proportional to $P_\xi(t-r) + P_\gamma(t-r)$. Their difference accounts for the field energy density stored in the space for the configuration we have studied.

In summary, the results in this section tell us that at late times the observer will not measure any net energy flow associated with the radiation emitted from the atom driven by the squeezed quantum field, but the observer can still detect a constant radiation energy density, which is related to the squeeze parameter. However, a restrictive condition is that the residual radiation energy density is of the near-field nature, and it falls off with the distance to the atom similar to $1/r^3$. Thus, its detection can be challenging at large distances, unless the squeeze parameter is large.

Even with these difficulties, one can still locally identify the squeezing via the response of the atom interacting with the squeezed field.

In the next Section, we discuss how the squeeze parameter depends on the parametric process the quantum field has experienced, such that one may acquire certain information about the process once one has measured the squeeze parameter.

5. Functional Dependence of the Squeeze Parameter on the Parametric Process

Now we turn to how one may possibly extract from the behavior of the squeeze parameter the information of the parametric process that occurs earlier. To be specific, consider the simple case that a massless quantum Klein–Gordon field in flat spacetime undergoes a parametric process such that the frequency of mode k transits smoothly from one constant value ω_i for $0 \leq t \leq t_a$ to another constant ω_f for $t \geq t_b > t_a$.

We would first like to show that the quantum field in the out-region behaves as if it is in its squeezed thermal state with a time-independent squeeze parameter, ζ_k, if the initial state of the field at $t = 0$ is a thermal state. Then, we express the squeeze parameter in terms of the Bogoliubov coefficients, a common tool to treat the dynamics of the quantum field in a parametric (time-varying) process. We also link the squeeze parameter to the fundamental solutions of the equation of motion of the field, in which information of the parametric process is embedded.

We work with the Heisenberg picture, so the field state remains in its initial state at $t = 0$. Formally, we can expand the field operator $\hat{\phi}(x)$ in terms of different sets of mode functions. In the out-region, two convenient choices are

$$u_k^{\text{IN}}(x) = \frac{1}{\sqrt{2\omega_i}} e^{ik\cdot x} \left[d_k^{(1)}(t) - i\omega_i d_k^{(2)}(t) \right], \qquad u_k^{\text{OUT}}(x) = \frac{1}{\sqrt{2\omega_f}} e^{ik\cdot x} e^{-i\omega_f t}, \qquad (57)$$

where $d_k^{(i)}(t)$ satisfies the equation $\ddot{d}_k^{(i)} + \omega^2(t) d_k^{(i)}(t) = 0$, and $u_k^{\text{OUT}}(x)$ is the standard plane-wave mode function in the out-region, while $u_k^{\text{IN}}(x)$ represents the mode function which evolves from the plane-wave mode function in the in-region. Thus, the field operator may have the expansions,

$$\hat{\phi}(x) = \begin{cases} \sum_k \hat{a}_k u_k^{\text{IN}}(x) + \hat{a}_k^\dagger u_k^{\text{IN}*}(x), \\ \sum_k \hat{b}_k u_k^{\text{OUT}}(x) + \hat{b}_k^\dagger u_k^{\text{OUT}*}(x), \end{cases} \qquad \sum_k = \int \frac{d^3 k}{(2\pi)^{\frac{3}{2}}}, \qquad (58)$$

in the out-region.

We further suppose that $(\hat{b}_k, \hat{b}_k^\dagger)$ are related to $(\hat{a}_k, \hat{a}_k^\dagger)$ by

$$\hat{b}_{+k} = \alpha_{+k} \hat{a}_{+k} + \beta_{-k}^* \hat{a}_{-k}^\dagger, \qquad (59)$$

whence the completeness condition $|\alpha_k|^2 - |\beta_k|^2 = 1$ implies that the Bogoliubov coefficients, α_k and β_k, can be parametrized by

$$\alpha_k = \cosh \eta_k, \qquad \beta_{-k}^* = -e^{i\theta_k} \sinh \eta_k, \qquad (60)$$

such that

$$\hat{b}_{+k} = \cosh \eta_k \hat{a}_{+k} - e^{i\theta_k} \sinh \eta_k \hat{a}_{-k}^\dagger = \hat{S}_2^\dagger(\zeta_k) \hat{a}_{+k} \hat{S}_2(\zeta_k). \qquad (61)$$

The parametrization (60) can be alternatively implemented by the two-mode squeeze operator $\hat{S}_2^\dagger(\zeta_k)$. We summarize its properties in Appendix A.

Then, a quantity of the field similar to the Hadamard function in the out-region can be cast into

$$\frac{1}{2}\langle\text{IN}|\{\hat{\phi}(x),\hat{\phi}(x')\}|\text{IN}\rangle$$
$$=\sum_k \frac{1}{2\omega_f}\left[\frac{1}{2}\langle\text{IN}|\{\hat{b}_k,\hat{b}_k\}|\text{IN}\rangle e^{ik\cdot(x+x')}e^{-i\omega_f(t+t')}+\frac{1}{2}\langle\text{IN}|\{\hat{b}_k,\hat{b}_{-k}\}|\text{IN}\rangle e^{ik\cdot(x-x')}e^{-i\omega_f(t+t')}\right.$$
$$\left.+\frac{1}{2}\langle\text{IN}|\{\hat{b}_k,\hat{b}_k^\dagger\}|\text{IN}\rangle e^{ik\cdot(x-x')}e^{-i\omega_f(t-t')}+\frac{1}{2}\langle\text{IN}|\{\hat{b}_k,\hat{b}_{-k}^\dagger\}|\text{IN}\rangle e^{ik\cdot(x+x')}e^{-i\omega_f(t-t')}+\text{C.C.}\right]$$
$$=\sum_k \frac{1}{2\omega_f}e^{ik\cdot(x-x')}\left[\frac{1}{2}\langle\zeta_{k,\text{IN}}^{\text{TMSQ}}|\{\hat{a}_k,\hat{a}_{-k}\}|\zeta_{k,\text{IN}}^{\text{TMSQ}}\rangle e^{-i\omega_f(t+t')}+\frac{1}{2}\langle\zeta_{k,\text{IN}}^{\text{TMSQ}}|\{\hat{a}_k,\hat{a}_k^\dagger\}|\zeta_{k,\text{IN}}^{\text{TMSQ}}\rangle e^{-i\omega_f(t-t')}+\text{C.C.}\right], \quad (62)$$

where $|\zeta_{k,\text{IN}}^{\text{TMSQ}}\rangle = \hat{S}_2(\zeta_k)|\text{IN}\rangle$ is the two-mode squeezed state of the initial in-state $|\text{IN}\rangle$. Thus, Equation (62) gives a Hadamard function in the two-mode squeezed in-state. Although here we use the pure state form, the result can be easily adapted for a mixed state.

On the other hand, the same Hadamard function can be expressed in terms of the in-mode functions, by the expansion

$$\frac{1}{2}\langle\text{IN}|\{\hat{\phi}(x),\hat{\phi}(x')\}|\text{IN}\rangle$$
$$=\sum_k \frac{1}{2\omega_i}e^{ik\cdot(x-x')}\left(N_k^{(\beta)}+\frac{1}{2}\right)\left[d_k^{(1)}(t)-i\omega_i d_k^{(2)}(t)\right]\left[d_k^{(1)}(t')+i\omega_i d_k^{(2)}(t')\right]+\text{C.C.} \quad (63)$$

If the in-state is a thermal state, then $\langle\text{IN}|\{\hat{a}_k,\hat{a}_k^\dagger\}|\text{IN}\rangle$ is understood in terms of the trace average, and gives $2N_k^{(\beta)}+1$, with $N_k^{(\beta)}$ being the average number density of the thermal state at temperature, β^{-1},

$$N_k^{(\beta)}=\frac{1}{e^{\beta\omega_i}-1}. \quad (64)$$

Observe the structural similarities between Equations (62) and (63).

To make them more revealing, we evaluate the expectation values on the right hand side of Equation (62),

$$\langle\text{IN}|\{\hat{b}_{+k},\hat{b}_{+k}\}|\text{IN}\rangle=0, \quad \langle\text{IN}|\{\hat{b}_{+k},\hat{b}_{-k}\}|\text{IN}\rangle=\left(\alpha_{+k}\beta_{+k}^*+\alpha_{+k}\beta_{-k}^*\right)\left(2N_k^{(\beta)}+1\right),$$
$$\langle\text{IN}|\{\hat{b}_{+k},\hat{b}_{-k}^\dagger\}|\text{IN}\rangle=0, \quad \langle\text{IN}|\{\hat{b}_{+k},\hat{b}_{+k}^\dagger\}|\text{IN}\rangle=\left(|\alpha_{+k}|^2+|\beta_{-k}|^2\right)\left(2N_k^{(\beta)}+1\right).$$

Thus, Equation (62) becomes

$$\frac{1}{2}\langle\text{IN}|\{\hat{\phi}(x),\hat{\phi}(x')\}|\text{IN}\rangle$$
$$=\sum_k \frac{1}{2\omega_f}e^{ik\cdot(x-x')}\left(N_k^{(\beta)}+\frac{1}{2}\right)\left[2\alpha_k\beta_k^* e^{-i\omega_f(t+t')}+\left(|\alpha_k|^2+|\beta_k|\right)e^{-i\omega_f(t-t')}+\text{C.C.}\right]. \quad (65)$$

Comparing this equation with Equation (63), we find that

$$|\alpha_k|^2+|\beta_k|^2=\frac{1}{2}\left[\frac{\omega_f}{\omega_i}d_k^{(1)2}(t_f)+\omega_f\omega_i d_k^{(2)2}(t_f)+\frac{1}{\omega_f\omega_i}\dot{d}_k^{(1)2}(t_f)+\frac{\omega_i}{\omega_f}\dot{d}_k^{(2)2}(t_f)\right], \quad (66)$$
$$2\alpha_k\beta_k^*=\frac{1}{2\omega_i\omega_f}\left[+i\omega_f d_k^{(1)}(t_f)+\omega_i\omega_f d_k^{(2)}(t_f)-\dot{d}_k^{(1)}(t_f)+i\omega_i \dot{d}_k^{(2)}(t_f)\right]$$
$$\times\left[-i\omega_f d_k^{(1)}(t_f)+\omega_i\omega_f d_k^{(2)}(t_f)+\dot{d}_k^{(1)}(t_f)+i\omega_i \dot{d}_k^{(2)}(t_f)\right]. \quad (67)$$

These results easily enable us to find the Bogoliubov coefficients α_k and β_k once we have the fundamental solutions $d_k^{(i)}(t)$ with $i=1,2$.

Following the same arguments leading to Equations (62) and (63), if we compute $\langle \hat{\phi}^2(x) \rangle$, $\langle \hat{\pi}^2(x) \rangle$ and $\langle \{\hat{\phi}(x), \hat{\pi}(x)\} \rangle$ in the out-region, one obtains that, for each mode,

$$\cosh 2\eta_k = \frac{1}{2}\left[\frac{1}{\omega_f \omega_i} d_k^{(1)2}(t) + \frac{\omega_i}{\omega_f} d_k^{(2)2}(t) + \frac{\omega_f}{\omega_i} d_k^{(1)2}(t) + \omega_f \omega_i d_k^{(2)2}(t)\right], \quad (68)$$

$$\cos(\theta_k - 2\omega_f t) \sinh 2\eta_k = \frac{1}{2}\left[\frac{1}{\omega_f \omega_i} d_k^{(1)2}(t) + \frac{\omega_i}{\omega_f} d_k^{(2)2}(t) - \frac{\omega_f}{\omega_i} d_k^{(1)2}(t) - \omega_f \omega_i d_k^{(2)2}(t)\right], \quad (69)$$

$$\sin(\theta_k - 2\omega_f t) \sinh 2\eta_k = -\left[\frac{1}{\omega_i} d_k^{(1)}(t) d_k^{(1)}(t) + \omega_i d_k^{(2)}(t) d_k^{(2)}(t)\right], \quad (70)$$

explicit relations occur between the squeeze parameters and the fundamental solutions. Here, $\hat{\pi}(x)$ is the canonical momentum conjugated to $\hat{\phi}(x)$. Equations (68)–(70) tell us how squeezing may dynamically arise from the parametric process of the field.

As a consistency check, we substitute the coefficients α_k and β_k on the right hand side by Equation (59), and obtain:

$$|\alpha_k|^2 + |\beta_k|^2 = \cosh 2\eta_k, \qquad 2\alpha_k \beta_k^* = -e^{i\theta_k} \sinh 2\eta_k. \quad (71)$$

Actually, Equations (66) and (67) only determine α_k, β_k up to a phase factor or a rotation, which we have been ignoring. For example, from (67), we may let

$$\alpha_k = \frac{1}{2\sqrt{\omega_i \omega_f}}\left[+i\omega_f d_k^{(1)}(t_f) + \omega_i \omega_f d_k^{(2)}(t_f) - d_k^{(1)}(t_f) + i\omega_i d_k^{(2)}(t_f)\right], \quad (72)$$

$$\beta_k = \frac{1}{2\sqrt{\omega_i \omega_f}}\left[-i\omega_f d_k^{(1)}(t_f) + \omega_i \omega_f d_k^{(2)}(t_f) + d_k^{(1)}(t_f) + i\omega_i d_k^{(2)}(t_f)\right], \quad (73)$$

and then we can directly show that Equation (66) is recovered and that

$$|\alpha_k|^2 - |\beta_k|^2 = d_k^{(1)}(t_f) d_k^{(2)}(t_f) - d_k^{(1)}(t_f) d_k^{(2)}(t_f) = 1. \quad (74)$$

However, in this case α_k is not real, as Equation (60) implies.

We may recover the missing phase by returning back the rotation operator that we have ignored all along. To be more specific, for example, let the rotation operator $\hat{R}$ be given by

$$\hat{R}(\Phi_i) = \exp\left[i\Phi_i\left(\hat{a}_i^\dagger \hat{a}_i + \frac{1}{2}\right)\right]. \quad (75)$$

We find $\hat{R}^\dagger(\Phi_i) \hat{a}_1 \hat{R}(\Phi_1) = \hat{a}_1 e^{i\Phi_1}$. Then, depending on the order of the rotation operator and the squeeze operator, we may have either

$$\hat{S}_2^\dagger(\zeta) \hat{R}^\dagger(\Phi_i) \hat{a}_1 \hat{R}(\Phi_i) \hat{S}_2(\zeta) = e^{i\Phi_1} \cosh \eta \, \hat{a}_1 - e^{i\Phi_1} e^{i\theta} \sinh \eta \, \hat{a}_2,$$

or

$$\hat{R}^\dagger(\Phi_i) \hat{S}_2^\dagger(\zeta) \hat{a}_1 \hat{S}_2(\zeta) \hat{R}(\Phi_i) = e^{i\Phi_1} \cosh \eta \, \hat{a}_1 - e^{i\Phi_2} e^{i\theta} \sinh \eta \, \hat{a}_2.$$

One can see that there is always a phase ambiguity. In both cases, from Equations (60) and (61), one finds that α_k will be complex in general, but actually, one may factor out the overall phase factor for each mode to render α_k real.

Before we proceed further with our analysis, let us examine a few illustrative examples.

5.1. Case 1

Consider the parametric process, in which the squared frequency $\omega^2(t)$ varies with time according to

$$\omega^2(t) = \begin{cases} \omega_i^2, & 0 \leq t \leq t_a, \\ \omega_i^2 + (\omega_f^2 - \omega_i^2)\dfrac{t-t_a}{t_b-t_a}, & t_a \leq t \leq t_b, \\ \omega_f^2, & t \geq t_b, \end{cases} \qquad (76)$$

That is, $\omega^2(t)$ is a piecewise-continuous function of time. The time evolution of $d_k^{(1)}(t)$, $d_k^{(2)}(t)$ are shown in Figure 2b,c, where one observes that the oscillation amplitudes of the two fundamental solutions change by different amounts, implying the occurrence of quantum squeezing. Notice in Figure 2d that, as $t \geq t_b$ the squeeze parameter η_k becomes a constant, squeezing is quite small because $\cosh 2\eta_k \sim 1$. This small squeezing results from the slow transition rate in the parametric process. The plots Figure 2e,f show oscillations of frequency $2\omega_f$, consistent with expectation. However, from these plots it is hard to tell whether θ_k is time-independent. We take another approach to show it in this section.

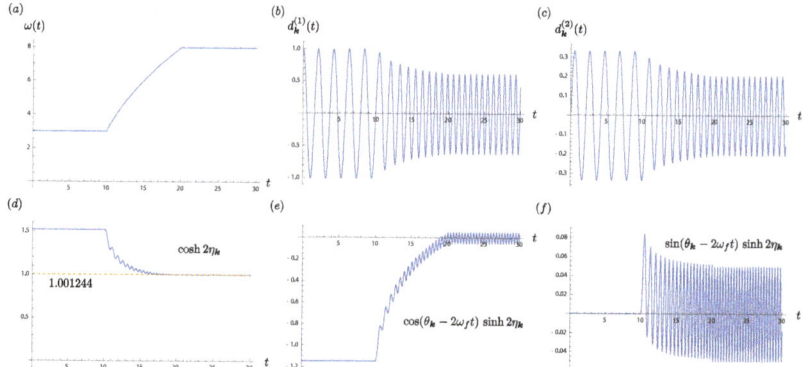

Figure 2. An example of a parametric process described by Equation (76): (**a**) the time dependence of the frequency ω; time evolutions of the two fundamental solutions, (**b**) $d_k^{(1)}$ and (**c**) $d_k^{(2)}$, and (**d**–**f**) the time dependence of the squeeze parameters. $\omega_i = 3$, $\omega_f = 8$, $t_a = 10$, and $t_b = 20$ are chosen.

5.2. Case 2

In the second example, we choose a nonmonotonic parametric process,

$$\omega^2(t) = \begin{cases} \omega_i^2, & 0 \leq t \leq t_a, \\ \omega_i^2 + (\omega_f^2 - \omega_i^2)\sin^2\left(\dfrac{t-t_a}{t_b-t_a}\dfrac{n\pi}{2}\right), & t_a \leq t \leq t_b, \\ \omega_f^2, & t \geq t_b. \end{cases} \qquad (77)$$

Here, n, an odd integer, gives the number of oscillations of the transition process; thus, the frequency variation is not monotonic. The corresponding plots are shown in Figure 3. Since both processes, though continuous, are not smooth, it allows us to see when the transitions start and end. Compared with case 1, the nonmonotonic transition introduces tumultuous behavior in the fundamental solutions $d_k^{(i)}$ during the transition, in Figure 3b,c, but right after $t = t_b$, the out mode immediately oscillates at frequency ω_f. Again, one can see the time-independence of η_k in the out-region.

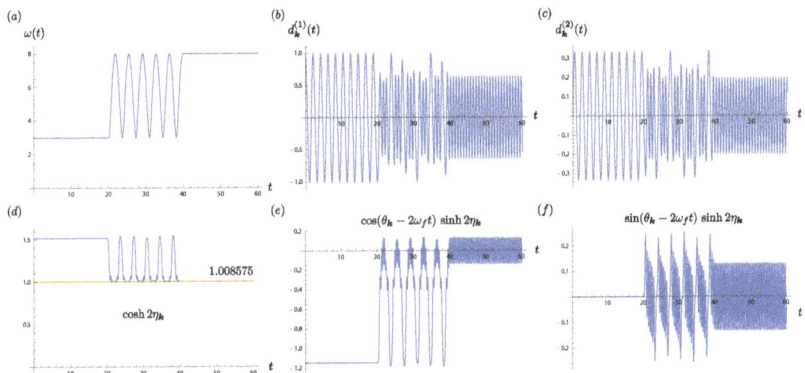

Figure 3. An example of a parametric process described by Equation (77): (**a–f**) same as in Figure 2. $\omega_i = 3$, $\omega_f = 8$, $t_a = 10$, $t_b = 20$, and $n = 11$ are chosen.

For the simple case we consider here, let us carry out some analysis about the time dependence of the squeeze parameter in the out-region. We may write $d_k^{(i)}(t)$ for $t \geq t_b$ as

$$d_k^{(1)}(t) = d_k^{(1)}(t_b) \cos \omega_f(t - t_b) + d_k^{(2)}(t_b) \frac{1}{\omega_f} \sin \omega_f(t - t_b), \tag{78}$$

$$d_k^{(2)}(t) = d_k^{(1)}(t_b) \frac{1}{\omega_f} \sin \omega_f(t - t_b) + d_k^{(2)}(t_b) \cos \omega_f(t - t_b). \tag{79}$$

The amplitudes of $d_k^{(i)}(t)$ in the out-region are determined by

$$d_k^{(1)}(t): \left[d_k^{(1)2}(t_b) + \frac{d_k^{(2)2}(t_b)}{\omega_f^2} \right]^{\frac{1}{2}}, \quad \text{and} \quad d_k^{(2)}(t): \left[\frac{d_k^{(1)2}(t_b)}{\omega_f^2} + d_k^{(2)2}(t_b) \right]^{\frac{1}{2}}, \tag{80}$$

which, in turn, are determined by the values of the fundamental solution at the end of the parametric process. We readily find that $d_k^{(1)}(t)\dot{d}_k^{(1)}(t) - \dot{d}_k^{(1)}(t)d_k^{(1)}(t) = 1$ for all t even for the parametric process via the Wronskian of the differential equation that the mode function satisfies.

According to Equation (67), we can show that

$$|\alpha_k(t)|^2 + |\beta_k(t)|^2 = |\alpha_k(t_b)|^2 + |\beta_k(t_b)|^2 \tag{81}$$

for all $t \geq t_b$. Since $|\alpha_k|^2 + |\beta_k|^2$ is proportional to $\cosh 2\eta_k$, Equation (81) then shows that η_k is a time-independent constant for all $t \geq t_b$. On the other hand, with the help of Equations (78) and (79), one finds:

$$\frac{1}{\omega_f \omega_i} d_k^{(1)2}(t) + \frac{\omega_i}{\omega_f} d_k^{(2)2}(t) - \frac{\omega_f}{\omega_i} d_k^{(1)2}(t) - \omega_f \omega_i d_k^{(2)2}(t) = A_k \sin(2\omega_f t_b - 2\omega_f t) + B_k \cos(2\omega_f t_b - 2\omega_f t)$$

$$= \sqrt{A_k^2 + B_k^2} \cos(\vartheta_k + 2\omega_f t_b - 2\omega_f t), \tag{82}$$

with

$$A_k = 2 \left[\frac{1}{\omega_i} d_k^{(1)}(t_b) \dot{d}_k^{(1)}(t_b) + \omega_i d_k^{(2)}(t) \dot{d}_k^{(2)}(t) \right], \tag{83}$$

$$B_k = \left[\frac{1}{\omega_f \omega_i} d_k^{(1)2}(t_b) + \frac{\omega_i}{\omega_f} d_k^{(2)2}(t_b) - \frac{\omega_f}{\omega_i} d_k^{(1)2}(t_b) - \omega_f \omega_i d_k^{(2)2}(t_b) \right]. \tag{84}$$

and

$$\cos \vartheta_k = \frac{B_k}{\sqrt{A_k^2 + B_k^2}}, \qquad \sin \vartheta_k = -\frac{A_k}{\sqrt{A_k^2 + B_k^2}}. \qquad (85)$$

From Equations (69) and (70), we see that $\sqrt{A_k^2 + B_k^2}$ is proportional to $\sinh 2\eta_k$, and thus is a time-independent constant. One can immediately identify ϑ_k in Equations (82)–(84) to be the same θ_k in Equations (68)–(70). Finally, we may note that Equation (82) still looks slightly different from the left hand sides of Equations (68)–(70) in the arguments of the trigonometric functions. It results from the choice of the out mode function in Equation (61). The reason for such a choice is that for an observer in the out-region, he has no reference to identify the origin of time coordinate. In addition, the choice of a fixed time origin at most amounts to an absolute phase, which in most cases is of no significance. However, in the current case we are comparing two formalisms; thus, for consistency's sake, we may define the origin of the time coordinate in the out-region at t_b rather than 0. By only shifting t in Equation (61) to $t \to t - t_b$, Equation (82) looks the same as those in Equations (68)–(70). At this point, we have shown that in the out-region an observer may report on having experienced a quantum field in a squeezed thermal state, with a time-independent squeeze parameter.

The squeeze parameter, η_k, we showed earlier is quite small because $\cosh 2\eta_k \sim 1$. This results from the fact that the parametric processes in the previous cases vary mildly. Now we show a case similar to that in Figure 3 but with a much sharper transition with $\omega_i = 1$, $\omega_f = 100$, $t_a = 3\pi/2$, $t_b = 3\pi/2 + 0.05$ and $n = 1$. The plot for $\cosh 2\eta_k$ is shown in Figure 4. We see in this case that $\cosh 2\eta_k = 25.62206$ in the out-region, much larger than the previous two cases. Therefore, it is consistent with the understanding that to generate large squeezing, the parametric process had better not be adiabatic. This is corroborated by our prior knowledge of cosmological particle production.

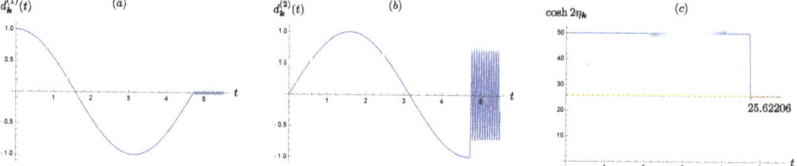

Figure 4. An example of a parametric process described by Equation (77): (**a**), (**b**), and (**c**) same as in Figure 3b, c, and d, respectively, but with $\omega_i = 1$, $\omega_f = 100$, $t_a = 3\pi/2$, $t_b = 3\pi/2 + 0.05$, and $n = 1$.

Finally, as a contrast to the previous piecewise-continuous parametric processes, we consider a sufficiently smooth $\omega^2(t)$, such as the thrice-differentiable function $\omega^2(t)$

$$\omega^2(t) = -\frac{(t - t_a)^4}{(t_b - t_a)^7} \left[20t^3 - 10(7t_b - t_a)\,t^2 + 4(21t_b^2 - 7t_b t_a + t_a^2)\,t - (35t_b^3 - 21t_b^2 t_a + 7t_b t_a^2 - t_a^3) \right], \qquad (86)$$

between $t = t_a$ and $t = t_b$. The corresponding results are shown in Figure 5. We do not discern any difference in the generic behavior, so for the quantities of interest in our present study, the piecewise-continuous $\omega^2(t)$ suffices.

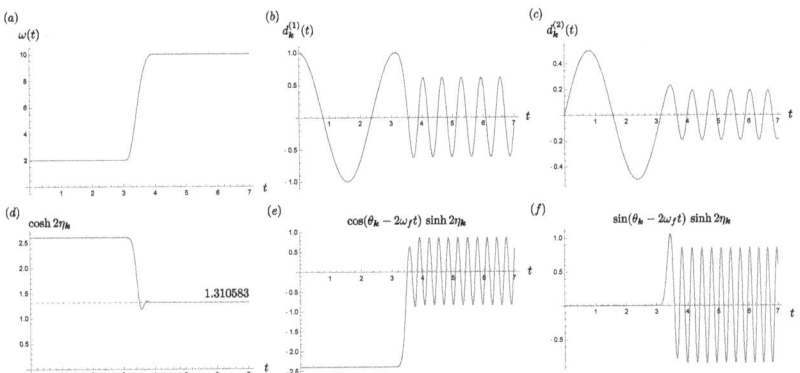

Figure 5. An example of a parametric process described by Equation (86): (**a–f**) same as in Figure 2. with $\omega_i = 2$, $\omega_f = 10$, $t_a = 3$, and $t_b = 4$ are chosen.

The fact that the squeeze parameter becomes time-independent in the out-region implies that this result is time-translation-invariant. To be more precise, if we shift the parametric process along the time axis, then as long as the conditions that (1) the observation time t is still in the out-region of the shifted process and (2) the initial time t_i remains in the in-region of the shifted process are satisfied, the observer at time t will still see the same squeeze parameter. In other words, the observer in the out-region cannot extract from the squeeze parameter any information about when the parametric process starts or ends. At first, it sounds odd that there exists such a time-translation invariance for a nonequilibrium, nonstationary process. We show this in Appendix D; however, after seeing these examples, this result may not appear that dubious. The full evolution from the initial time t_i to the observation time t is not invariant under time translation, but after the parametric process has ended at t_b, the mode functions in the out-region reverts to a sinusoidal time dependence. The time translation in a sinusoidal function of time amounts to a phase shift. Thus, if the quantity of interest is independent of this phase shift, it appears to possess time-translation invariance.

Although in the current configuration the detector in the out-region will sense the same squeeze parameter when the parametric process is shifted along the time axis, the measured results in the detector still depend on the duration $t_b - t_a$ and the functional form of the process. Actually, this can be expected from Equations (68)–(70), where the squeeze parameters are expressed in terms of the fundamental solutions, which in turn depend on the functional form of the parametric process in their equation of motion.

For illustration, Figure 6a shows the dependence of the squeeze parameter, η_k, on the duration of the parametric process, given by Equation (76). We fix the starting time, t_a, of the parametric process and the moment t the measurement in the out-region is performed. We find that the squeeze parameter η_k, defined in Equation (68), is oscillatory but decreases with increasing ending time, t_b. Except for the small oscillations, the curve in Figure 6a gives the general trend that the squeezing (as manifested, e.g., in particle pair production), is subdued with a slower transition rate or longer transition duration $t_b - t_a$. This example shows that this kind of parametric process gives a lower production of particles as it moves toward the adiabatic regime. The mild oscillations may be related to the kinks at t_a and t_b due to the nonsmoothness of $\omega(t)$. This may be seen from the observation that the oscillations, as shown in the blow-ups in Figure 6a, appear shallower with larger $t_b - t_a$ because the kinks are less abrupt. This argument also find its support from Figure 7b,d.

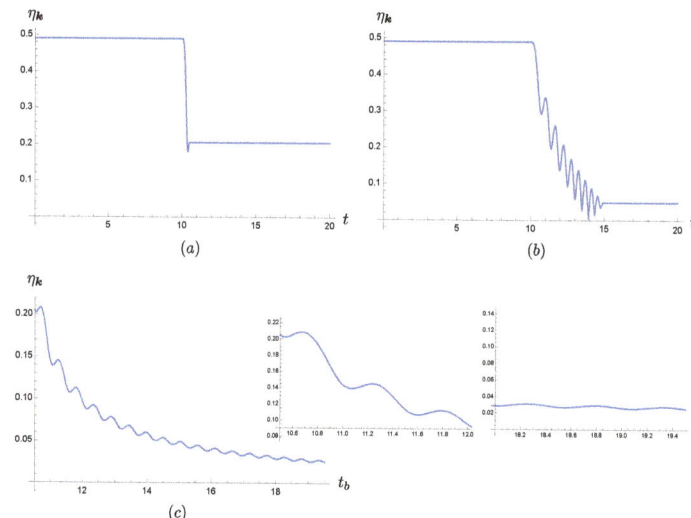

Figure 6. (**a**) The t_b-dependence of the squeeze parameter, η_k for the parametric process described by Equation (76) with $\omega_i = 3$, $\omega_f = 8$, $t_i = 0$, and $t_a = 10$. A fictitious detector is placed at $t = 20$, and t_b varies from 10.5 to 19.5. For comparison, the time dependence of η_k when (**b**) $t_b = 10.5$ and (**c**) $t_b = 15$ are shown. These two cases are highlighted in (**a**) by red circles.

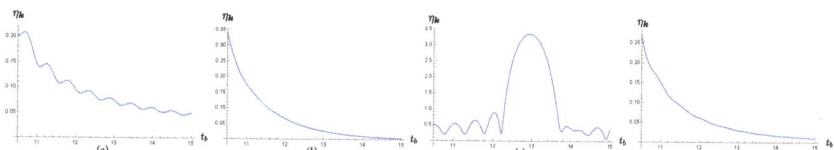

Figure 7. The dependence of the squeeze parameter η on the functional form of $\omega^2(t)$. The functional form of the parametric process is described by (**a**) Equation (76) and (**b**) Equation (86). The same Equation (77) is used with (**c**) $n = 11$ and (**d**) $n = 1$. The relevant parameters are $\omega_i = 3$, $\omega_f = 8$, $t_i = 0$, and $t_a = 10$. The measurement is carried out at $t = 20$, and t_b varies from 10.5 to 15.

In Figure 7, the dependence of the squeeze parameter η_k on the functional form of the parametric process is shown. We choose three different parametric processes of $\omega^2(t)$ for examples. In Figure 7a, the process has the piecewise-continuous form given by Equation (76), so it gives the same result as Figure 6a. Figure 7b is described by the smooth transition, Equation (86). We have the same process, Equation (77), for Figure 7c,d, but with different choices of n. We choose $n = 11$ for Figure 7c and $n = 1$ for Figure 7d. Thus, in Figure 7c, the transition does not monotonically vary with time.

Since both Figure 7b,d are associated with smooth parametric processes, when we prolong the duration of the parametric processes by fixing t_a and increasing t_b, we see that the value of η_k monotonically decreases without any ripples. Thus, together with the behavior of the curve in Figure 6a, they imply that the ripples in Figure 7a may originate from the nonsmoothness of the parametric process at the transition times t_a and t_b.

Among the plots in Figure 7, Figure 7c shows a more interesting behavior. The transition described by Equation (77) with $n = 11$ is sinusoidal, as shown in Figure 3a. The squeeze parameter η_k in Figure 7c then reveals more structures, and may signify the presence of resonance, shown by a large peak. It seems to imply that the particle production can be greatly enhanced if the transition duration is tuned to the right value for fixed ω_i and ω_f, as in parametric resonance. This may constitute a mechanism to generate stronger squeezing, in addition to the common one in a runaway setting [23,24]. Another unusual feature is that η_k in Figure 7c is not significantly reduced when we have a larger t_b.

Actually, this resonance feature can be traced back to parametric instability. We write the equation of motion for a parametric oscillator with the frequency given by Equation (77) into the canonical form

$$x''(\tau) + [a - 2q \cos(2\tau)] x(\tau) = 0, \qquad (87)$$

where a prime represents taking the derivative with respect to τ, $\Omega t = 2\tau$, and

$$a = \frac{4A^2}{\Omega^2}, \quad q = \frac{2B^2}{\Omega^2}, \quad \Omega = \frac{n\pi}{t_b - t_a}, \quad A^2 = \frac{\omega_f^2 + \omega_i^2}{2}, \quad B^2 = \frac{\omega_f^2 - \omega_i^2}{2}. \qquad (88)$$

Now, comparing the stability diagram Figure 8a of Equation (87) with Figure 7c, one immediately notices that the portion of the curve in Figure 8b that corresponds to exceptionally high squeezing is essentially located within the unstable region (white area) of the stability diagram in Figure 8a. Thus, the large squeezing in the mode driven by a nonmonotonic $\omega(t)$ is caused by instability in parametric resonance.

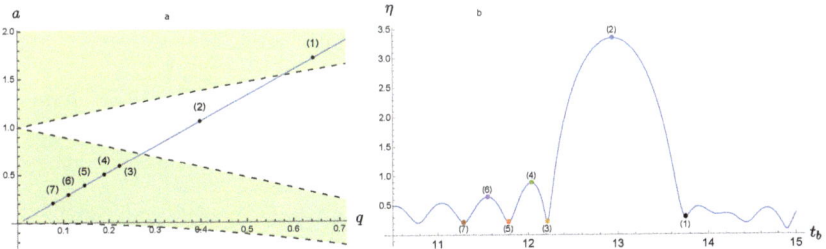

Figure 8. Comparison between the stability diagram of Equation (87) (**a**) with Figure 7c (**b**). The motion of the parametric oscillator (87) is stable in color-shaded regions. We highlight a few peaks and valleys of the curve in (**b**), and the corresponding points are shown in the stability diagram (**a**). The straight line in (**a**) with varying hues of red, yellow, green, cyan, blue, and magenta corresponds to the time $t_b = 10.5$ to 15 in (**b**). The parameters in this plot are the same as in Figure 7.

The above examples illustrate a point that the dependence of the squeeze parameter η_k on the duration of the parametric process does reflect qualitative features of the parametric process. Actually, we can apply the same idea to the spectral dependence of the squeeze parameter for various functional forms of the parametric processes. These will serve as templates from which we may extract qualitative information about the parametric process that the quantum field has experienced.

6. Summary

A major task of theoretical cosmology is to find ways to deduce the state of the early universe from relics observed today such as features of primordial radiation and matter contents. In addition to particle creation at the Planck time and structure formation after the GUT (grand unified theory) time, we add here another fundamental quantum process: quantum radiation from atoms induced by of quantum fields.

A massless quantum field in Minkowski space is subjected to a parametric process which varies the frequencies of the normal modes of the field from one constant value to another over a finite time interval. The initial (in-) state of the field will become squeezed after the process. In the out-region we couple an atom to this field in the out-state; the response of the atom certainly depends on this squeezing. Meanwhile, the radiation generated by the interacting atom will send out signals outwards, such that an observer or a probe at a much later time may still be able to identify the squeezing via the signatures in the radiation.

Once we have identified the squeezing, locally or remotely, we then ask how the squeezing may depend on the details of the parametric process the field had been subjected to. If we can work out templates that relate these quantities, then measurements of certain features of squeezing can tell us something about the parametric process.

The salient points in our results are summarized below. In the first part of this paper we showed the following:

- The radiation field at a location far away from the atom looks stationary; its nonstationary component decays with time exponentially fast.
- The net energy flow cancels at late times, similar to the case discussed in Ref. [2]. These features are of particular interest considering that the atom that emits this radiation is coupled to a squeezed field, which is nonstationary by nature. However, they are consistent with the fact that the atom's internal dynamics relaxes in time.
- This implies that we are unable to measure the extent of squeezing by measuring the net radiation energy flow at a location far away from the atom.
- On the other hand, one can receive residual radiation energy density at late times, which is a time-independent constant and is related to the squeeze parameter. However, it is of a near-field nature, so the observer cannot be located too far away from the atom.

Note that there is always an ambient free component of the same squeezed quantum field everywhere, in addition to the above radiation component generated by the atom's internal motion.

In the second part of this paper we focused on the dependence of the squeeze parameter on the parametric process and obtained these results:

- Formally it can be shown that the squeeze parameter depends on the evolution of the field in the parametric process.
- In the current configuration, for a given parametric process, the squeeze parameter depends only on the duration of the process; it does not depend on the starting or the ending time of the process.
- In general, for a monotonically varying process, the value of the squeeze parameter decreases with increasing duration of the process.
- This implies that, for an adiabatic parametric process, the squeezing tends to be quite small, but it can be quite significant for nonadiabatic parametric processes. These results are consistent with studies of cosmological particle creation in the 1970s as parametric amplification of vacuum fluctuations.
- If the parametric process changes with time sinusoidally, then the dependence of the squeeze parameter on the duration of the process shows interesting additional structures. For certain lengths of the process, the squeeze parameter can have unusually large values.
- This nonmonotonic behavior turns out to be related to parametric instability. The resulting large squeeze parameter is caused by the choice of the parameter that falls within the unstable regime of the parametric process.

Our results in this second part indicate that not much information about a monotonic parametric process can be gleaned from the squeeze parameter. On the other hand, for a nonmonotonic parametric process, the squeeze parameter shows nontrivial dependence on the duration of the process. The latter is expected to be the typical case in nature in which various scales are involved in a transition. Moreover, to conform to realistic measurements, we can apply similar considerations to examine the mode dependence of the squeeze parameter. Thus, the current scenario seems to offer a viable means to extract partial information about the parametric process a quantum field has undergone.

As a final remark, we comment on applying the discussions in Section 4 to the case of the atom coupled to quantized electromagnetic fields. To begin with, in conventional sense, such a configuration has supra-Ohmic, Markovian dynamics. It is inherently unstable, as discussed in Ref. [25]. It has runaway solutions, so it makes no sense to discuss the late-time behavior, relaxation, energy balance, or the fluctuation–dissipation relation of such a system. Although we may render such a system to have stable dynamics by resorting to order reduction [26] via the critical manifold argument, this approach is not satisfactory from the viewpoint of open systems because order reduction only asymmetrically changes the behavior of the dissipation on the atom side; it does not accordingly modify the fluctuation noise force on the atom. Thus, the resulting nonequilibrium fluctuation–dissipation relation associated with the atom's internal dynamics will not take the elegant form we usually see for the Ohmic, Markovian dynamics. Furthermore, this relation will depend on the parameters of the reduced system, and thus loses its universality for interacting linear systems and for some nonlinear systems.

To restore the stable dynamics of the atom's internal dynamics [25] and the beauty of the associated nonequilibrium fluctuation–dissipation relation, it is probably easier if we generalize the Markovian spectrum of the quantized electromagnetic field to the non-Markovian one. However, this introduces additional complexity to the analysis presented in Section 4 because the radiation field will not have a simple local (apart from retardation) form similar to Equation (47). Thus, it is not clear yet whether the rather general properties we expounded in Section 4 for a quantum scalar field also convey to a quantized electromagnetic field; this is a topic saved for future investigation.

Author Contributions: Conceptualization, J.-T.H. and B.-L.H.; methodology, M.B. and J.-T.H.; formal analysis, M.B. and J.-T.H.; investigation, M.B. and J.-T.H.; writing—original draft preparation, M.B., J.-T.H. and B.-L.H.; writing—review and editing, M.B., J.-T.H. and B.-L.H.; visualization, J.-T.H.; funding acquisition, J.-T.H. All authors have read and agreed to the published version of the manuscript.

Funding: J.-T.H. is supported by the Ministry of Science and Technology of Taiwan, R.O.C., under Grant No. MOST 111-2811-M-008-022.

Data Availability Statement: Not applicable.

Acknowledgments: B.-L.H. appreciates the kind hospitality of Chong-Sun Chu of the National Center for Theoretical Sciences and Hsiang-nan Li of the Academia Sinica in the finishing stage of this paper.

Conflicts of Interest: The authors declare no conflict of interest.

Appendix A. Two-Mode Squeezed State

Here, we outline the properties of the two-mode squeezed operator $\hat{S}_2(\zeta)$ associated mode 1 and 2,

$$\hat{S}_2(\zeta) = \exp\left[\zeta^* \hat{a}_1 \hat{a}_2 - \zeta\, \hat{a}_1^\dagger \hat{a}_2^\dagger\right], \tag{A1}$$

such that the two-mode squeezed thermal state is defined by

$$\hat{\rho}_{\text{TMST}} = \hat{S}_2(\zeta)\, \hat{\rho}_\beta\, \hat{S}_2^\dagger(\zeta). \tag{A2}$$

where $\hat{\rho}_\beta$ is a thermal state. The creation and annihilation operators satisfy the standard commutation relation $[\hat{a}, \hat{a}^\dagger] = 1$. It is convenient to place the squeeze parameter ζ into a polar form $\zeta = \eta\, e^{i\theta}$, with $\eta \in \mathbb{R}^+$ and $0 \leq \theta < 2\pi$. We thus have

$$\hat{S}_2^\dagger \hat{a}_1 \hat{S}_2 = \cosh\eta\, \hat{a}_1 - e^{+i\theta} \sinh\eta\, \hat{a}_2^\dagger, \qquad \hat{S}_2^\dagger \hat{a}_2 \hat{S}_2 = \cosh\eta\, \hat{a}_2 - e^{+i\theta} \sinh\eta\, \hat{a}_1^\dagger \tag{A3}$$

such that

$$\langle \hat{a}_i^2 \rangle_{\text{TMST}} = \text{Tr}\left\{\rho_\beta\, \hat{S}_2^\dagger \hat{a}_i^2\, \hat{S}_2\right\} = 0, \qquad \langle \hat{a}_i^{\dagger 2} \rangle_{\text{TMST}} = 0, \tag{A4}$$

$$\langle \hat{a}_1 \hat{a}_1^\dagger \rangle_{\text{TMST}} = (\bar{n}_1 + 1)\cosh^2\eta + \bar{n}_2 \sinh^2\eta, \qquad \langle \hat{a}_1^\dagger \hat{a}_1 \rangle_{\text{TMST}} = \bar{n}_1 \cosh^2\eta + (\bar{n}_2 + 1)\sinh^2\eta, \tag{A5}$$

$$\langle \hat{a}_2 \hat{a}_2^\dagger \rangle_{\text{TMST}} = (\bar{n}_2 + 1)\cosh^2\eta + \bar{n}_1 \sinh^2\eta, \qquad \langle \hat{a}_2^\dagger \hat{a}_2 \rangle_{\text{TMST}} = \bar{n}_2 \cosh^2\eta + (\bar{n}_1 + 1)\sinh^2\eta, \tag{A6}$$

$$\langle \hat{a}_1 \hat{a}_2 \rangle_{\text{TMST}} = -\frac{e^{+i\phi}}{2}(\bar{n}_1 + \bar{n}_2 + 1)\sinh 2\eta, \qquad \langle \hat{a}_1^\dagger \hat{a}_2^\dagger \rangle_{\text{TMST}} = -\frac{e^{-i\phi}}{2}(\bar{n}_1 + \bar{n}_2 + 1)\sinh 2\eta, \tag{A7}$$

where $\bar{n}_i$ is the mean particle number of the thermal state associated with mode i,

$$\bar{n}_i = \text{Tr}\{\hat{\rho}_\beta \hat{a}_i^\dagger \hat{a}_i\} \tag{A8}$$

It is interesting to compare with the single-mode squeezed thermal state $\hat{\rho}_{ST}$, defined by

$$\hat{\rho}_{ST} = \hat{S}(\zeta)\,\hat{\rho}_\beta\,\hat{S}^\dagger(\zeta). \tag{A9}$$

Here, $\hat{S}(\zeta)$ is the squeezed operator,

$$\hat{S}(\zeta) = \exp\left[\frac{1}{2}\zeta^*\hat{a}^2 - \frac{1}{2}\zeta\hat{a}^{\dagger 2}\right]. \tag{A10}$$

We then find:

$$\langle \hat{a}\rangle_{ST} = 0 = \langle a^\dagger\rangle_{ST}, \tag{A11}$$

$$\langle \hat{a}^2\rangle_{ST} = -e^{+i\theta}\sinh 2\eta\left(\bar{n}+\frac{1}{2}\right), \qquad \langle a^{\dagger 2}\rangle_{ST} = -e^{-i\theta}\sinh 2\eta\left(\bar{n}+\frac{1}{2}\right), \tag{A12}$$

$$\langle \hat{a}^\dagger \hat{a}\rangle_{ST} = \cosh 2\eta\left(\bar{n}+\frac{1}{2}\right) - \frac{1}{2} = \cosh 2\eta\,\bar{n} + \sinh^2\eta, \tag{A13}$$

where $\bar{n}$ is the mean particle number of the thermal state. Both the two single-mode squeezed state and the two-squeezed state are connected. We first write two modes in terms of their normal modes $i = \pm$,

$$\hat{a}_1 = \frac{\hat{a}_+ + \hat{a}_-}{\sqrt{2}}, \qquad \hat{a}_2 = \frac{\hat{a}_+ - \hat{a}_-}{\sqrt{2}}, \tag{A14}$$

with $[\hat{a}_+, \hat{a}_-] = 0$. Then we can write the two-mode squeeze operator, $\hat{S}_2$ in (A1), as

$$\begin{aligned}
\hat{S}_2 &= \exp\left[\frac{\zeta^*}{2}(\hat{a}_+ + \hat{a}_-)(\hat{a}_+ - \hat{a}_-) - \frac{\zeta}{2}(\hat{a}_+^\dagger + \hat{a}_-^\dagger)(\hat{a}_+^\dagger - \hat{a}_-^\dagger)\right] \\
&= \exp\left[\frac{\zeta^*}{2}\hat{a}_+^2 - \frac{\zeta}{2}\hat{a}_+^{\dagger 2}\right] \times \exp\left[-\frac{\zeta^*}{2}\hat{a}_-^2 + \frac{\zeta}{2}\hat{a}_-^{\dagger 2}\right] \\
&= \hat{S}_{\hat{a}_+}(\zeta) \times \hat{S}_{\hat{a}_-}(-\zeta).
\end{aligned} \tag{A15}$$

That is, in terms of the normal modes, it can be decomposed into a product of two single-mode squeezed operators.

In the current setting, the connection is even closer because the pair of modes in $\hat{S}_2$ have the opposite momenta $\pm k$. Since the two-mode squeeze operator, $\hat{S}_2$ in (A1) symmetrically contains the annihilation and the creation operators of both modes, during the mode counting, contributions from each pair $(k, -k)$ will appear twice, and the resulting expressions may look similar to that given by two single-mode squeezed state. For example, let us compute the Hadamard function of the scalar field ϕ in the two-mode squeezed thermal state in spatially isotropic spacetime,

$$\begin{aligned}
G_{H,0}^{(\phi)}(x,x') &= \int\frac{d^3k}{(2\pi)^{\frac{3}{2}}}\frac{1}{\sqrt{2\omega}}\int\frac{d^3k'}{(2\pi)^{\frac{3}{2}}}\frac{1}{\sqrt{2\omega'}}\left[\frac{1}{2}\langle\{\hat{a}_k,\hat{a}_{k'}\}\rangle_{\text{TMST}}\,e^{ik\cdot x+ik'\cdot x'}e^{-i\omega t-i\omega't'}\right. \\
&\qquad \left. + \frac{1}{2}\langle\{\hat{a}_k,\hat{a}_{k'}^\dagger\}\rangle_{\text{TMST}}\,e^{ik\cdot x-ik'\cdot x'}e^{-i\omega t+i\omega't'} + \text{H.C.}\right] \\
&= \int\frac{d^3k}{(2\pi)^3}\frac{1}{2\omega}\left[\frac{1}{2}\langle\{\hat{a}_k,\hat{a}_k\}\rangle_{\text{TMST}}\,e^{ik\cdot(x+x')}e^{-i\omega(t+t')} + \frac{1}{2}\langle\{\hat{a}_k,\hat{a}_{-k}\}\rangle_{\text{TMST}}\,e^{ik\cdot(x-x')}e^{-i\omega(t+t')}\right. \\
&\qquad \left. + \frac{1}{2}\langle\{\hat{a}_k,\hat{a}_k^\dagger\}\rangle_{\text{TMST}}\,e^{ik\cdot(x-x')}e^{-i\omega(t-t')} + \frac{1}{2}\langle\{\hat{a}_k,\hat{a}_{-k}^\dagger\}\rangle_{\text{TMST}}\,e^{ik\cdot(x+x')}e^{-i\omega(t-t')} + \text{H.C.}\right] \\
&= \int\frac{d^3k}{(2\pi)^3}\frac{1}{2\omega}\,e^{ik\cdot(x-x')}\left[\left(n_\omega+\frac{1}{2}\right)\cosh 2\eta_\omega\,e^{-i\omega(t-t')} - e^{i\theta_\omega}\left(n_\omega+\frac{1}{2}\right)\sinh 2\eta_\omega\,e^{-i\omega(t+t')} + \text{H.C.}\right],
\end{aligned} \tag{A16}$$

with $\omega = |\mathbf{k}|$ and the help of Equations (A4)–(A7). This is the same as the Hadamard function of the field in the squeezed thermal state, consistent with Equation (A15).

Appendix B. Late-Time Behavior of $\langle \Delta \hat{T}_{\mu\nu} \rangle$

Appendix B.1. Late-Time Energy Flux Density $\langle \Delta \hat{T}_{tr} \rangle$

Here, we examine the late-time net radiation flux of the field at distance r sufficiently far away from the atom, so that only the dominant contribution, independent r, is of our interest. The energy flux density is given by

$$\langle \Delta \hat{T}_{rt}(x) \rangle = \lim_{x' \to x} \frac{\partial^2}{\partial r \partial t'} \left[G_H^{(\phi)}(x,x') - G_{H,0}^{(\phi)}(x,x') \right]. \tag{A17}$$

Let us first look at the stationary component.

Appendix B.1.1. Stationary Component $\langle \Delta \hat{T}_{tr} \rangle_{\text{ST}}$

We assume that $t \gg r$, but place no other restrictions on r for the moment. Following the decomposition in Equation (24), the stationary part of the contribution purely from the radiation field is given by

$$\lim_{x' \to x} \frac{\partial^2}{\partial r \partial t'} \frac{1}{2} \langle \{\hat{\phi}_{\text{BR}}(x), \hat{\phi}_{\text{BR}}(x')\} \rangle_{\text{ST}} = -e^2 \int_{-\infty}^{\infty} \frac{d\omega}{2\pi} \, \omega^2 \, \widetilde{G}_H^{(\chi)}(\omega) \, |\widetilde{G}_{R,0}^{(\phi)}(x;\omega)|^2, \tag{A18}$$

with the help of

$$\frac{\partial}{\partial r} \widetilde{G}_{R,0}^{(\phi)}(x;\omega) = \left(i\omega - \frac{1}{r}\right) \widetilde{G}_{R,0}^{(\phi)}(x;\omega), \tag{A19}$$

$$\frac{\partial}{\partial r} \widetilde{G}_{H,0}^{(\phi),\text{ST}}(x;\omega) = \cosh 2\eta \, \coth \frac{\beta \omega}{2} \left[\omega \, \text{Re} \, \widetilde{G}_{R,0}^{(\phi)}(x;\omega) - \frac{1}{r} \, \text{Im} \, \widetilde{G}_{R,0}^{(\phi)}(x;\omega) \right], \tag{A20}$$

where the stationary part of the free-field Hadamard function is

$$\widetilde{G}_{H,0}^{(\phi),\text{ST}}(x,0;\omega) \equiv \widetilde{G}_{H,0}^{(\phi),\text{ST}}(x;\omega) = \cosh 2\eta \, \coth \frac{\beta \omega}{2} \frac{\sin \omega r}{4\pi r}. \tag{A21}$$

To arrive at Equation (A18), we observe that the term proportional to r^{-1} is odd in ω, thus, vanishing, and that $\text{Re} \, \widetilde{G}_R^{(\chi)}(\omega)$ is an even function of ω. At time greater than the relaxation time scale we may make the identification

$$\widetilde{G}_H^{(\chi)}(\omega) = \cosh 2\eta \, \coth \frac{\beta \omega}{2} \, \text{Im} \, \widetilde{G}_R^{(\chi)}(\omega), \tag{A22}$$

by the nonequilibrium fluctuation–dissipation relation [18,19,22] of the internal dynamics of the atom if it is initially coupled to a squeezed thermal field.

On the other hand, for the cross-term, containing the correlation between the radiation field and the free field at large distance from the atom, its stationary part is

$$\lim_{x' \to x} \frac{\partial^2}{\partial r \partial t'} \frac{1}{2} \left[\langle \{\hat{\phi}_h(x), \hat{\phi}_{\text{BR}}(x')\} \rangle_{\text{ST}} + \langle \{\hat{\phi}_{\text{BR}}(x), \hat{\phi}_h(x')\} \rangle_{\text{ST}} \right]$$
$$= -i e^2 \int_{-\infty}^{\infty} \frac{d\omega}{2\pi} \, \omega^2 \, \rho^{(\text{ST})}(\omega) \, \widetilde{G}_R^{(\chi)}(\omega) \, |\widetilde{G}_{R,0}^{(\phi)}(x;\omega)|^2. \tag{A23}$$

Hereafter, it is convenient to introduce the shorthand notations

$$\rho^{(\text{ST})}(\omega) = \cosh 2\eta \, \coth \frac{\beta \omega}{2}, \quad \text{and} \quad \rho^{(\text{NS})}(\omega) = \sinh 2\eta \, \coth \frac{\beta \omega}{2}. \tag{A24}$$

Since the contribution from $\text{Re}\,\widetilde{G}_R^{(\chi)}(\omega)$ is an even function of ω, one immediately notices that it is the opposite of Equation (A18) by Equation (A22), and thus cancels with Equation (A18). We already know from the case of the unsqueezed thermal state that the corresponding contributions of Equations (A18) and (A23) cancel with one another. Therefore, here, we showed that at late times the stationary part of the net energy flux density far away from the atom vanishes even though the field is initially in the squeezed thermal state, a nonstationary configuration.

Here, it is worth emphasizing the significance of vanishing stationary component of the net radiation flux. This is a unique feature of self-consistent quantum dynamics. The cancellation between Equations (A18) and (A23) is not possible if the radiated field $\phi_R(x)$ is not correlated with the free field $\phi_h(x)$. This correlation is established because the atom that sends out the radiation field is driven by a free field, as seen by the Langevin equation (7). It still holds when the field state is a vacuum state [2], so a pure classical system will not have such correlation. Secondly, our system is distinct from the classical driven dipole because the latter has a net outward radiation flux supplied by the external driving agent and is then independent of the field state. This is best seen from the component (A18) of the flux that is solely composed of the radiation field. This flux is field-state-dependent because the internal dynamics of the atom are driven by the quantum fluctuations of the field, rather than an external agent.

Appendix B.1.2. Nonstationary Component $\langle \Delta \hat{T}_{tr} \rangle_{\text{NS}}$

Now we turn to the nonstationary contribution in $\langle \Delta \hat{T}_{rt}(x) \rangle$. The component purely from the radiation field is given by

$$\lim_{x' \to x} \frac{\partial^2}{\partial r \partial t'} \frac{1}{2} \langle \{\hat{\phi}_{BR}(x), \hat{\phi}_{BR}(x')\} \rangle_{\text{NS}}$$
$$= i e^4 \int_0^\infty \frac{d\omega}{2\pi} \frac{\omega^2}{4\pi} \rho^{(\text{NS})}(\omega) \left(i\omega - \frac{1}{r} \right) \left[\widetilde{G}_R^{(\chi)}(\omega) \widetilde{G}_{R,0}^{(\phi)}(x;\omega) \right]^2 e^{-i2\omega t + i\theta} + \text{C.C.}, \quad (A25)$$

while the counterpart from the cross-term is

$$\lim_{x' \to x} \frac{\partial^2}{\partial r \partial t'} \frac{1}{2} \left[\langle \{\hat{\phi}_h(x), \hat{\phi}_{BR}(x')\} \rangle_{\text{NS}} + \langle \{\hat{\phi}_{BR}(x), \hat{\phi}_h(x')\} \rangle_{\text{NS}} \right] \quad (A26)$$
$$= i e^2 \int_0^\infty \frac{d\omega}{2\pi} \omega \rho^{(\text{NS})}(\omega) \widetilde{G}_R^{(\chi)}(\omega) \widetilde{G}_{R,0}^{(\phi)}(x;\omega) \left[\omega \widetilde{G}_{R,0}^{(\phi)}(x;\omega) - \frac{2}{r} \text{Im}\,\widetilde{G}_{R,0}^{(\phi)}(x;\omega) \right] e^{-i2\omega t + i\theta} + \text{C.C.}.$$

Contrary to the stationary contribution, in general, Equation (A25) does not cancel Equation (A26). This can be explicitly shown by examining the sum of the expressions which are proportional to ω inside the curly brackets in both equations. We find that these terms together give

$$i e^2 \int_0^\infty \frac{d\omega}{2\pi} \omega^2 \rho^{(\text{NS})}(\omega) \widetilde{G}_R^{(\chi)}(\omega) \left[\widetilde{G}_{R,0}^{(\phi)}(x;\omega) \right]^2 \left[i e^2 \frac{\omega}{4\pi} \widetilde{G}_R^{(\chi)}(\omega) + 1 \right] e^{-i2\omega t + i\theta} + \text{C.C.}$$
$$= i e^2 \int_0^\infty \frac{d\omega}{2\pi} \omega^2 \rho^{(\text{NS})}(\omega) \,\text{Re}\,\widetilde{G}_R^{(\chi)}(\omega) \frac{\widetilde{G}_R^{(\chi)}(\omega)}{\widetilde{G}_R^{(\chi)*}(\omega)} \left[\widetilde{G}_{R,0}^{(\phi)}(x;\omega) \right]^2 e^{-i2\omega t + i\theta} + \text{C.C.}, \quad (A27)$$

where we use the fact that $e^2 = 8\pi \gamma m$ and

$$\text{Im}\,\widetilde{G}_R^{(\chi)}(\omega) = 2m\gamma\omega \,|\widetilde{G}_R^{(\chi)}(\omega)|^2. \quad (A28)$$

Apparently, Equation (A27) does not vanish in general. If the squeeze angle θ is really a mode- and time-independent constant, we may set $\theta = 0$ without loss of generality and further simplify the nonstationary contribution (A27), in $\langle \Delta \hat{T}_{rt}(x) \rangle$ to

$$i\,\mathrm{e}^2 \int_0^\infty \frac{d\omega}{2\pi}\, \omega^2 \rho^{(\mathrm{NS})}(\omega) \left[\widetilde{G}_{R,0}^{(\phi)}(\boldsymbol{x};\omega)\right]^2 \frac{\widetilde{G}_R^{(\chi)}(\omega)}{\widetilde{G}_R^{(\chi)*}(\omega)}\, \mathrm{Re}\, \widetilde{G}_R^{(\chi)}(\omega)\, e^{-i2\omega t}. \tag{A29}$$

Equation (A27), or the special case Equation (A29), in general, is time-dependent and is not identically zero.

For completeness, let us write down the sum of the terms proportional to $1/r$ in the square brackets in Equations (A25) and (A26), which is

$$-i\frac{\mathrm{e}^2}{r}\int_0^\infty \frac{d\omega}{2\pi}\, \omega \rho^{(\mathrm{NS})}(\omega)\, \widetilde{G}_R^{(\chi)}(\omega)\, \widetilde{G}_{R,0}^{(\phi)}(\boldsymbol{x};\omega)\left[\mathrm{e}^2 \frac{\omega}{4\pi}\widetilde{G}_R^{(\chi)}(\omega)\widetilde{G}_{R,0}^{(\phi)}(\boldsymbol{x};\omega) + 2\,\mathrm{Im}\,\widetilde{G}_{R,0}^{(\phi)}(\boldsymbol{x};\omega)\right] e^{-i2\omega t + i\theta}$$

$$= -\frac{\mathrm{e}^2}{r}\int_0^\infty \frac{d\omega}{2\pi}\, \omega \rho^{(\mathrm{NS})}(\omega)\left\{\left[\widetilde{G}_{R,0}^{(\phi)}(\boldsymbol{x};\omega)\right]^2 \frac{\widetilde{G}_R^{(\chi)}(\omega)}{\widetilde{G}_R^{(\chi)*}(\omega)}\,\mathrm{Re}\,\widetilde{G}_R^{(\chi)}(\omega) + \widetilde{G}_R^{(\chi)}(\omega)|\widetilde{G}_{R,0}^{(\phi)}(\boldsymbol{x};\omega)|^2\right\} e^{-i2\omega t + i\theta}, \tag{A30}$$

plus its complex conjugate. Here, again, we used Equation (A28) to recast the expression into a more compact form. This will be useful later to check the consistency via the continuity equation.

The sum of Equations (A25) and (A26) gives the only contribution of the net radiation flux that is likely to survive after the relaxation time scale. However, in Appendix C, we can argue that although Equations (A25) and (A26) do not cancel in general, their sum still decays with time and, furthermore, it actually falls off to zero exponentially fast, proportional to $e^{-2\gamma t}$.

Thus, combining the discussions about the stationary component, we show that at late times no net radiation flux exudes to spatial infinity from a static atom even though the internal degree of freedom of the atom is driven by the quantum fluctuations of the squeezed field caused by its parametric process. At late times it leaves no trace of detectable signal to a detector at distance much greater than the typical scales in the atom's internal dynamics, except for the case of extremely large squeezing where we might have the chance to see a trace of the net flux since it will take a longer time to decay. However, the information of the squeeze parameter η can still be read out from the asymptotic equilibrium state of the atom's internal dynamics.

Appendix B.2. Late-Time Field Energy Density $\langle \Delta \hat{T}_{tt} \rangle$

The energy density is given by

$$\langle \Delta \hat{T}_{tt}(x) \rangle = \lim_{x' \to x} \frac{1}{2}\left(\frac{\partial^2}{\partial t \partial t'} + \frac{\partial^2}{\partial r \partial r'}\right)\left[G_H^{(\phi)}(x,x') - G_{H,0}^{(\phi)}(x,x')\right]. \tag{A31}$$

That is, we subtract off the energy density of the free field in the squeezed state, and Equation (A31) tells us the change in energy density due to the atom–field interaction.

Appendix B.2.1. Stationary Component $\langle \Delta \hat{T}_{tt} \rangle_{\mathrm{ST}}$

We first consider the stationary component of Equation (A31). Following the decomposition (24), the part purely caused by the radiation field is given by

$$\lim_{x' \to x} \frac{1}{2}\left(\frac{\partial^2}{\partial t \partial t'} + \frac{\partial^2}{\partial r \partial r'}\right)\frac{1}{2}\langle\{\hat{\phi}_{\mathrm{BR}}(x), \hat{\phi}_{\mathrm{BR}}(x')\}\rangle_{\mathrm{ST}}$$

$$= \mathrm{e}^2 \int_{-\infty}^\infty \frac{d\omega}{2\pi}\, \rho^{(\mathrm{ST})}(\omega)\left(\omega^2 + \frac{1}{2r^2}\right)\mathrm{Im}\,\widetilde{G}_R^{(\chi)}(\omega)\,|\widetilde{G}_{R,0}^{(\phi)}(\boldsymbol{x};\omega)|^2. \tag{A32}$$

The contribution from the cross-term is then

$$\lim_{x'\to x} \frac{1}{2}\left(\frac{\partial^2}{\partial t \partial t'} + \frac{\partial^2}{\partial r \partial r'}\right) \frac{1}{2}\left[\langle\{\hat{\phi}_h(x), \hat{\phi}_{BR}(x')\}\rangle_{ST} + \langle\{\hat{\phi}_{BR}(x), \hat{\phi}_h(x')\}\rangle_{ST}\right] \quad (A33)$$

$$= -e^2 \int_{-\infty}^{\infty} \frac{d\omega}{2\pi} \rho^{(ST)}(\omega) \left\{\left(\omega^2 + \frac{1}{2r^2}\right) |\widetilde{G}_{R,0}^{(\phi)}(x;\omega)|^2 \operatorname{Im}\widetilde{G}_R^{(\chi)}(\omega) + \frac{\omega}{r} \operatorname{Re}\left[\widetilde{G}_R^{(\chi)}(\omega) \widetilde{G}_{R,0}^{(\phi)\,2}(x;\omega)\right]\right.$$

$$\left. - \frac{1}{2r^2} \operatorname{Im}\left[\widetilde{G}_R^{(\chi)}(\omega) \widetilde{G}_{R,0}^{(\phi)\,2}(x;\omega)\right]\right\},$$

with the help of Equations (A19) and (A20) to re-express the term

$$2\operatorname{Re}\left[\widetilde{G}_R^{(\chi)}(\omega) \widetilde{G}_{R,0}^{(\phi)}(x;\omega)\right] \operatorname{Im}\widetilde{G}_{R,0}^{(\phi)}(x;\omega) = -|\widetilde{G}_{R,0}^{(\phi)}(x;\omega)|^2 \operatorname{Im}\widetilde{G}_R^{(\chi)}(\omega) + \operatorname{Im}\left[\widetilde{G}_R^{(\chi)}(\omega) \widetilde{G}_{R,0}^{(\phi)\,2}(x;\omega)\right].$$

Adding both contributions together gives

$$\langle \Delta \hat{T}_{tt}\rangle_{ST} = -e^2 \int_{-\infty}^{\infty} \frac{d\omega}{2\pi} \rho^{(ST)}(\omega) \left\{\frac{\omega}{r} \operatorname{Re}\left[\widetilde{G}_R^{(\chi)}(\omega) \widetilde{G}_{R,0}^{(\phi)\,2}(x;\omega)\right] - \frac{1}{2r^2} \operatorname{Im}\left[\widetilde{G}_R^{(\chi)}(\omega) \widetilde{G}_{R,0}^{(\phi)\,2}(x;\omega)\right]\right\}, \quad (A34)$$

at late times and at large distance away from the atom. The dominant contribution of the stationary component of the local field energy density vanishes, so Equation (A34) is subdominant since it is proportional to $1/r^3$. It comes from the correlation between the radiation field and the free field at the location far away from the atom, and it decays faster with r. Note that there are two type of $1/r^3$ contributions in the cross-terms (A33) but one of them cancels with its counterpart in Equation (A32). The residual energy density (A34) has a relatively short range by nature. We note that it is proportional to e^2, which is proportional to the damping constant, γ. Since this is a time-independent constant, it means that this may be an unambiguous aftereffect of the transient peregrinating radiation field.

Next, we turn to the nonstationary component.

Appendix B.2.2. Nonstationary Component $\langle \Delta \hat{T}_{tt}\rangle_{NS}$

We first examine the late-time contribution purely from the radiation field,

$$\lim_{x'\to x} \frac{1}{2}\left(\frac{\partial^2}{\partial t \partial t'} + \frac{\partial^2}{\partial r \partial r'}\right) \frac{1}{2}\langle\{\hat{\psi}_{BR}(x), \hat{\psi}_{BR}(x')\}\rangle_{NS}$$

$$= e^4 \int_0^{\infty} \frac{d\omega}{2\pi} \frac{\omega}{4\pi} \rho^{(NS)}(\omega) \left(\omega^2 + i\frac{\omega}{r} - \frac{1}{2r^2}\right) \left[\widetilde{G}_R^{(\chi)}(\omega) \widetilde{G}_{R,0}^{(\phi)}(x;\omega)\right]^2 e^{-i2\omega t + i\theta} + \text{C.C.}. \quad (A35)$$

On the other hand, the corresponding contribution from the cross-term gives

$$\lim_{x'\to x} \frac{1}{2}\left(\frac{\partial^2}{\partial t \partial t'} + \frac{\partial^2}{\partial r \partial r'}\right) \frac{1}{2}\left[\langle\{\hat{\phi}_h(x), \hat{\phi}_{BR}(x')\}\rangle_{NS} + \langle\{\hat{\phi}_{BR}(x), \hat{\phi}_h(x')\}\rangle_{NS}\right]$$

$$= -e^2 \int_0^{\infty} \frac{d\omega}{2\pi} \rho^{(NS)}(\omega) \left\{i\left(\omega^2 + i\frac{\omega}{r}\right) \widetilde{G}_R^{(\chi)}(\omega) \left[\widetilde{G}_{R,0}^{(\phi)}(x;\omega)\right]^2\right.$$

$$\left. + \frac{1}{r^2} \widetilde{G}_R^{(\chi)}(\omega) \widetilde{G}_{R,0}^{(\phi)}(x;\omega) \operatorname{Im}\widetilde{G}_{R,0}^{(\phi)}(x;\omega)\right\} e^{-i2\omega t + i\theta} + \text{C.C.}, \quad (A36)$$

Now we use the same trick (A28) to combine both contributions, and we obtain:

$$\langle \Delta \hat{T}_{tt}\rangle_{NS} = -ie^2 \int_0^{\infty} \frac{d\omega}{2\pi} \rho^{(NS)}(\omega) \left\{-i\left(\omega^2 + i\frac{\omega}{r} - \frac{1}{2r^2}\right) \left[\widetilde{G}_{R,0}^{(\phi)}(x;\omega)\right]^2 \frac{\widetilde{G}_R^{(\chi)}(\omega)}{\widetilde{G}_R^{(\chi)*}(\omega)} \operatorname{Re}\widetilde{G}_R^{(\chi)}(\omega)\right.$$

$$\left. + \frac{1}{2r^2} \widetilde{G}_R^{(\chi)}(\omega) |\widetilde{G}_{R,0}^{(\phi)}(x;\omega)|^2\right\} e^{-i2\omega t + i\theta} + \text{C.C.}. \quad (A37)$$

The dominant term in Equation (A37) has the same form as Equation (A27), and thus will vanish at late times at distance far away from the atom.

Therefore, at late times, $\langle \Delta \hat{T}_{tt} \rangle$ settles down to a constant value whose value decays similar to $1/r^3$ away from the atom.

Appendix B.3. Continuity Equation

Here, in passing, we would like to examine the consistency of our results of the nonstationary components of $\langle \Delta \hat{T}_{\mu t} \rangle$ in terms of the continuity equation $\partial^\mu T_{\mu t} = 0$. It turns out convenient to list the previous results for the relevant components

$$\langle \hat{T}_{rt} \rangle_{\text{NS}}^{(\text{RR})} = i e^4 \int_0^\infty \frac{d\omega}{2\pi} \frac{\omega^2}{4\pi} \rho^{(\text{NS})}(\omega) \left(i\omega - \frac{1}{r} \right) \left[\widetilde{G}_R^{(\chi)}(\omega) \widetilde{G}_{R,0}^{(\phi)}(x;\omega) \right]^2 e^{-i2\omega t + i\theta} + \text{C.C.},$$

$$\langle \hat{T}_{rt} \rangle_{\text{NS}}^{(\text{HR})} = i e^2 \int_0^\infty \frac{d\omega}{2\pi} \omega \rho^{(\text{NS})}(\omega) \widetilde{G}_R^{(\chi)}(\omega) \widetilde{G}_{R,0}^{(\phi)}(x;\omega) \left[\omega \widetilde{G}_{R,0}^{(\phi)}(x;\omega) - \frac{2}{r} \operatorname{Im} \widetilde{G}_{R,0}^{(\phi)}(x;\omega) \right] e^{-i2\omega t + i\theta} + \text{C.C.},$$

$$\langle \hat{T}_{tt} \rangle_{\text{NS}}^{(\text{RR})} = e^4 \int_0^\infty \frac{d\omega}{2\pi} \frac{\omega}{4\pi} \rho^{(\text{NS})}(\omega) \left(\omega^2 + i\frac{\omega}{r} - \frac{1}{2r^2} \right) \left[\widetilde{G}_R^{(\chi)}(\omega) \widetilde{G}_{R,0}^{(\phi)}(x;\omega) \right]^2 e^{-i2\omega t + i\theta} + \text{C.C.},$$

$$\langle \hat{T}_{tt} \rangle_{\text{NS}}^{(\text{HR})} = -e^2 \int_0^\infty \frac{d\omega}{2\pi} \rho^{(\text{NS})}(\omega) \widetilde{G}_R^{(\chi)}(\omega) \left\{ i \left(\omega^2 + i\frac{\omega}{r} \right) \left[\widetilde{G}_{R,0}^{(\phi)}(x;\omega) \right]^2 \right.$$
$$\left. + \frac{1}{r^2} \widetilde{G}_{R,0}^{(\phi)}(x;\omega) \operatorname{Im} \widetilde{G}_{R,0}^{(\phi)}(x;\omega) \right\} e^{-i2\omega t + i\theta} + \text{C.C.}.$$

The superscript "RR" represents the contributions purely from the radiation field such as Equations (A25) and (A35), while the superscript "HR" denotes those from the cross-terms such as Equations (A26), and (A36). For our configuration, we do not have $T_{\vartheta t}$ and $T_{\varphi t}$ where ϑ and φ are, respectively, the polar angle, and azimuthal angle, so, explicitly, the continuity equation can be written in the form

$$\partial^\mu \hat{T}_{\mu t} = 0 = \partial_t T_{tt} - \frac{1}{r^2} \partial_r \left(r^2 T_{rt} \right). \tag{A38}$$

Thus, we find:

$$\frac{1}{r^2} \partial_r \left(r^2 \langle \hat{T}_{rt} \rangle_{\text{NS}}^{(\text{RR})} \right) = -2i e^4 \int_0^\infty \frac{d\omega}{2\pi} \frac{\omega^2}{4\pi} \rho^{(\text{NS})}(\omega) \left(\omega^2 + i\frac{\omega}{r} - \frac{1}{2r^2} \right) \left[\widetilde{G}_R^{(\chi)}(\omega) \widetilde{G}_{R,0}^{(\phi)}(x;\omega) \right]^2 e^{-i2\omega t + i\theta} + \text{C.C.},$$

which is equal to $\partial_t \langle \hat{T}_{tt} \rangle_{\text{NS}}^{(\text{RR})}$, and

$$\frac{1}{r^2} \partial_r \left(r^2 \langle \hat{T}_{rt} \rangle_{\text{NS}}^{(\text{HR})} \right) = 2e^2 \int_0^\infty \frac{d\omega}{2\pi} \rho^{(\text{NS})}(\omega) \widetilde{G}_R^{(\chi)}(\omega) \left\{ i \left(\omega^2 + i\frac{\omega}{r} \right) \left[\widetilde{G}_{R,0}^{(\phi)}(x;\omega) \right]^2 \right.$$
$$\left. + \frac{1}{r^2} \widetilde{G}_{R,0}^{(\phi)}(x;\omega) \operatorname{Im} \widetilde{G}_{R,0}^{(\phi)}(x;\omega) \right\} e^{-i2\omega t + i\theta} + \text{C.C.}.$$

It is exactly $\partial_t \langle \hat{T}_{tt} \rangle_{\text{NS}}^{(\text{HR})}$. Thus, the continuity equation is satisfied.

Appendix C. Late-Time Behavior of the Nonstationary Contribution in $\langle \Delta \hat{T}_{rt} \rangle$

It turns out that the nonstationary contribution in $\langle \Delta \hat{T}_{rt}(x) \rangle$ still decays with time. To begin with, it proves useful to quote the late-time expressions of the energy exchange between the atom and the surrounding field [2,22]. For times greater than the relation time, the nonstationary contribution of the energy flow into the atom from the field has the form,

$$P_\xi^{(\text{NS})}(t) = i e^2 \int_0^\infty \frac{d\omega}{2\pi} \frac{\omega^2}{4\pi} \sinh 2\eta \coth \frac{\beta\omega}{2} \left\{ \widetilde{G}_R^{(\chi)}(\omega) e^{-i2\omega t + i\theta} - \text{C.C.} \right\}, \tag{A39}$$

and the corresponding contribution of the energy flow out of the atom due to friction is

$$P_\gamma^{(\text{NS})}(t) = -e^4 \int_0^\infty \frac{d\omega}{2\pi} \frac{\omega^3}{(4\pi)^2} \sinh 2\eta \coth \frac{\beta\omega}{2} \left\{ \left[\widetilde{G}_R^{(\chi)}(\omega) \right]^2 e^{-i2\omega t + i\theta} + \text{C.C.} \right\}. \tag{A40}$$

For comparison, let us place the terms proportional to ω inside of the curly brackets of (A26) and (A25) here,

$$(A26): \quad i e^2 \int_0^\infty \frac{d\omega}{2\pi} \frac{\omega^2}{(4\pi)^2 r^2} \sinh 2\eta \coth \frac{\beta \omega}{2} \left\{ \widetilde{G}_R^{(x)}(\omega) e^{-i2\omega(t-r)+i\theta} - \text{C.C.} \right\}, \qquad (A41)$$

$$(A25): \quad -e^4 \int_0^\infty \frac{d\omega}{2\pi} \frac{\omega^3}{(4\pi)^3 r^2} \sinh 2\eta \coth \frac{\beta \omega}{2} \left\{ \left[\widetilde{G}_R^{(x)}(\omega)\right]^2 e^{-i2\omega(t-r)+i\theta} + \text{C.C.} \right\}. \qquad (A42)$$

One can see correspondence between Equations (A39) and (A41), as well as between Equations (A40) and (A42). For example, the integrand of Equation (A41) is $\dfrac{e^{i2\omega r}}{4\pi r^2}$ times the integrand of Equation (A39). The same applies to the pair of Equations (A40)–(A42). For a fix r and at times $t \gg r$, we may shift or redefine t in Equations (A41) and (A42) by $t - r \to t$ such that the integrals in both equations are essentially proportional to those in Equations (A39) and (A40). We numerically showed in Ref. [19] that Equations (A39) and (A40) vanish as $t \to \infty$, so we conclude that the terms proportional to ω inside of the curly brackets of Equations (A25) and (A26) will also vanish at late times. For example, as shown in Figure A1, generated by numerical calculations, the contribution in Equation (A25) actually decays exponentially fast to zero.

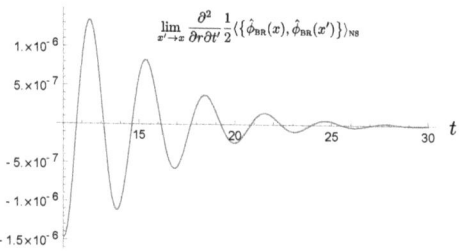

Figure A1. The time-dependence of Equation (A25), one of the nonstationary components of the radiation flux. The curve falls off to zero exponentially fast. Here, $\beta^{-1} = 0$, $\gamma = 0.2$, $r = 10$, and $\omega_R = 1$ are chosen.

For terms that are proportional to $1/r$ inside of the curly brackets of Equations (A26) and (A25), we immediately see that if we take the time derivative of Equation (A25), we obtain an expression which is $2/r$ times Equation (A42). Likewise, the time derivative of Equation (A26) gives

$$\frac{2}{r} \times i e^2 \int_0^\infty \frac{d\omega}{2\pi} \frac{\omega^2}{(4\pi)^2 r^2} \sinh 2\eta \coth \frac{\beta \omega}{2} \left\{ \widetilde{G}_R^{(x)}(\omega) \left[e^{-i2\omega(t-r)+i\theta} - e^{-i2\omega t + i\theta}\right] - \text{C.C.} \right\}. \qquad (A43)$$

Following the previous arguments, for a fixed r, it vanishes in the limit $t \to \infty$ too. Thus, we have demonstrated that the time derivatives of the terms proportional to $1/r$ inside of the curly brackets of Equations (A26) and (A25) vanish, so these terms must be constants for sufficiently large time. On the other hand, these terms are time-dependent and are not sign-definite, so the asymptotic constants must be zero.

Placing these results together, we reach the conclusion that the nonstationary terms of the net radiation flux vanish eventually at large distance away from the atom. However, this does tell us how slowly the nonstationary terms decay. Actually, at least for the zero-temperature limit $\beta \to \infty$, we can carry out the above integrals analytically, and the brute-force calculations show that for fixed r, these integrals give results that decay exponentially fast to zero in a form proportional to $\sinh 2\eta \, e^{-2\gamma t}$. Thus, at late times, no net radiation energy flux leaks to the spatial infinity from the atom even though the internal degree of freedom of the atom is coupled to a nonstationary squeezed quantum field.

Appendix D. Time-Translational Invariance of the Squeeze Parameter in the Out-Region

Next we argue that if we shift the parametric process by Δ along the time axis, as shown in Figure A2, then we have

$$\mathsf{d}_i^{(\mathrm{II})}(t+\Delta, t_a+\Delta) = d_i^{(\mathrm{II})}(t, t_a) \tag{A44}$$

where $i = 1, 2$, and $\mathsf{d}_i^{(\mathrm{II})}(t+\Delta, t_a+\Delta)$, in sans serif style, is the fundamental solution in-region II for the shifted case described by the orange curve in Figure A2.

Since the fundamental solutions $\mathsf{d}_i^{(\mathrm{II})}(y, y_a)$ in-region II for the orange curve satisfy the equation of motion

$$\ddot{\mathsf{d}}_i^{(\mathrm{II})}(y, y_a) + \omega^2(y)\, \mathsf{d}_i^{(\mathrm{II})}(y, y_a) = 0, \tag{A45}$$

with the standard initial conditions given at $y_a = t_a + \Delta$, while the fundamental solutions $d_i^{(\mathrm{II})}(y, y_a)$ in-region II for the blue curve satisfy the equation of motion

$$\ddot{d}_i^{(\mathrm{II})}(t, t_a) + \omega^2(t)\, d_i^{(\mathrm{II})}(t, t_a) = 0, \tag{A46}$$

with the standard initial conditions given at t_a, one can immediately notice that both cases are related by $y = t + \Delta$, and thus we arrive at Equation (A44), and in particular,

$$\mathsf{d}_i^{(\mathrm{II})}(t_b+\Delta, t_a+\Delta) = d_i^{(\mathrm{II})}(t_b, t_a). \tag{A47}$$

Here below, we use this result.

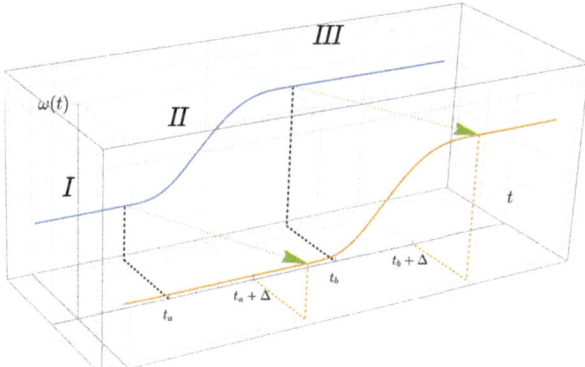

Figure A2. The parametric process described by a piecewise continuous function. The transition begins at t_a, and ends at t_b, region II. Between t_i and t_a, region I, the frequency $\omega(t)$ takes on a constant value ω_i and is given by another constant ω_f after $t \geq t_b$, region III. We consider a shift Δ of the parametric process along the timeline while keeping the functional form, the duration $t_b - t_a$ of the process, and the initial time t_i of the motion unchanged. We displace these two cases by a horizontal displacement for discernibility. See text for details.

Now we would like to explicitly show that a detector, in the out-region with $t > t_b + \Delta$, will find the same squeeze parameter for both cases described in Figure A2, independent of the shift Δ, even if it carries out the measurement at the same time t. Note that in both cases the motion starts at the same $t_i = 0$. We focus on the moment $t = t_b + \Delta$, since afterward, the squeeze parameter is a constant. Before we start, let us first note that $d_i^{(\mathrm{I})}(t)$ is essentially the same as $\mathsf{d}_i^{(1)}(t)$, except that the former applies to the time interval $0 = t_i \leq t \leq t_a + \Delta$ while the latter only to $0 = t_i \leq t \leq t_a$. With this recognition, we can write $\mathsf{d}_i^{(1)}(t_a + \Delta)$ as

$$\mathsf{d}_i^{(1)}(t_a + \Delta) = d_1^{(1)}(\Delta)\, \mathsf{d}_i^{(1)}(t_a) + d_2^{(1)}(\Delta)\, \dot{\mathsf{d}}_i^{(1)}(t_a)\,. \tag{A48}$$

Now we write $\mathsf{d}_i(t_b + \Delta, 0)$ in terms $d_i^{(X)}$ with X = I, II, and III of the unshifted process; see Figure A2. We show the calculations explicitly for $\mathsf{d}_1(t_b + \Delta)$, and the result for $\mathsf{d}_2(t_b + \Delta)$ follows similarly. Pretty straightforwardly, we have:

$$\begin{aligned}\mathsf{d}_1(t_b + \Delta, 0) &= \mathsf{d}_1^{(II)}(t_b + \Delta, t_a + \Delta)\, \mathsf{d}_1^{(1)}(t_a + \Delta) + \mathsf{d}_2^{(II)}(t_b + \Delta, t_a + \Delta)\, \dot{\mathsf{d}}_1^{(1)}(t_a + \Delta) \\ &= \mathsf{d}_1^{(II)}(t_b, t_a)\left[d_1^{(1)}(\Delta)\, \mathsf{d}_1^{(1)}(t_a) + d_2^{(1)}(\Delta)\, \dot{\mathsf{d}}_1^{(1)}(t_a)\right] + \mathsf{d}_2^{(II)}(t_b, t_a)\left[\dot{d}_1^{(1)}(\Delta)\, \mathsf{d}_1^{(1)}(t_a) + \dot{d}_2^{(1)}(\Delta)\, \dot{\mathsf{d}}_1^{(1)}(t_a)\right]\end{aligned} \tag{A49}$$

with the help of Equations (A44) and (A48). We use a trick that only applies to the undamped harmonic oscillator. For in-region I, we have the identities,

$$\dot{d}_2^{(1)}(t) = d_1^{(1)}(t)\,, \qquad \Rightarrow \qquad \dot{d}_2^{(1)}(\Delta)\, \mathsf{d}_1^{(1)}(t_a) = d_2^{(1)}(t_a)\, \dot{d}_1^{(1)}(\Delta)\,, \tag{A50}$$

and it allows us to rewrite Equation (A49),

$$\begin{aligned}\mathsf{d}_1(t_b + \Delta, 0) &= \mathsf{d}_1^{(II)}(t_b, t_a)\left[d_1^{(1)}(\Delta)\, \mathsf{d}_1^{(1)}(t_a) + \dot{d}_1^{(1)}(\Delta)\, d_2^{(1)}(t_a)\right] + \mathsf{d}_2^{(II)}(t_b, t_a)\left[\dot{d}_1^{(1)}(\Delta)\, \mathsf{d}_1^{(1)}(t_a) + d_1^{(1)}(\Delta)\, \dot{\mathsf{d}}_1^{(1)}(t_a)\right] \\ &= \left[\mathsf{d}_1^{(II)}(t_b, t_a)\, \mathsf{d}_1^{(1)}(t_a) + \mathsf{d}_2^{(II)}(t_b, t_a)\, \dot{\mathsf{d}}_1^{(1)}(t_a)\right] d_1^{(1)}(\Delta) + \left[\mathsf{d}_1^{(II)}(t_b, t_a)\, d_2^{(1)}(t_a) + \mathsf{d}_2^{(II)}(t_b, t_a)\, \mathsf{d}_1^{(1)}(t_a)\right] \dot{d}_1^{(1)}(\Delta) \\ &= \mathsf{d}_1(t_b, 0)\, d_1^{(1)}(\Delta) + \mathsf{d}_2(t_b, 0)\, \dot{d}_1^{(1)}(\Delta)\,.\end{aligned} \tag{A51}$$

Similarly, for $\mathsf{d}_2(t_b + \Delta, 0)$, we have

$$\mathsf{d}_2(t_b + \Delta, 0) = \mathsf{d}_1(t_b, 0)\, d_2^{(1)}(\Delta) + \mathsf{d}_2(t_b, 0)\, \dot{d}_2^{(1)}(\Delta)\,. \tag{A52}$$

Now we plug these results into the expressions in Equations (68)–(70) to find the corresponding squeeze parameter in out-region III,

$$\frac{1}{\omega_f \omega_i}\, \dot{\mathsf{d}}_k^{(1)2}(t_b + \Delta) + \frac{\omega_i}{\omega_f}\, \mathsf{d}_k^{(2)2}(t_b + \Delta) + \frac{\omega_f}{\omega_i}\, \mathsf{d}_k^{(1)2}(t_b + \Delta) + \omega_f \omega_i\, \mathsf{d}_k^{(2)2}(t_b + \Delta)$$

$$= \frac{1}{\omega_f \omega_i}\, \dot{\mathsf{d}}_k^{(1)2}(t_b) + \frac{\omega_i}{\omega_f}\, \mathsf{d}_k^{(2)2}(t_b) + \frac{\omega_f}{\omega_i}\, \mathsf{d}_k^{(1)2}(t_b) + \omega_f \omega_i\, \mathsf{d}_k^{(2)2}(t_b)\,, \tag{A53}$$

$$\frac{1}{\omega_f \omega_i}\, \dot{\mathsf{d}}_k^{(1)2}(t_b + \Delta) + \frac{\omega_i}{\omega_f}\, \mathsf{d}_k^{(2)2}(t_b + \Delta) - \frac{\omega_f}{\omega_i}\, \mathsf{d}_k^{(1)2}(t_b + \Delta) - \omega_f \omega_i\, \mathsf{d}_k^{(2)2}(t_b + \Delta)$$

$$= \frac{1}{\omega_f \omega_i}\, \dot{\mathsf{d}}_k^{(1)2}(t_b) + \frac{\omega_i}{\omega_f}\, \mathsf{d}_k^{(2)2}(t_b) - \frac{\omega_f}{\omega_i}\, \mathsf{d}_k^{(1)2}(t_b) - \omega_f \omega_i\, \mathsf{d}_k^{(2)2}(t_b)\,, \tag{A54}$$

$$\frac{1}{\omega_i}\, \mathsf{d}_k^{(1)}(t_b + \Delta)\dot{\mathsf{d}}_k^{(1)}(t_b + \Delta) + \omega_i\, \mathsf{d}_k^{(2)}(t_b + \Delta)\dot{\mathsf{d}}_k^{(2)}(t_b + \Delta)$$

$$= \frac{1}{\omega_i}\, \mathsf{d}_k^{(1)}(t_b)\dot{\mathsf{d}}_k^{(1)}(t_b) + \omega_i\, \mathsf{d}_k^{(2)}(t_b)\dot{\mathsf{d}}_k^{(2)}(t_b)\,, \tag{A55}$$

due to

$$\dot{d}_1^{(1)}(\Delta)\, d_1^{(1)}(\Delta) = -\omega_i^2 d_2^{(1)}(\Delta)\, \dot{d}_2^{(1)}(\Delta)\,, \qquad d_1^{(1)2}(\Delta) + \omega_i^2 d_2^{(1)2}(\Delta) = 1\,, \qquad \frac{1}{\omega_i^2}\, \dot{d}_1^{(1)2}(\Delta) + \dot{d}_2^{(1)2}(\Delta) = 1\,.$$

The results for the shifted process are the same as those for the unshifted process. Thus, both cases generate the same squeezing at $t \geq t_b + \Delta$.

References

1. Unruh, W.G. Notes on black-hole evaporation. *Phys. Rev. D* **1976**, *14*, 870–892. [CrossRef]
2. Hsiang, J.-T.; Hu, B.L. Atom-field interaction: From vacuum fluctuations to quantum radiation and quantum dissipation or radiation reaction. *Physics* **2019**, *1*, 430–444. [CrossRef]
3. DeWitt, B.S. Quantum gravity: The new synthesis. In *General Relativity: An Einstein Centenary Survey*; Hawking, S.W., Israel, W., Eds.; Cambridge University Press: New York, NY, USA, 1979; pp. 680–745.
4. Ackerhalt, J.R.; Knight, P.L.; Eberly, J.H. Radiation reaction and radiative frequency shifts. *Phys. Rev. Lett.* **1973**, *30*, 456–460. [CrossRef]
5. Milonni, P.W.; Smith, W.A. Radiation reaction and vacuum fluctuations in spontaneous emission. *Phys. Rev. A* **1975**, *11*, 814–824. [CrossRef]
6. Dalibard, J.; Dupont-Roc, J.; Cohen-Tannodji, C. Vacuum fluctuations and radiation reaction: Identification of their respective contributions. *J. Phys. France* **1982**, *43*, 1617–1638. [CrossRef]
7. Dalibard, J.; Dupont-Roc, J.; Cohen-Tannodji, C. Dynamics of a small system coupled to a reservoir: Reservoir fluctuations and self-reaction. *J. Phys. France* **1984**, *45*, 637–656. [CrossRef]
8. Jackson, J.D. *Classical Electrodynamics*; John Wiley & Sons, Inc.: New York, NY, USA, 1975. Available online: https://archive.org/details/ClassicalElectrodynamics2nd/page/n5/mode/2up (accessed on 5 May 2023).
9. Rohrlich, F. *Classical Charged Particles: Foundation of Their Theories*; Routledge/Taylor & Francis Group: New York, NY, USA, 2019. [CrossRef]
10. Johnson, P.R.; Hu, B.L. Stochastic theory of relativistic particles moving in a quantum field: Scalar Abraham-Lorentz-Dirac-Langevin equation, radiation reaction, and vacuum fluctuations. *Phys. Rev. D* **2002**, *65*, 065015. [CrossRef]
11. Johnson, P.R.; Hu, B.L. Unruh effect in a uniformly accelerated charge: From quantum fluctuations to classical radiation. *Found. Phys.* **2005**, *35*, 1117–1147. [CrossRef]
12. Hsiang, J.-T.; Hu, B.L. Quantum radiation and dissipation in relation to classical radiation and radiation reaction. *Phys. Rev. D* **2022**, *106*, 045002. [CrossRef]
13. Drummond, P.D.; Ficek, Z. *Quantum Squeezing*; Springer: Berlin/Heidelberg, Germany, 2004. [CrossRef]
14. Grishchuk, L.P.; Sidorov, Y.V. Squeezed quantum states of relic gravitons and primordial density fluctuations. *Phys. Rev. D* **1990**, *42*, 3413–3421. [CrossRef]
15. Hu, B.L.; Kang, G.; Matacz, A. Squeezed vacua and the quantum statistics of cosmological particle creation. *Int. J. Mod. Phys. A* **1994**, *9*, 991–1007. [CrossRef]
16. Weinberg, S. *Cosmology*; Oxford University Press: New York, NY, USA, 2008.
17. Hsiang, J.T.; Hu, B.L. NonMarkovianity in cosmology: Memories kept in a quantum field. *Ann. Phys.* **2021**, *434*, 168656. [CrossRef]
18. Hsiang, J.T.; Hu, B.L. Fluctuation-dissipation relation for a quantum Brownian oscillator in a parametrically squeezed thermal field. *Ann. Phys.* **2021**, *433*, 168594. [CrossRef]
19. Arısoy, O.; Hsiang, J.T.; Hu, B.L. Quantum-parametric-oscillator heat engines in squeezed thermal baths: Foundational theoretical issues. *Phys. Rev. E* **2022**, *105*, 014108. [CrossRef]
20. Parker, L. Quantized fields and particle creation in expanding universes. I. *Phys. Rev.* **1969**, *183*, 1057–1068. [CrossRef]
21. Zel'dovich, Y.B. Particle production in cosmology. *JETP Lett.* **1970**, *12*, 307–311. Available online: http://jetpletters.ru/ps/1734/article_26352.shtml (accessed on 5 March 2023).
22. Hsiang, J.T.; Chou, C.H.; Subaşı, Y.; Hu, B.L. Quantum thermodynamics from the nonequilibrium dynamics of open systems: Energy, heat capacity, and the third law. *Phys. Rev. E* **2018**, *97*, 012135. [CrossRef]
23. Guth, A.H.; Pi, S.-Y. Quantum mechanics of the scalar field in the new inflationary universe. *Phys. Rev. D* **1985**, *32*, 1899–1920. [CrossRef]
24. Hsiang, J.T.; Hu, B.L. No intrinsic decoherence of inflationary cosmological perturbations. *Universe* **2022**, *8*, 27. [CrossRef]
25. Hsiang, J.T.; Hu, B.L. Non-Markovian Abraham-Lorenz-Dirac equation: Radiation reaction without pathology. *Phys. Rev. D* **2022**, *106*, 125108. [CrossRef]
26. Spohn, H. The critical manifold of the Lorentz–Dirac equation. *Europhys. Lett. (EPL)* **2000**, *49*, 287–292. [CrossRef]

Disclaimer/Publisher's Note: The statements, opinions and data contained in all publications are solely those of the individual author(s) and contributor(s) and not of MDPI and/or the editor(s). MDPI and/or the editor(s) disclaim responsibility for any injury to people or property resulting from any ideas, methods, instructions or products referred to in the content.

Article

Zero-Point Energy Density at the Origin of the Vacuum Permittivity and Photon Propagation Time Fluctuation

Christophe Hugon [1] and Vladimir Kulikovskiy [2,*]

[1] Research & Development on Marseilles, 65 Bd de la Libération, 13001 Marseille, France; chr.hugon@protonmail.com
[2] Istituto Nazionale di Fisica Nucleare, INFN Sezione di Genova, Via Dodecaneso 33, 16146 Genova, Italy
* Correspondence: vladimir.kulikovskiy@ge.infn.it

Abstract: We give a vacuum description with zero-point density for virtual fluctuations. One of the goals is to explain the origin of the vacuum permittivity and permeability and to calculate their values. In particular, we improve on existing calculations by avoiding assumptions on the volume occupied by virtual fluctuations. We propose testing of the models that assume a finite lifetime of virtual fluctuation. If during its propagation, the photon is stochastically trapped and released by virtual pairs, the propagation velocity may fluctuate. The propagation time fluctuation is estimated for several existing models. The obtained values are measurable with available technologies involving ultra-short laser pulses, and some of the models are already in conflict with the existing astronomical observations. The phase velocity is not affected significantly, which is consistent with the interferometric measurements.

Keywords: virtual pair; virtual fluctuations; light velocity fluctuation

Citation: Hugon, C.; Kulikovskiy, V. Zero-Point Energy Density at the Origin of the Vacuum Permittivity and Photon Propagation Time Fluctuation. *Physics* **2024**, *6*, 94–107. https://doi.org/10.3390/physics6010007

Received: 5 October 2023
Revised: 1 December 2023
Accepted: 5 December 2023
Published: 10 January 2024

Copyright: © 2024 by the authors. Licensee MDPI, Basel, Switzerland. This article is an open access article distributed under the terms and conditions of the Creative Commons Attribution (CC BY) license (https://creativecommons.org/licenses/by/4.0/).

1. Introduction

With the concept of virtual fluctuations composed of two photons, several effects can be introduced and numerically estimated. This includes the known Lamb shift measured in the Lamb–Retherford experiment, the measured Casimir effect [1], the observed dynamic Casimir effect [2], and the predicted Unruh effect, as well as Hawking radiation. There is no general consensus, however, whether the virtual fluctuations are rather a physical phenomenon than just a useful mathematical tool [3].

Quantum mechanics (QM) is based on postulated equations which do not have an intuitive introduction despite and because of the more than a dozen quite contradictory interpretations available [4,5]. The attempts to derive some of the principles of QM through classical stochastic processes are ongoing in order to provide a deeper understanding of the (experimental) phenomena from a wider point of view and context. In particular, in stochastic electrodynamics (SED), the interaction of the zero-point field (ZPF) with real particles is evaluated. This interaction may explain several, if not all, of the quantum phenomena (for one of the most recent papers; see [4]). The energy density of the ZPF, $w(\omega) \propto \omega^3$, can be derived from the condition that there is no average force of the ZPF acting on any physical harmonic oscillator with a frequency ω (Einstein–Hopf formula [6] (Appendix)). Thus, this energy density has the same form as ZPF in quantum electrodynamics (QED). Some of the quantum relations, such as the canonical commutation relation, $[\hat{x}; \hat{p}] = i\hbar$, with $\hat{x}$ and $\hat{p}$ the coordinate and momentum operators, respectively, and $\hbar$ the reduced Planck's constant, $h/(2\pi)$, can be obtained in SED only if the interaction with ZPF is treated in a non-perturbative way [5,7]. This is true in particular for free particles: the transition from a classical deterministic behaviour to an indeterministic quantum one happens in SED once the interaction with ZPF takes the leading role and the information on the initial condition is lost [8].

In QED, the virtual fluctuations or ZPF manifest themselves as an additional $1/2\hbar\omega$ term of the total energy stored in a single oscillation mode. That term appears for each of the light modes for the photons in the box, and the number of modes becomes continuous once the infinite box size limit is considered for Planck's law derivation [9]. Conventionally, the $1/2\hbar\omega$ term is omitted in order to evaluate the energy difference with respect to the so-called zero-point level. This is sufficient for most applications, where only the energy difference matters. Nevertheless, the zero-point energy is not null even in the absence of real particles (photons), and it affects gravity at the cosmological scales. Moreover, the mode's energy density is $w(\varepsilon) \propto \varepsilon^3$, infinite for $\varepsilon \to \infty$. In the framework of quantum field theory (QFT), generally, the energy upper limit at the Planck scale, Λ, is hypothesized. As a result, the zero-point energy density $w \propto \Lambda^4$ has up to 120 orders of difference with the observed energy density—an issue known as the cosmological constant problem.

In the present study, the concept of virtual fluctuations composed by virtual fermion–antifermion pairs is explored. The motivations to enrich virtual fluctuations with fermion–antifermion pairs are not new, and overviews can be found in Refs. [10–12]. Here, we use virtual fluctuations as a synonym for virtual pairs and mostly consider only fermions and antifermions.

In Refs. [10,12], it is assumed that the virtual pairs may appear for a short lifetime connected with their energy by the Heisenberg uncertainty relation. The pairs are CP-symmetric, what permits them to have zero values for the total angular momentum, colour, and spin. In the presence of the electromagnetic field, the pairs should polarise, and thus the vacuum has dielectric behaviour. The vacuum's dielectric properties, i.e., the permittivity, ϵ_0, and the permeability, μ_0, are already axiomatically postulated in Maxwell's equations. One can, however, take a step back and reconsider this by looking at the following equations of the electric displacement field, $\vec{D}$, and the magnetic field strength, $\vec{H}$, for dielectrics:

$$\vec{D} = \epsilon_0 \vec{E} + \vec{P}, \tag{1}$$

$$\vec{H} = \frac{1}{\mu_0}\vec{B} - \vec{M}, \tag{2}$$

with $\vec{E}$ and $\vec{B}$ being electric field strength and magnetic flux density, respectively.

Polarisation, P, and magnetisation, M, induced by the external field at the microscopic level correspond to the sum of electric and magnetic moments in a unit of volume, respectively. Following Refs. [10,12], one can consider that the first terms in Equations (1) and (2) are due to the vacuum polarisation and magnetisation, i.e., these terms can also be estimated as the sums of electric and magnetic moments of the virtual fermion pairs.

In order to calculate ϵ_0 and $1/\mu_0$, the moments generated by a virtual pair are needed, along with the volume occupied by each pair. For the moments' estimation, the calculations in Refs. [11–13] take a shortcut by assuming the oscillation model for the virtual pair with two states separated by $2mc^2$ with m the particle mass and c the speed light in the vacuum. The calculations in Ref. [10] start from the dipole moment of a pair with opposite charges and fermion magnetic moment and then assume that the pair lifetime is modified in the presence of the field. For the volume estimation, the typical size of a Compton length is commonly involved. In particular, in [12], this is motivated by the assumption that in order for the virtual pair to interact with the external field, the energy conservation should be violated locally by $\Delta\varepsilon \gtrsim 2mc^2$, which is non-detectable in a period shorter than $\hbar/(2mc^2)$. If the speed of particles is at the maximum, i.e., at the speed of light, c, the pair must remain separated by a distance $\lambda_C = \hbar/mc$. In order to obtain a measured value of ϵ_0, the average volume of $V \simeq 0.41\lambda_C^3$ is required, and it is equally occupied by the virtual fermions of all known types.

In the study presented here, we address the assumptions on the volume occupied by virtual fluctuations in the above mentioned models. We consider that a self-consistent and intuitive way to introduce the occupied volume is to go back to the assumed origin of virtual fluctuations, i.e., ZPF which appears in solutions of QED equations or which is introduced from the beginning in SED. One already knows the ZPF energy density and the energy per mode. Thus, the ZPF itself already provides the modes' density, $\rho(\varepsilon)$. The mode's density can be used in order to obtain any property density if the expression of the property per mode is known. The assumption of the infinity of the modes and a distinct energy associated with each mode is incompatible with the assumption that virtual fluctuations appear only with total energy, $\varepsilon = 2mc^2$. Another feature is that by introducing ZPF density, one does not need to further assume that the virtual pairs become real for the time related to their energy following the Heisenberg uncertainty. The only assumption needed is that at the ground state the energy of virtual fluctuation is $\varepsilon = 1/2\hbar\omega$, while after interaction, it becomes $\varepsilon = (1/2 + n)\hbar\omega$. In QED, this means that before the interaction, no real particles were present, while after the interaction, n particles were produced. In SED, the virtual fluctuations are part of the ZPF, which is real even before or in the absence of an interaction.

In Section 2, we review the Planck's law derivation in order to introduce the density of virtual fermions and to explain their mathematical origin. We show how ϵ_0 can be estimated using vacuum density in Section 3.

One may further assume that the photon propagation speed is actually finite only because the photon is delayed at each interaction with virtual fluctuations by their annihilation time. The real photon propagation then fluctuates, without affecting the phase velocity measured by interferometers [14], and the arrival time spread can be estimated by knowing the average lifetime of a virtual pair and the distance travelled. This is performed for several theories in Section 4, and such spread presents a new prediction, measurable with the available technologies.

2. Statistics for Virtual Pairs

The conventional derivation of the Planck's law is briefly summarised in this Section with the explicit assumptions needed. In many textbooks, these calculations are oriented to provide a measurable energy density, so the vacuum-related terms are often hidden and the physical origins behind some assumptions are not reported properly.

2.1. Density of Modes for a Particle in a Box

A particle in a box may only have modes respecting the boundary conditions $\psi(0) = \psi(L) = 0$, where L is the size of a box. This is satisfied for standing waves which have half-wavelength ($\lambda/2$) multiplets equal to a box size:

$$l_i \frac{\lambda_i}{2} = L_i, \tag{3}$$

for every space coordinate i. This condition shows that states are discrete for a box with a limited size. Each set of non-negative integer numbers l_i defines a mode. The wavelength is connected with momentum p_i as:

$$\lambda_i = h/p_i. \tag{4}$$

So, the number of states in the mode space corresponds to the number of the states in the momentum space as follows:

$$dl_x dl_y dl_z = \frac{8L_x L_y L_z}{h^3} dp_x dp_y dp_z = \frac{8V}{h^3} d^3 p, \tag{5}$$

for a box with volume V.

For a relativistic particle,

$$\varepsilon^2 = (mc^2)^2 + (p_x c)^2 + (p_y c)^2 + (p_z c)^2. \tag{6}$$

Thus, the number of states (discrete points) in a sphere of radius ε corresponds to the number of states inside an ellipsoid with volume $4/3\pi p_x p_y p_z$. The non-zero mass only shifts the ellipsoid's centre without affecting its volume. To determine a mode's space with positive numbers l_i and momentums p_i, only one octant should be considered, so the number of modes can be evaluated as:

$$dN = 1/8 dl_x dl_y dl_z = \frac{L_x L_y L_z}{h^3} dp_x dp_y dp_z = \frac{V}{h^3} 4\pi p^2 dp. \quad (7)$$

Thus, the mode density becomes:

$$d\rho = \frac{4\pi p^2 dp}{h^3}. \quad (8)$$

This is valid for massive and massless particles in a general (relativistic) case even for an infinite box size. In the case of $L \to \infty$, the mode distribution becomes continuous.

This result is also valid if a boundary condition $\psi(0) = \psi(L) = 0$ of an isolated box is relaxed to a continuity condition $\psi(0) = \psi(L)$. In such a condition, negative values of l_i are allowed, and $\vec{p}$ becomes a vector with values in a full space (and not just one octant). However, only wavelength multiplets being equal to the size of the box are now allowed (opposite to the half-wavelength multiplets). This gives a $(1/2)^3$ reduction in the number of modes that is numerically equal to one octant condition.

2.2. Mode Energy

The electromagnetic field can be quantized starting from Maxwell's equations and performing Fourier analysis for modes in the box with periodic conditions [15]. For such systems, the Hamiltonian becomes equivalent to an infinite set of oscillators. The mode energy levels are, as for a harmonic oscillator:

$$E = \hbar\omega\left(\frac{1}{2} + n\right), \quad (9)$$

where n is the number of quanta with energy $\hbar\omega$—the smallest value of energy that can be taken or added to the mode with ω frequency. For the electromagnetic field, a quantum is associated with a single photon; thus, $\varepsilon = \hbar\omega = pc$.

2.3. Statistics

For this part, it is worth mentioning that conventionally and historically, black-body radiation is described for a closed box in thermodynamical equilibrium with the temperature T and the volume V being fixed. In statistical mechanics, this is, however, ambiguous since it may correspond to any of the following.

- The grand canonical ensemble (T, V, μ fixed), where μ is the chemical potential: The system can exchange energy and particles with a reservoir, so that various possible states of the system can differ in both their total energy and the total number of particles.
- The canonical ensemble (T, V, N fixed): The system can exchange energy with the heat bath, so that the states of the system will differ in total energy; the number of particles is fixed.

In the equilibrium, the probability of a state with energy distribution E_i is described by the following relations. For canonical ensemble,

$$p_i = \frac{e^{-E_i/(kT)}}{Z}, \quad (10)$$

where Z is the partition function assuring probability normalisation,

$$Z = \Sigma_i e^{-E_i/(kT)}, \quad (11)$$

and k is the Boltzmann constant.

The most powerful and general way to obtain this probability is using the information-theoretic Jaynesian maximum entropy approach [16,17].

Similarly, for the grand canonical ensemble:

$$p_i = \frac{e^{(N_i\mu - E_i)/(kT)}}{Z}, \quad (12)$$

$$Z = \Sigma_i e^{(N_i\mu - E_i)/(kT)}. \quad (13)$$

For the black-body radiation, the number of photons is not conserved, so it is considered that the grand canonical ensemble is the right choice, although even in some recent studies the opposite choice is made (see critiques in Ref. [18], for example). The confusion is supported by the feature that for black-body radiation, $\mu = 0$ is assumed, so both probabilities become identical. The choice of $\mu = 0$ is a consequence of the property that the black-body radiation in a closed cell should be completely defined by only two macroscopic parameters: T and V. As a result, the system pressure is defined only by temperature, and the Gibbs free energy $G = \mu N$ becomes 0 (actually, not well-defined), so for a system with a non-zero N, it is required to have $\mu = 0$ [19].

In a case of non-interacting bosons, each available single-particle level (mode) forms a separate thermodynamic system in contact with the reservoir. Thus, the analysis of the system behaviour can be conducted within a single mode, and then the properties of the whole system can be obtained by integrating it over the modes with their densities. For a single mode, the grand canonical partition function becomes (omitting μ term):

$$Z = \Sigma_{n=0}^{\infty} e^{-\hbar\omega(n+1/2)/(kT)} = e^{-\hbar\omega/(2kT)} \frac{1}{1 - e^{-\hbar\omega/(kT)}}. \quad (14)$$

And each state probability within a mode is:

$$p_i = p(n) = \frac{e^{-\hbar\omega(n+1/2)/(kT)}}{Z} = e^{-\hbar\omega n/(kT)}(1 - e^{-\hbar\omega/(kT)}). \quad (15)$$

Interestingly, the probability does not contain the zero-point energy term, i.e., the probability would be identical if this term was omitted from the beginning. For the average energy evaluation, it is convenient to use the following property of the partition function:

$$\langle E \rangle = \frac{\Sigma_{n=0}^{\infty} E_n e^{-E_n\beta}}{Z} = \frac{dZ}{d\beta}\frac{1}{Z} = -\frac{d(\log Z)}{d\beta} = \hbar\omega\left(\frac{1}{2} + \frac{1}{e^{\hbar\omega/(kT)} - 1}\right), \quad (16)$$

where $\beta = 1/(kT)$ is introduced for simplicity. The average number of particles can be obtained from the average property:

$$\langle E \rangle = \left\langle \hbar\omega\left(\frac{1}{2} + n\right) \right\rangle = \hbar\omega\left(\frac{1}{2} + \langle n \rangle\right), \quad (17)$$

and by comparing Equations (16) and (17), it can be expressed as follows:

$$\langle n \rangle = \frac{1}{e^{\hbar\omega/(kT)} - 1}. \quad (18)$$

For the canonical ensemble of bosons, the number of particles, N, is fixed (the system cannot exchange particles with a reservoir) and the modes cannot be considered as independent thermodynamical systems. In the limit of $n_i \to \infty$ (the number of particles in each mode is extremely large), one can, however, obtain a surprisingly similar average number of particles for each mode (energy level) as in the case of the canonical ensemble [20].

2.4. Degeneracy

In the above considerations, the feature that each mode for the electromagnetic field has a degeneracy $g = 2$ corresponding to the two polarisations was omitted. This, effectively, makes the mode density twice as high.

2.5. Energy Density

Combining the results from Section 2, one obtains the Planck's law:

$$w(p)dp = 2\frac{4\pi p^2 dp}{h^3}\varepsilon\left(\frac{1}{2} + \frac{1}{e^{\varepsilon/(kT)} - 1}\right), \tag{19}$$

from where the vacuum density can be deduced:

$$\rho(p)dp = \frac{4\pi p^2 dp}{h^3}. \tag{20}$$

Following the derivation, let us summarise some features.

- The vacuum density is numerically equal to the density of the modes—its $1/2$ factor from the zero-level is compensated by the degeneracy $g = 2$;
- Modes and their zero-point energies appear for the electromagnetic field after Fourier transformation, so, strictly speaking, assuming a mode occupies some volume (even on average) is improper;
- Derivation of the statistical distribution for modes assumes they are independent and that the number of quanta in each mode is high: this may leave some speculation whether the vacuum density is described by Equation (20) in cases when there are few or no quanta (associated with real photons);
- The vacuum density is temperature- and energy-independent (non-thermal, linear); in current derivation, this appears as a consequence of several features: modes are independent and each mode's microstate energy has a zero-level energy that is proportional to the mode quantum's energy with the same factor for each mode;
- The vacuum density and its energy density are infinite if there is no cutoff at high energies/frequencies; the introduction of such an ultraviolet cutoff is a viable option, for example, in doubly special relativity [18].

Similarly to photons, one may apply the discussed derivation to fermions. The mode density, as discussed in Section 2.1 above, is the same for massive and massless particles. The degeneracy for fermions is $g = 2$. Quark fermions have additional factor 3 degeneracy due to their colour. The vacuum density should also follow Equation (20) since Bose–Enstein or Fermi–Dirac statistical terms are not relevant. The quantum energy of each mode is actually $((mc^2)^2 + (pc)^2)^{1/2}$, where m is the mass of a fermion. Assuming now that virtual fluctuations should appear as fermion–antifermion pairs with opposite spin and momentum, their density should also be described by the same equation, in which p always refers to the momentum of a single fermion or antifermion. Finally, let us note that for boson pairs (W^+W^-) the vacuum density expression is also the same as for fermions since the statistical term is relevant only for real quanta; however, degeneracy $g = 3$ for a spin of 1 should be used.

3. The Vacuum Permeability and Permittivity
3.1. Calculation with ZPF Density Using Oscillator Model

In order to perform vacuum permittivity calculations, we consider some of the assumptions of Ref. [12], in particular:

- each virtual fermion–antifermion pair behaves as a harmonic oscillator with the levels separated by $\hbar\omega$; specifically, this is the energy gap between the ground state of a virtual pair and the excited state where both particles become real.

This assumption is compatible with the mode energy (9) of a quantised electromagnetic field. We further try to make the calculation in Ref. [12] more consistent with modes and their quanta. This brings the following assumptions:

- a virtual fermion–antifermion pair becomes real if $\varepsilon = 2((mc^2)^2 + (pc)^2)^{1/2}$ is added to the mode so the oscillator frequency is actually $\hbar \omega = \varepsilon(p)$ instead of $2mc^2$;
- we use the vacuum (mode) density for the estimation of ϵ_0 instead of the average volume following the alternative proposed at the end of [12].

Following the study in Ref. [12], if one arranges the two-level approximation, so that only transitions between the ground and the first excited states of the oscillator are relevant, one finds that the maximum possible induced dipole moment becomes:

$$d_{max} = q_e \langle \psi_1 | \hat{x} | \psi_0 \rangle = q_e \int_{-\infty}^{\infty} \left(\frac{m\omega}{\pi \hbar}\right)^{1/4} e^{-\frac{m\omega x^2}{2\hbar}} \left(\frac{m\omega}{\pi \hbar}\right)^{1/4} e^{-\frac{m\omega x^2}{2\hbar}} \sqrt{2} \left(\frac{m\omega}{\hbar}\right)^{1/2} x \, dx$$

$$= q_e \int_{-\infty}^{\infty} \sqrt{\frac{2}{\pi}} \left(\frac{m\omega}{\hbar}\right) e^{-\frac{m\omega x^2}{\hbar}} x^2 = q_e \sqrt{\frac{2}{\pi}} \sqrt{\frac{\hbar}{m\omega}} \int_{-\infty}^{\infty} e^{-z^2} z^2 dz = q_e \sqrt{\frac{\hbar}{2m\omega}}. \quad (21)$$

Here, q_e denotes the electron charge.

The time-averaged induced dipole moments following calculations in Ref. [12] (Equation (A3)) can be expressed as follows:

$$d = \frac{2d_{max}^2}{\hbar \omega} E = \frac{q_e^2}{m\omega^2} E. \quad (22)$$

The expressions for d_{max} and d here differ from Equations (A2) and (A3) of Ref. [12] by factors $1/\sqrt{2}$ and $1/2$, respectively, as soon as $\hbar \omega = 2mc^2$ is assumed. Equation (22) is same as Equation (25) obtained in Ref. [13] within a semiclassical treatment.

The vacuum permittivity calculation then can be estimated as follows, while assuming only one type of fermion so far:

$$\epsilon_0 = \int_0^{p_{max}} \frac{4\pi p^2 \, dp}{h^3} \frac{q_e^2}{m\omega^2} = \frac{q_e^2}{4\pi \hbar c} \frac{1}{2\pi} \frac{1}{mc^2} \int_0^A \frac{(pc)^2 d(pc)}{(pc)^2 + (mc^2)^2}, \quad (23)$$

where the cutoff on pc at A is introduced in order to have a finite integral. The introduction of the cutoff is just instrumental here, and its possible physical origin is discussed below in this Section and in Section 5. One can rewrite the above equation by using fine-structure constant, α:

$$\frac{1}{\alpha} = \frac{1}{2\pi} \frac{1}{mc^2} \int_0^A \frac{(pc)^2 d(pc)}{(pc)^2 + (mc^2)^2} = \frac{1}{2\pi} \left[\frac{A}{mc^2} - \tan^{-1}\left(\frac{A}{mc^2}\right)\right]. \quad (24)$$

One can see that in order to obtain $1/\alpha \approx 137$, a value of A should be on the order of $A \approx (2\pi/\alpha)mc^2 \approx 861 mc^2$, where the factor $2\pi/\alpha$ resembles the inverted QED correction needed for the electron magnetic moment (one-loop result for the anomalous magnetic moment). If all known charged elementary particles are considered, the ϵ_0 and $1/\alpha$ expressions contain a sum over charged fermions, or more generally, over all charged elementary particles:

$$\frac{1}{\alpha} = \frac{1}{2\pi} \sum_i Q_i^2 c_i \frac{g_i}{2} \left[\frac{A}{m_i c^2} - \tan^{-1}\left(\frac{A}{m_i c^2}\right)\right], \quad (25)$$

where charge scale $Q_i = 1$ for leptons and W-bosons, $Q_i = -1/3$ for d, s, b quarks, and $Q_i = 2/3$ for u, c, t quarks; the degeneracy, g_i, due to a spin is 2 for fermions and 3 for $W^- W^+$ pairs, while the degeneracy, c_i, due to a colour is 3 for quarks and 1 for the other types. If one keeps the upper energy value the same for each type, a value of the threshold $A \approx 292$ MeV is needed to reach the measured value of ϵ_0 or α. The most important contribution is coming from electrons and the second contribution from u quarks, for which $m_u = 1.5$ MeV was used. Other contributions are at the percent level and below. The threshold, A, is much higher than $m_e c^2$, and the greatest contribution is coming from the ultra-relativistic pairs since, for $A = m_e c^2$, only 0.1%ϵ_0 is reached. This poses doubt if the used non-relativistic harmonic oscillator model is reasonable. One possibility to lower the threshold, A, and to avoid relativistic pairs is to assume that instead of the real charges, one should use bigger, unscreened charges.

The obtained value of A is above or on a level of the chiral symmetry breaking value ($\approx$100 MeV or pion mass), the energy above which the quark condensate should disappear [21]. One should investigate how to incorporate this effect. A possible solution could be to consider that for the quarks the A threshold is set to this level. Since the quark contribution to $1/\alpha$ is sub-dominant, this should not strongly affect the evaluation in Equation (25). The current threshold is three orders of magnitude below the electroweak unification (246 GeV). Such a threshold could be a physically motivated choice for the leptonic virtual pairs; however, ϵ_0 would not be matched with the measured value.

It is possible to make the above calculations with vacuum density consistent with the calculations in Ref. [12], which exploit the concept of the average volume occupied by virtual pairs. In this paper, we can numerically estimate the average virtual pair volume as the inverted density. In order to make both calculations compatible, one needs to assume that the separation of energy levels is energy independent, and it is equal to $2mc^2$, as in Ref. [12]. This allows one to split the constant multiplication in Equation (23) from the integral, which becomes just the vacuum density, $\rho = \int_0^{A/c} 4\pi p^2 dp / h^3$. In order the average volume to be proportional to $\lambda_C^3 = (\hbar/(mc))^3$, one needs to set the limit, A, proportional to mc^2: $A = amc^2$, i.e., to assume a different threshold for each elementary particle type.

With $A = amc^2$ thresholds and $\hbar\omega = 2mc^2$, the expression for $1/\alpha$ reads:

$$\frac{1}{\alpha} = \frac{1}{2\pi} \sum_i Q_i^2 c_i \frac{g_i}{2} \frac{1}{(mc^2)^3} \int_0^{amc^2} (pc)^2 d(pc) = \frac{1}{2\pi} \sum_i Q_i^2 c_i \frac{g_i}{2} \frac{a^3}{3}. \qquad (26)$$

With this threshold, one obtains a factor $a = \sqrt[3]{6\pi/(\sum_i Q_i^2 c_i (g_i/2)\alpha)} \approx 6.5$ (we consider $g_i = 3$ for W^+W^-, so $\sum_i Q_i c_i (g_i/2)^2 = 9.5$, and not 9 as in Ref. [12]) and the average volume reads:

$$\langle V \rangle = 1/n = 1/\int_0^{A/c} \frac{4\pi p^2 dp}{h^3} = \frac{6\pi^2}{a^3}\left(\frac{\hbar}{mc}\right)^3 \approx 0.22\left(\frac{\hbar}{mc}\right)^3, \qquad (27)$$

what is consistent with the calculations in Ref. [12], considering the difference in the dipole moment by a factor of 2 and the differences in g_i for W-bosons.

3.2. Relations between Magnetic and Electric Dipole Moments of Virtual Pairs

Here, we reiterate to avoid the oscillator approximation approach and to start from a magnetic moment of each fermion in the pair along with an electric dipole moment of a pair in order to arrive at the evaluation of ϵ_0 and μ_0. We provide below an ansatz for calculating magnetic and electric dipole moments for virtual pairs that keeps the c, ϵ_0, and μ_0 interrelation.

First, it is worth mentioning that in some models of vacuum description [10] and for real fermion gas (Equations (59.4) and (59.12) in Ref. [22]), there are the following dependences for permittivity and permeability:

$$1/\mu_0 = \sum_i f(\beta_i^2), \qquad (28)$$

$$\epsilon_0 = \sum_i f((d_i/2)^2), \qquad (29)$$

where β is a magnetic moment of a single fermion, d is an electric dipole moment of a fermion–antifermion pair, and function f is of the same dependence in Equations (28) and (29) as soon as the corresponding potentials are of the same expression, namely, $U = -\vec{d}\vec{E}$ and $U = -\vec{\beta}\vec{B}$. Given $c = 1/\sqrt{\epsilon_0 \mu_0}$, one finds that this leads to the following relationship between β and d:

$$\beta = (d/2)c. \tag{30}$$

The magnetic moment can be obtained by comparing relativistic energy for an electron in a magnetic field [23] (in SI units):

$$E_M = \sqrt{m^2 c^4 + p_z^2 c^2 + 2 q_e \hbar c^2 B(n + 1/2 - gs/2)} \approx \varepsilon_f + \frac{q_e \hbar c^2 B}{2 \varepsilon_f}(2n + 1 - gs), \tag{31}$$

where we consider the first order approximation as valid for common fields: $\beta_B B \ll m_e c^2$ (here, β_B is the Bohr magneton).

This approximation can be compared to the non-relativistic energy levels in Ref. [22] (in SI units):

$$E_M^{NR} = \frac{p_z^2}{2m} + \frac{q_e \hbar B}{2m}(2n + 1 - gs). \tag{32}$$

Comparing relativistic and non-relativistic expressions, one can see that they match if mc^2 is substituted with fermion energy, ε_f. Thus, the following relativistic magnetic moment can be used instead of the Bohr magneton, $\beta_B = q_e \hbar/(2m)$:

$$\beta = \frac{q_e \hbar}{2 \varepsilon_f / c^2}. \tag{33}$$

For electric dipole moments, one can use the following classical expression:

$$d = Q q_e x, \tag{34}$$

where x is considered a "distance" between fermions.

By comparing expressions (33) and (34), one can realise that Equation (30) is satisfied if:

- the magnetic moment, β, for fermions with charge $Q q_e$ has an additional factor Q;
- for electric dipole moment, d, the "distance", x, is defined as $x = \hbar/(\varepsilon_f/c)$.

If the last point is seen as the Heisenberg uncertainty, this requires that the fermions travel at a speed of light or they are of zero mass.

3.3. Other Considerations for the Virtual Pair Models

If a virtual electron–positron pair is seen as a positronium [11] it is worth establishing a connection with an up-to-date description of the positronium states. In the classical description of such a system, similarly to the electron in hydrogen, it is assumed that the electron and positron are orbiting around the centre of mass. Thus, the energy levels are quantised as in the Bohr model but using a reduced mass equal to $m_e/2$. The precise calculation of the bound states comes from the Bethe–Salpeter equation. Noticeably, there are two solutions with the orthogonal states: one in which particles can be bound at atomic-like distances, similarly to the Bohr model, and the other with the nuclear-like quantised distances [24]. The latter states have zero energy, and thus they could be very promising for describing virtual pairs.

It is worth mentioning that the optical properties, namely, polarisation and nonquantum entanglement [25], can be described using the mechanical model analogy [26]. This mechanical model for a photon or a 2-dimensional (2D) beam consists of two masses, each representing the eigenvalue of a polarisation coherence matrix. The polarisation and entanglement are then quantitatively associated with mechanical concepts of the centre of mass and the moment of inertia through the Huygens–Steiner theorem for rigid body rotation. Although it may be a mere coincidence of this analogy, one may want to search for a deeper physical meaning.

4. Photon Propagation and Propagation Time Dispersion

Most theories explaining electromagnetic vacuum properties with virtual fluctuations assume that particle–antiparticle pairs are continuously appearing in the vacuum and their lifetimes are limited by the Heisenberg uncertainty. We detail here several theories and the lifetimes assumed:

- $\tau \approx \hbar/(2mc^2)$, where this time, in particular, serves to define the size/volume of the virtual pair [12];
- $\tau = \hbar/(K \times 2mc^2)$ with the best fit for $K \approx 31.9$, where the lifetime modified in the presence of electromagnetic field serves to evaluate ϵ_0 and μ_0 [10];
- $\tau = \hbar/(\alpha^5 mc^2)$ since, after interaction with the photon, the virtual pair forms a quasi-stationary state in Ref. [11].

In a bare vacuum, the dielectric and magnetic moments of the virtual fermion pairs are absent, and thus ϵ_0 and $1/\mu_0$ become 0. The speed of light in the Maxwell equations in a vacuum is $c = [(1/\mu_0)/\epsilon_0]^{1/2}]$ and becomes not well defined [27]. This indicates that the photon propagation is tightly connected with the presence of virtual fluctuations. It is quite natural then, to assume that the photon propagation speed is actually finite only because the photon is delayed at each interaction with virtual fluctuations by their annihilation time, i.e., there is no additional propagation delay in the bare vacuum. By knowing the average lifetime of a virtual pair, τ, and the total time $T = L/c$ needed to cover a distance L, one can straightforwardly estimate the total number of interactions, $N = T/\tau$, and its fluctuation, $N^{1/2}$, which gives an approximate fluctuation time estimate:

$$\sigma_T = N^{1/2}\tau = \sqrt{\frac{\tau}{c}}\sqrt{L}. \tag{35}$$

Thus, for the three theories mentioned above, this fluctuation time becomes:

- $\sigma_T \approx 1.5\,\text{fs}\sqrt{L[\text{m}]}$ for consideration in Ref. [12];
- $\sigma_T \approx 0.26\,\text{fs}\sqrt{L[\text{m}]}$ for for consideration in Ref. [10] (this can be compared with a more precise estimation at $\sigma_T \approx 0.05\,\text{fs}\sqrt{L[\text{m}]}$ given in Ref. [10]);
- $\sigma_T \approx 0.46\,\text{ns}\sqrt{L[\text{m}]}$ for for consideration in Ref. [11].

Let us note that the above calculations are meant to provide an order of magnitude. For a better estimate, one may assume that the photon is delayed at each interaction only by a portion of a lifetime since the photon interaction can occur at a random moment of the pair's appearance from the vacuum, and one can consider a smooth distribution for the possible lifetimes. The fluctuation in propagation time can actually be removed completely if one assumes that during each interaction with time τ_i, a photon propagates for a distance of $c\tau_i$, which may look less intuitive. Indeed, one would need to explain the photon energy transfer in space inside a virtual pair that absorbs the photon.

Nowadays, the strongest constraints on the photon propagation time fluctuation are established by astrophysical observations, mainly GRBs (gamma ray bursts) and pulsars [28,29]. The current limits are at $0.2 - 0.3\,\text{fs}\,\text{m}^{-1/2}$. This means that if the estimate presented here is correct, at least at the order of magnitude, the model in Ref. [11] to be excluded, while the other two considered models are still viable.

The dependence of the fluctuations as $L^{1/2}$ shows that for the measurement, the time resolution has a stronger impact compared to the photon path length. Actually, the astrophysical measurements are based on the observation of events with a duration of the order of 10^{-3} s at distances of megaparsecs, and these measurements are hardly improvable due to the finite size of the Universe and the intrinsic event durations. Instead, the current state of the art of laser technologies allows the generation of femtosecond and even attosecond light pulses, and a pulse duration evaluation is conducted using the autocorrelation, i.e., probing the pulse by its own copy, so the precision is better than the pulse width.

An experiment able to reach the required sensitivity can be realised with femtosecond laser pulses propagating over a few kilometres in a vacuum tunnel without reflections. In the end of the tunnel, the beam is measured with the autocorrelation system. A minimal realisation of the autocorrelation system can consist of a beam splitter, beam routing with three unprotected, gold, single-layer mirrors per split beam and a second harmonic crystal. The mirror system generates a variable delay between the two beam parts. Using a commercially available laser source with $\mathcal{T} = 2$ fs pulse width (~5 fs full-width half-maximum, FWHM), the vacuum optical path of L = 10 km (such tunnels are routinely used in modern gravitational wave detectors) and a pulse duration measurement device with $\sigma_\mathcal{T} = 1\%$ precision, one could already set a limit that is two orders of magnitude lower compared to the existing ones, as is estimated below:

$$\sqrt{\mathcal{T}^2 + \sigma^2 L} - \mathcal{T} = \sigma_\mathcal{T} \mathcal{T}, \qquad (36)$$

$$\sigma = \mathcal{T}\sqrt{\frac{\sigma_\mathcal{T}(2 + \sigma_\mathcal{T})}{L}} \approx 2\sqrt{\frac{0.02}{10^4}} \text{ fs m}^{-1/2} \approx 0.003 \text{ fs m}^{-1/2}. \qquad (37)$$

If one considers the laser with an attosecond pulse duration, the generated pulse would be broadened up to a 16-as duration after only a 1-cm propagation with $\sigma_\mathcal{T} \approx 0.05$ fs $\sqrt{L[\text{m}]}$. Interestingly, the world's shortest pulse generated is measured at the FWHM of 43 ± 1 as [30]. Such a pulse already after 1 cm would be at the FWHM of 57 as or, alternatively, a pulse with a negligible duration would have a duration of 43 as after 13 cm. If fluctuations at such scales are real, currently measured attosecond pulses could be already close to the limit of the generation of shortest pulses and the measurement of their duration. Setting limits to the fluctuations in the speed of light with the best available attosecond laser and a 1–50 cm path can be a viable alternative for such measurement. Finally, let us note that since the fluctuation in propagation time is frequency-independent, the frequency-resolved optical gating pulse characterisation in the time and frequency domains should provide robust effect measurements and enable its discrimination from the matter effects. This effect also has distinct dependence from the distance, so performing additional measurements with a shorter path is another way to control the systematic effects.

5. Conclusions and Discussions

There is certainly an interest in the physics community concerning the idea of virtual fluctuations being at the origin of vacuum electromagnetic properties. As it is shown, similarly to $\vec{P}$, polarisation of the medium, $\epsilon_0 \vec{E}$, can be associated with the polarisation of the vacuum, and similarly, $1/\mu_0 \vec{B}$ can be associated with vacuum magnetisation. This view can clarify the historical controversy between $\vec{H}$ versus $\vec{B}$ and $\vec{D}$ versus $\vec{E}$ [31]. In the absence of matter, $\vec{H}$ becomes the response of the vacuum to the external field $\vec{B}$, while in the presence of matter, it is a combination of the vacuum response to the external field and the matter's magnetisation (permanent, or induced by the field $\vec{B}$). This follows Faraday and Maxwell, which makes $\vec{H}$ the cause of $\vec{B}$ and similarly for $\vec{D}$ and $\vec{E}$.

The authors of [12] show how one can evaluate vacuum polarisation, namely ϵ_0 and $1/\mu_0$, within QED from the interactions of photons with electron pairs using the Feyman diagrams description. In this approach, one should deal with a high momentum cutoff, as this prevents the exact calculation of ϵ_0 without the introduction of a new variable.

A tempting approach to avoid infinite integrals that require cutoffs or renormalisations is to use a constant or an average volume occupied by virtual pairs along with average magnetic and electric moments. Here, we revisit the mathematical origin of zero-point energy appearing in Planck's law derivation in order to stress the feature that this energy appears in the mode space associated with particle momentum. Actually, it is impossible to estimate the volume occupied by a virtual pair with a known momentum due to the

Heisenberg uncertainty. The evaluation of ϵ_0 and μ_0 can be performed using vacuum density in the form $\rho(p) = 4\pi p^2/h^3$ instead of using an averaged pair volume. Moreover, from the mathematical expression of Plank's law, each mode has its own frequency, ω, or energy quanta, $\hbar\omega$; thus, it is more logical to assume that the zero-point energy, $\frac{1}{2}\hbar\omega$, should generate virtual pairs of fermions with non-zero momentum per fermion. Here, we show how these calculations can be performed starting from the ideas in Ref. [12]. In order to obtain a finite value for ϵ_0, a threshold on the maximum fermion momentum is needed. The value of this threshold in our calculation does not provide any intriguing coincidence; however, the calculation shows that most of the fermions should be relativistic. Thus, the use of a non-relativistic quantum harmonic oscillator as a model could be argued. One could use, however, bigger, unscreened charges in order to lower this threshold. We do not propose any physical explanation of the obtained threshold at 292 MeV. A physically motivated choice for such threshold for quarks could be a chiral symmetry breaking energy ($\approx$100 MeV), while for leptons, it could be the electroweak unification energy ($\approx$256 GeV). This indicates an incompleteness of our model. An understanding of how to incorporate the breaking of symmetry and the generation of mass, i.e., the Higgs mechanism, could be the key to explaining the threshold. We note here for completeness that calculating ϵ_0 within QFT requires knowledge of the Landau pole energy. Its value, obtained in [12], does not provide any intriguing coincidence either.

As it is shown, dealing with infinite integrals in the calculations involving virtual pair density takes two distinct paths. In one approach, explored in Ref. [12], it is considered that the volume occupied by virtual pairs has a finite, momentum-independent value, and the vacuum properties are evaluated using this volume scale as a parameter. In [12], it is demonstrated that this approach is consistent with a calculation within QFT, which, in turn, requires the Landau pole energy as a parameter. In another approach, in this study, the density of virtual pairs is used, following its zero-point field origin. The calculations require a cutoff at a high momentum value, which becomes a parameter. The average volume per pair can still be estimated as the inverted density. In a particular case of momentum-independent transition between the virtual fluctuation states, the calculation with the average volume and the density become equivalent. The predictive power of both approaches should be quantified by the number of properties that can be estimated based on such parameters (pair volume or momentum cutoff) and a consistency with the existing observations. The properties may include the vacuum permeability, the vacuum density, and the cosmological vacuum energy density. Meanwhile, the observations should include quantum phenomena such as the interference patterns in double slit experiments and the stability of atoms and their energy levels. In particular, the presence of virtual fluctuations in the vacuum must have an impact on the cosmological scales.

The introduction of the negative gravitational mass for the antiparticles makes each massive virtual pair a gravitational dipole. This model solves several cosmological issues at once, including effects ascribed to dark matter and the cosmological constant problem [32]. The recent results from the ALPHA-g experiment on antihydrogen atoms show that antimatter accelerates towards the Earth with 0.75 ± 0.13 g (statistical+systematic) ± 0.16 g (simulation) [33], so the theories involving antigravity for antimatter are hardly viable.

Several theories that describe photon interactions with virtual pairs include the pair lifetime [10–12]. We speculate that one of the consequences of this assumption is that at each interaction, the photon is trapped until the pair is annihilated, and this may give rise to a fluctuation in the photon propagation time. This fluctuation can be estimated for the mentioned theories. In particular, the theory in Ref. [11] could already be in conflict with the available measurements. For the fluctuations predicted for theories in Ref. [10,12], one needs measurements to be made with available ultrafast laser technologies. The observation of the fluctuation in photon propagation time would certainly be incompatible with quantum field theory.

Author Contributions: Conceptualization, C.H. and V.K.; methodology, C.H. and V.K.; formal analysis, V.K.; writing—original draft preparation, V.K.; writing—review and editing, C.H. and V.K.; funding acquisition, V.K. All authors have read and agreed to the published version of the manuscript.

Funding: The APC and the study was funded by the Istituto Nazionale di Fisica Nucleare, Italy (INFN), including the grant for training activities 195393 attributed to the project ReWOLF-Cub, Italy.

Data Availability Statement: Calculation notebook is available at https://doi.org/10.5281/zenodo.10457877.

Conflicts of Interest: The authors declare no conflict of interest.

Abbreviations

The following abbreviations are used in this manuscript:

GRB	Gamma-ray burst
FWHM	Full-width half-maximum
SED	Stochastic electrodynamics
QED	Quantum electrodynamics
QFT	Quantum field theory
QM	Quantum mechanics
ZPF	Zero-point fluctuations

References

1. Lamoreaux, S.K. Demonstration of the Casimir force in the 0.6 to 6 μm range. *Phys. Rev. Lett.* **1997**, *78*, 5–8. [CrossRef]
2. Wilson, C. Observation of the dynamical Casimir effect in a superconducting circuit. *Nature* **2011**, *479*, 376–379. [CrossRef]
3. Jaeger, G. Are virtual particles less real? *Entropy* **2009**, *21*, 141. [CrossRef]
4. Cavalleri, G. A quantitative assessment of stochastic electrodynamics with spin (SEDS): Physical principles and novel applications. *Front. Phys. China* **2010**, *5*, 107–122. [CrossRef]
5. Cetto, A.M.; de la Peña, L. The electromagnetic vacuum field as an essential hidden ingredient of the quantum-mechanical ontology. *Entropy* **2022**, *24*, 1717. [CrossRef]
6. Boyer, T.G. Derivation of the blackbody radiation spectrum without quantum assumptions. *Phys. Rev.* **1969**, *182*, 1374–1383. [CrossRef]
7. Huang, W.C.-W.; Batelaan, H. Testing quantum coherence in stochastic electrodynamics with squeezed Schrödinger cat states. *Atoms* **2019**, *7*, 42. [CrossRef]
8. Cetto, A.M.; de la Peña, L.L. Role of the electromagnetic vacuum in the transition from classical to quantum mechanics. *Found. Phys.* **2022**, *52*, 84. [CrossRef]
9. Planck, M. Über die Begründung des Gesetzes der schwarzen Strahlung. *Ann. Phys.* **1912**, *342*, 642–656. [CrossRef]
10. Urban, M.; Couchot, F.; Sarazin, X.; Djannati-Ataï, A. The quantum vacuum as the origin of the speed of light. *Eur. Phys. J. D* **2013**, *67*, 58. [CrossRef]
11. Mainland, G.B.; Mulligan, B. Polarization of vacuum fluctuations: Source of the vacuum permittivity and speed of light. *Found. Phys.* **2020**, *50*, 457–480. [CrossRef]
12. Leuchs, G.; Hawton, M.; Sánchez-Soto, L.L. Physical mechanisms underpinning the vacuum permittivity. *Physics* **2023**, *5*, 179–192. [CrossRef]
13. Margan, E. *Some Intriguing Consequences of the Quantum Vacuum Fluctuations in the Semi-Classical Formalism*; Technical Report; Jozef Stefan Institute: Ljubljana, Slovenia, 2011. Available online: https://www-f9.ijs.si/~margan/Articles/SomeConsequences.pdf (accessed on 3 December 2023).
14. Urban, M.; Couchot, F.; Sarazin, X.; Djannati-Ataï, A. Reply to the comment on: The quantum vacuum as the origin of the speed of light. *Eur. Phys. J. D* **2013**, *67*, 220. [CrossRef]
15. Mandl, F.; Shaw, G. *Quantum Field Theory*; John Wiley & Sons, Ltd.: Chichester, UK, 2010; pp. 1–9.
16. Jaynes, E.T. Information theory and statistical mechanics. *Phys. Rev.* **1957**, *106*, 620–630. [CrossRef]
17. Jaynes, E.T. Information theory and statistical mechanics. II. *Phys. Rev.* **1957**, *108*, 171–190. [CrossRef]
18. Mishra, D.K.; Chandra, N.; Vaibhav, V. Equilibrium properties of blackbody radiation with an ultraviolet energy cut-off. *Ann. Phys.* **2017**, *385*, 605–662. [CrossRef]
19. Kelly, R.E. Thermodynamics of black body radiation. *Am. J. Phys.* **1981**, *49*, 714–719. [CrossRef]
20. Müller-Kirsten, H.J.W. *Basics of Statistical Physics*; World Scientific Publishing Co. Pte. Ltd.: Singapore, 2022. [CrossRef]
21. Rugh, S.E.; Zinkernagel, H. The quantum vacuum and the cosmological constant problem. *Stud. Hist. Philos. Sci. B Stud. Hist. Philos. Mod. Phys.* **2002**, *33*, 663–705. [CrossRef]
22. Landau, L.D.; Lifshitz, E.M. *Statistical Physics*; Elsevier Ltd.: Oxford, UK, 2013. [CrossRef]

23. Steinmetz, A.; Formanek, M.; Rafelski, J. Magnetic dipole moment in relativistic quantum mechanics. *Eur. Phys. J. A* **2019**, *55*, 40. [CrossRef]
24. Patterson, C.W. Properties of the anomalous states of positronium. *Phys. Rev. A* **2023**, *107*, 042816. [CrossRef]
25. Simon, B.N.; Gori, S.F.; Santarsiero, M.; Mukunda, B.N.; Simon, R. Nonquantum entanglement resolves a basic issue in polarization optics. *Phys. Rev. Lett.* **2010**, *104*, 023901. [CrossRef] [PubMed]
26. Qian, X.-F.; Izadi, M. Bridging coherence optics and classical mechanics: A generic light polarization-entanglement complementary relation. *Phys. Rev. Res.* **2023**, *5*, 033110. [CrossRef]
27. Leuchs, G.; Villar, A.S.; Sanchez-Soto, L.L. The quantum vacuum at the foundations of classical electrodynamics. *Appl. Phys. B* **2010**, *100*, 9–13. [CrossRef]
28. Abdo, A.A.; Ackermann, M.; Ajello, M.; Asano, K.; Atwood, W.B.; Axelsson, M.; Baldini, L.; Ballet, J.; Barbiellini, G.; Baring, M.G.; et al. A limit on the variation of the speed of light arising from quantum gravity effects. *Nature* **2009**, *462*, 331–334. [CrossRef]
29. Crossley, J.H.; Eilek, J.A.; Hankis, T.H.; Kern, J.S. Short-lived radio bursts from the Crab pulsar. *Astrophys. J.* **2010**, *722*, 1908–1920. [CrossRef]
30. Gaumnitz, T.; Jain, A.; Pertot, Y.; Huppert, M.; Jordan, I.; Ardana-Lamas, F.; Wörner, H.J. Streaking of 43-attosecond soft-X-ray pulses generated by a passively CEP-stable mid-infrared driver. *Opt. Express* **2017**, *25*, 27506–27518. [CrossRef]
31. Roche, J.J. B and H, the intensity vectors of magnetism: A new approach to resolving a century-old controversy. *Am. J. Phys.* **2000**, *68*, 438–449. [CrossRef]
32. Hajdukovic, D.S. Antimatter gravity and the Universe. *Mod. Phys. Lett. A* **2020**, *35*, 2030001. [CrossRef]
33. Anderson, E.K.; Baker, C.J.; Bertsche, W.; Bhatt, N.M.; Bonomi, G.; Capra, A.; Carli, I.; Cesar, C.L.; Charlton, M.; Christensen, A.; et al. Observation of the effect of gravity on the motion of antimatter. *Nature* **2023**, *621*, 716–722. [CrossRef]

Disclaimer/Publisher's Note: The statements, opinions and data contained in all publications are solely those of the individual author(s) and contributor(s) and not of MDPI and/or the editor(s). MDPI and/or the editor(s) disclaim responsibility for any injury to people or property resulting from any ideas, methods, instructions or products referred to in the content.

MDPI
Grosspeteranlage 5
4052 Basel
Switzerland
Tel.: +41 61 683 77 34

Physics Editorial Office
E-mail: physics@mdpi.com
www.mdpi.com/journal/physics

Disclaimer/Publisher's Note: The statements, opinions and data contained in all publications are solely those of the individual author(s) and contributor(s) and not of MDPI and/or the editor(s). MDPI and/or the editor(s) disclaim responsibility for any injury to people or property resulting from any ideas, methods, instructions or products referred to in the content.

www.ingramcontent.com/pod-product-compliance
Lightning Source LLC
LaVergne TN
LVHW070712100526
838202LV00013B/1079